Nature and Nature's Laws

A volume
in
THE DOCUMENTARY HISTORY
of
WESTERN CIVILIZATION

NATURE AND NATURE'S LAWS
Documents of the Scientific Revolution

edited by
MARIE BOAS HALL

WALKER AND COMPANY
New York

NATURE AND NATURE'S LAWS
Introduction, editorial notes, translations by the editor, and compilation
Copyright © 1970 by Marie Boas Hall.

Library of Congress Catalog Card Number: 69-15560.
ISBN: 0-8027-2016-1

Printed in the United States of America.

*Illustrations 3-12 inclusive are used by courtesy of the
Trustees of the British Museum.*

Published in the United States of America in
1970 by the Walker Publishing Company, Inc.

Published simultaneously in Canada by
The Ryerson Press, Toronto.

Volumes in this series are published in association
with Harper & Row, Publishers, Inc., from
whom paperback editions are available in Harper
Torchbooks.

Contents

Illustrations

Preface

IT IS impossible to document the scientific revolution fully: no one book can adequately sample even the major writings of some two hundred and fifty prolific years. Scientists of the sixteenth to eighteenth centuries had leisure, literary interest, and an audience which needed to be convinced. The scientific paper, the brief announcement of work done, intended for fellow scientists, was indeed invented during the period, but most men wrote books—long books. Since every attempt is made in this series to avoid very short excerpts which cannot give the flavor of the whole, it has been necessary to be severely restrictive. Preference has been given either to the most influential writers—Galileo, Descartes, Newton —or to those available in contemporary translation. In some cases (especially Galileo and Newton) the selections have been truncated on the grounds that there are readily available modern editions in paperback (cited in the bibliography). In other cases documents mentioned in the introductory commentary have been omitted for lack of space or lack of suitable translation or because they have often been reproduced.

As far as possible, language has been preserved intact, though spelling and punctuation have frequently been modernized, especially in the earlier selections, so that the reader need not be separated from the author by the barrier of obsolete usage. Very occasionally (as in the selection from Digges) the word order has also been modified to conform to modern usage.

Where not otherwise specified the translations have been made specially for this volume. I have to thank A. Rupert Hall for the translation in Document 5, and for permission to use the translations in Documents 4 and 33. I am grateful to Columbia University Press for permission to use the translation in Document 14; and to A. E. Gunther for permission to use the translation in

Document 11. Illustrations 3 through 12 are used by courtesy of the Trustees of the British Museum.

Marie Boas Hall

London
September, 1967

Introduction:
The Scientific Revolution

HISTORIANS, unlike geologists, tend to see the past at first view not as a long, continuous development, but as a series of tenuously connected jumps, which they call revolutions. Only later does the critic move in to insist that every revolution has a slow period of development that precedes it, or at least to argue that *your* revolution is better characterized as the culmination of slow growth, although *his* may be genuine. Nineteenth century historians were interested in art and literature and secular rationalism, and so *their* revolution—which catapulted Europe, they thought, into the modern world—was the Renaissance. Twentieth-century historians first criticized this by finding earlier examples of "rebirth"—the Renaissance of the twelfth century, the Carolingian and Ottonian Renaissances of the ninth and tenth centuries—and then denied that there was any such thing as a "rebirth" of anything, emphasizing that all human history has been continuous development and change. (In support of older historians it must not be forgotten that the men of Renaissance Europe thought that they were indeed reviving and re-creating the best of the remote past.)

The concept of the scientific revolution is much more recent that the concept of the Renaissance, much more solidly based, and, so far, not successfully attacked. It places the central period for the development of modern science in the seventeenth century, which has long been recognized as an extraordinarily fertile and creative period. The Age of Newton, like the Age of Archimedes or the Age of Einstein, is bound to seem extremely creative if only because of the presence in it of one of the few truly great scientific minds of all time. The term "scientific revolution" to characterize this period, and above all to characterize the development of the methods of early modern science, so recognizably akin to the science of our own day, was, I think, first used by Herbert Butterfield in 1948 in the lectures which became the *The Origins of Modern Science* (1949). This book greatly influenced a whole generation of young historians of science just after World War II, and in paperback is doubtless continuing its influence. The term was given further publicity through its adoption by A. Rupert Hall as the title of a book—his *Scientific Revolution* (1954). It almost immediately passed' into current usage, and although individual attributes of the age have been traced to earlier periods, and individual scientists have been shown to have drawn much from their medieval predecessors, on the whole the revisionists have found the position too strongly supported by a wealth of evidence to be able to destroy it. Further study only shows it to have been more complex, not less. It is true that T. S. Kuhn in *The Structure of Scientific Revolutions* has argued that the seventeenth-century scientific revolution is only one among many; but, whether his view is accepted or not, *the* scientific revolution remains. Further, the seventeenth-century-oriented revolution easily embraces many minor revolutions, such as "the Copernican revolution" in astronomy, the sixteenth-century "revolution in anatomy," or "the chemical revolution" of the eighteenth century; for we all follow Butterfield in allowing the scientific revolution to begin soon after 1500 and end about 1790 (thereby following the precedent of French historians, who have always allowed the seventeenth century to end with the death of Louis XIV in 1715).

When it comes to the point of beginning and ending all historians agree: Copernicus and Vesalius accidentally published simultaneously in 1543 two great, though very different, works, which are both at once the culmination of an old tradition and the beginnings of a new scientific approach. The *De Revolutionibus* is very like Ptolemy's *Almagest* of thirteen hundred years before stood on its head, because it is the earth that moves in Copernicus' book, not the sun and the celestial spheres. Vesalius' beautifully illustrated work *De Humani Corporis Fabrica (On the Fabric of the Human Body)* is much as Galen might have written it about thirteen hundred years earlier if he had been able to dissect a few human bodies. But each, while obviously conceived in terms of the work of classical antiquity, was consciously novel in approach and each proved to be revolutionary in its implications. (Ironically, more so in the case of Copernicus, who tried to minimize his radical departures from tradition, than in the case of Vesalius, who overpraised his own novelty.)

General awareness of Copernicus' work among the educated public was far greater than of Vesalius', and the reason for both is significant. Astronomy was the best understood, most widely studied, and most controversial science of its day for many good reasons; while there was great lay interest in medical remedies, and most educated men had some private list of drugs to recommend to their friends, medical science was studied only by would-be physicians, and anatomy could be studied only in connection with medical schools. Anatomical dissection hardly came within the range of experience of laymen, and not even of scientists unless they had access to the public dissections performed for the benefit of student physicians or surgeons. But astronomy impinged upon everyone, as it had for centuries. Cosmological astronomy was useful in telling the time of day; even public clocks were uncommon before the eighteenth century. The introduction of Huygens' pendulum clocks after the 1650's slowly made domestic clocks possible, but few were built before the last decade of the century. Astronomy was essential for calendrical computation, whether in calculating the date of Easter and other Church festivals (one of the points mentioned by Copernicus as an important indication of the need for the reform of established planetary theory) or in merely following the details of a printed calendar. Astronomy became of great use to sailors and hence to cartographers in the fifteenth and sixteenth centuries: as they sailed into uncharted waters out of sight of land, or came to hitherto unknown and unexplored lands, sailors needed methods of determining their latitude and longitude. Astronomy, it was hoped, could provide both; indeed land-based mathematical astronomers quickly taught sailors how to use simplified astronomical instruments to determine their latitude. In the course of the sixteenth century both methods and instruments were improved, and the development of "Mercator's projection" provided useful charts. In the same period suggestions for longitude determination were numerous. Two of these, the method of lunar distances (involving fairly accurate stellar observations and complex tables) and the method of clocks (involving the development of technically refined chronometers), were both exploited in the seventeenth century and rendered useful in the eighteenth; chronometers became universal only in the mid-nineteenth century.

Finally, astronomy was useful in predicting the future. Ever since antiquity mathematical astronomers had regarded astrology as the legitimate and practical application of astronomical knowledge to human needs. Copernicus was the first astronomer since Ptolemy to try to calculate planetary positions for any but astrological use, yet the tables based upon his calculations were to be widely used by astrologers. Astrology was frowned upon in the sixteenth century as magical, mystic, and doubtfully legitimate. Excessive use of personal horoscopes was even illegal. But the milder forms of astrology were socially and legally acceptable, and medical astrology was taught in the universities, although (as critics of astrology had already pointed out) the doubtful accuracy of most of their data made the work of astrologers of questionable value. It was only the pursuit of the rational and rejection of the occult that, during the scientific revolution, ensured the banishment of astrology to the realm of pure quackery and mysticism.

Hence the work of Copernicus was of general interest, particularly because his specialized, highly mathematical work was preceded by simpler little works of his own (the manuscript *Commentariolus*) and of his disciple Rheitcus. For a half a century Copernicus was more talked of than read. Then, through the telescopic discoveries and polemic publications of Galileo, Copernicanism was publicized in a highly dramatic manner. As a result of Galileo's association of Copernicanism with his own discoveries in mechanics, as well as of his trial and condemnation, acceptance of the Copernican system (though now much modified and probably too radical for Copernicus himself) became a moral issue. On the one hand, Copernicanism was an essential part of the new philosophy—rational, consciously radical, profoundly anti-Aristotelian and anti-Ptolemaic. On the other hand, Copernicanism was a symbol of the struggle for free, secular, unconventional intellectual development. After 1532 to support Ptolemy wholeheartedly was to be obscurantist, even reactionary; a modified Tychonianism was the only way out for those not prepared to accept a moving earth.

All this discussion, together with the invention of the astronomical telescope and the great step forward in mechanics—both by Galileo—made the mathematical and physical sciences stand well in advance of the biological sciences. Galileo's fruitful mind, indeed, posed numerous physical questions, which his successors saw how to unravel; and acoustics, mechanics, optics, pneumatics all owe much to Galileo and his followers. With Bacon, he became the patron saint of the new philosophy in England, where the scientific revolution reached its finest stage in the work of the Royal Society and especially of its most celebrated Fellow, Isaac Newton. The best scientists even of the Royal Society were primarily concerned with the physical sciences, not the biological sciences; this may in part explain why the biological sciences advanced more slowly. Perhaps, too, Auguste Comte was right when he said that the biological sciences can only develop when the mathematical sciences are fully matured. Be that as it may, the biological sciences indeed advanced less rapidly, though they were not without their achievements. The work of the sixteenth-century anatomists had involved much physiological discussion, but no startling advances had been made. The great achievement of the scientific revolution was to elucidate the structure and function of the blood vessels—heart, veins, and arteries—to demonstrate their interconnection, and to establish further the effect of respiration on the blood supply. The great achievement was Harvey's conception of the circulation and his convincing experimental, observational, and rational proof of its existence. What loose ends he left were tidied up by two very able anatomists, Malpighi and Richard Lower. Lower was also interested in experimental physiology, an interest which such physical scientists as Hooke, Boyle, and Wren shared. There were numerous experiments at Oxford and before the Royal Society in London on the effect of injecting substances into the veins. Medical men in Italy and Germany also experimented with this interesting problem, but little emerged. In the 1660's came daring and naive experiments on blood transfusion. They ceased promptly with the inevitable death of the poor madman in Paris into whose veins sheep's blood was introduced in the hope of calming his frenzy. Otherwise there was much for medical men still to do in pure anatomy, as work like that of De Graaf on the human reproductive system showed.

The microscope offered a new world to the anatomist. Harvey had worked without one, but his successors employed it effectively. This new instrument was a fascinating means of discovery, as Henry Power (physician and amateur scientist) and Hooke (professional scientist and a physicist) showed in their respective books. Power's descriptions were the more popular; any amateur could, with him, marvel at the appearance of a butterfly's wing. Hooke's were more exact, and his observations incessantly led him into digressive discussions of physical science, so that *Micrographia* was most influential, and is best remembered, because of Hooke's treatment of optics and combustion. Soon, however, biologists, especially those concerned with the small, found in the microscope a superb instrument of discovery. Witness Malpighi's work on the silkworm; Swammerdam's elucidation of the life cycle of insects; Leeuwenhoek's exploration of omnipresent "animals" (microorganisms) so small as to be invisible to the unaided eye. Where the microscope of the day could not see, careful experiment might demonstrate, as with Redi's clever proof that maggots on decaying meat came out of eggs laid by flies, not out of the putrefaction itself. The smallest of living creatures were thus detected and studied in detail, and they had to be fitted into the newly developing science of taxonomy. In the hands of a series of men from John Ray in the later seventeenth century to Linnaeus in the mid-eighteenth century, taxonomy was to become an absorbing passion as zoologists and above all botanist strove to cram the whole range of living things into some scheme of classification, to fathom, as they thought, the plan that God had had in mind at the Creation. They could not know that it was doomed to failure because of the existence of the complex relations developed in the course of slow evolutionary processes. Instruments played a great role in developing various sciences. Like the telescope and the microscope, the barometer and, even more, the air pump opened a whole new world of exploration. Experiments *in vacuo* were not easy to devise or perform: far and away the best were those made by Boyle with the aid of his specially designed air pumps. All previous conjectures about the physical nature of air were put by him to the test and proved or disproved. He measured the elasticity of the air, giving us Boyle's Law; explored the relations of air and respiration and air and combustion; made the first step toward the discovery of the production of air in fermentation; and produced dozens of fascinating new experiments and new problems associated with strange or anomalous conditions. Boyle and his air pump became the symbols of the new experimental science.

An extremely important ingredient of the scientific revolution was a deep concern with methodology. With success in discovery and invention came a profound belief that continued success was possible through the further discovery of a scientific method, which was presumed to be unique and at the same time universal. The most important men who wrote specifically on this subject were two very different philosophers, the French mathematician René Descartes and the English lawyer, historian, and writer, Francis Bacon. Descartes represented and codified the tradition of formal rigorous philosophic thought. Bacon was a looser, less consistent figure. Both had a vision of the potentialities of contemporary science that was enormously influential.

Both Bacon and Descartes saw natural philosophy as the most urgent and profitable field for investigation; both saw it yielding its secrets rapidly if the proper methods of research were applied; both insisted that the world is utterly rational, composed of nothing but matter and motion, and that only rational, antimystical methods of thought could therefore comprehend it. Otherwise they differed profoundly in their views on science and scientific method. Descartes' vision was of a philosophy—and a natural philosophy—constructed and understood on the basis of reasoned conviction. He had a high view of the potentialities of the human mind, even the untrained human mind, for recognizing what is logical and reasonable and rejecting what is not. He had arrived at his own convictions by searching for the simplest, most self-evident principles (as the fact of his own existence, and hence of the existence of God). From such principles, step by logical step, he examined each for its logical connection with the previous one, building up a series of ever more complex metaphysical, physical, cosmological, even biological conclusions, until (as in the suppressed treatise *Le Monde*) he had fabricated a universe which was very like, perhaps even identical with, the real universe. Descartes, laying stress upon rational, deductive argument, established the method of thought known to most seventeenth-century philosophers (Spinoza, for example) as "mathematical"; this meant imitating the logic and the rigor of mathematics, not its symbolism or numerical method. There was a place for sense experience in Descartes' system, but it necessarily was a relatively small one. Sense experience could help produce the conviction that some logically derived facts were true; sense experience was a determining factor when two equally logical but differing hypotheses were in question; and sense experience discovered new facts, thereby opening up new fields for logical investigation. But sense experience at best could confirm it could never prove or even aid substantially in proving any scientific or philosophic theory.

To Bacon, could he have known of it, Descartes' method would have seemed what it was for the later Baconians—the method of *a priori* hypothesis, speculative building of systems on the insubstantial basis of mere logical thought. Bacon, very curiously for a university-educated lawyer, saw empiricism as the only hope, perhaps because he was not interested in mathematics. Where Descartes began with logical conviction, Bacon began with empirically determined fact. Though apparently less radical than Descartes, and outwardly ready to retain more of past learning, Bacon was really more novel in his approach: never before had anyone argued that the best way to arrive at truth was through the fallible, readily confused senses. But Bacon perceived that empiricism need not consist of crude, casual sense experience. He saw the possibility of controlled, well-thought-out experiment, through which the secrets of nature could be revealed to the inquiring and thoughtful man. Bacon did not reject hypotheses. He valued the fact-gathering ant higher than the metaphysical, web-spinning spider, for he thought that empirically determined fact must come first. He rejected *a priori* hypotheses—those not grounded on empirical evidence—in favor of hypotheses based on sense experience. And further, he insisted that all hypotheses before acceptance must be tested by specially devised, "crucial" experiments, by which they should stand or fall. Bacon had an exalted view of the potentialities of fact; he thought it would be a very useful (though lowly) occupation for men everywhere to gather "histories of nature," which could then be sorted out and used by others. He had (not unnaturally, for he was no scientist) little conception of the complexities of the true experimental method. But he had an inspired dream of its possibilities in an age when few men did so, and he inspired later generations of both French and English scientists. Above all, his works encouraged men to popularize science, or rather to render the discoveries of scientists familiar to educated men.

Bacon died in 1626. Twenty-five years later young men in London were meeting to seek relief from civil strife in the discussion of scientific ideas, meetings rendered momentous by the fact that it was many of these same men who decided in 1660 after the restoration of King Charles II to organize the Royal Society of

London, the oldest continuous scientific society in the world, whose earliest members were much concerned with the methodology of science. These men, as John Wallis the mathematician remembered it, saw themselves as having been inspired jointly by Bacon and Galileo. Bacon inspired them to a profound belief in experiment; Galileo to a conviction of the importance of mathematical physics, mechanics, observational Copernicanism, and perhaps experiment.

It is impossible to tell now how far these English mathematicians, astronomers, medical men, and virtuosi (as educated amateurs were then called) thought of Galileo as an experimentalist. Indeed it is a point upon which modern historians are by no means agreed. At one time, taking Galileo literally, it was assumed that he had arrived at the truth of the law of falling bodies through experiment. Later, especially through the influence of the great French historian of science, Alexandre Koyré, writing in the 1940's and 1950's, the view that Galileo performed more thought experiments than real experiments became accepted, and it was realized that he could not have derived the law of falling bodies empirically. Lately it has been shown that Galileo's description of experiments is accurate and correct, so that there is a tendency once again to argue that he was something of an empiricist. But Pascal a few decades later was quite capable of accurately describing experiments that he did not and could not perform (since they might involve a man's sitting in ten feet of water for some considerable length of time); so it appears not impossible that Galileo, who clearly would have been a clever experimentalist if he had been an empiricist, could also have described accurately the conditions of experiments he never performed. For a man like Galileo, who believes that the book of nature is written in mathematical language, is not likely to lay great emphasis upon experiment as a key to discovery. It is more than probable that the first man to use experiment to gather data for a mathematically expressed law was a true empiricist, Robert Boyle, who perhaps did not quite realize what a mathematical expression his law truly is.

If Wallis and his friends meeting at Gresham College in the late 1640's thought of Galileo as a kind of mathematical Baconian it would not, indeed, be surprising. Certainly they valued his telescopic observations very highly as providing important support for the Copernican system (they do not offer any *proof*), though they valued his work on dynamics even more highly. They did not share

Descartes' view that Galileo should have looked for the cause of gravity before he described how bodies behaved under its influence. Indeed, they tended to distrust Descartes' tendency to seek total explanations (or "final causes"). Perhaps they saw also that although Descartes sought to be as rigorous in physics as in mathematics, and although his description of body as extension meant that physics was a kind of geometry, he did not, in fact, present a mathematical physics: his *Principia Philosophiae* was purely descriptive. It was rather Galileo who showed how to describe mathematically an empirically verifiable and qualitative world. Yet Descartes had a profound influence in England as well as in France, especially among the younger men. Boyle, Hooke, and Newton successively saw much to admire and emulate in Descartes, though all rejected many of his ideas. Hooke, the greatest empiricist of the three, came closest to accepting Descartes' detailed views, especially his views on the nature of matter, the existence of the ether, and the conviction that a truly rational explanation cannot be overthrown by "mere" empirical evidence. Huygens, the Dutch scientist—who, as the leading member of the French Academy of Science after 1666, shaped French ideas and activities—was very like Hooke in his combination of Baconianism and Cartesianism, though a far abler mathematician. He was by no means out of place as an early Fellow of the Royal Society, for he believed in and practiced, experiment, and though he was not an empiricist he was not a blind Cartesian either. He also made decisive contributions to mathematical physics; for that very reason his work is not readily accessible to the nonscientific reader.

The best-known early Fellow of the Royal Society, Robert Boyle, had no pretensions to being a mathematical physicist. Though he frequently applied quantitative methods, his real interest was in experimental science. He was far more of an empiricist than Hooke or Huygens, for he truly believed that experiment could prove as well as confirm or falsify a hypothesis or theory. Yet Boyle too was profoundly influenced by Descartes, and it was his reading of Descartes that led him to accept and extend the mechanical philosophy. Boyle devoted much of his life to trying to prove that the universe really was a vast, divinely created machine in which the particles of matter, following the laws of motion, alone accounted for the observed properties of the external world as we perceive it. Boyle's idea of proof was to provide hundreds, even thousands, of examples of experimental evidence from the realm of chemistry (a subject doubtfully a part of natural philosophy until he took it up and made it so).

Newton, greatest of all the English scientists, was to read Descartes and Hooke and Boyle in his earliest creative years, and Galileo only a little later. From Descartes he drew the great vision of a mechanical universe based upon definitions of matter and laws of motion; from Boyle and Hooke he learned experimental fact; from Boyle a profound, Baconian distrust of *a priori* hypothesis and a predeliction for induction (the deriving of hypothesis and theory from empirical evidence); from Galileo he derived the concept of a mathematical universe to be properly described only in the language of mathematics. So Newton's *Principia* differs from Descartes'. It was *The Mathematical Principles of Natural Philosophy,* presented in mathematical language, but with every effort taken to ensure that it described the real world as it really existed and was no romance or dream, however enticingly rational. This is the Newtonian triumph.

The scientific revolution demonstrated to its successors that man could understand the universe because it was eminently rational, reasonable, nonmystic, mathematical, and investigable through the mechanism of empiricism. In their conviction that the methods of mathematical physics and astronomy were the correct ones, the natural philosophers of the seventeenth and eighteenth centuries often went astray. Descartes' mechanistic physiology and psychology—in which sensation derives from the effect on the mind of a mechanical stimulus to the nervous system—denied sensation to all animals except man. Borelli successfully applied mechanical concepts to the motions of the arms, legs, and wings but failed in his attempt to do so for the motion of the blood. Eighteenth-century physiologists produced wonderfully mechanical systems, but these did not increase their knowledge of the true workings of the human body. Similarly, early Newtonian chemists tried unsuccessfully to apply the inverse square law of attraction to explain chemical bodies. More useful were the tables of affinity drawn up by the French chemist Geoffroy, after a hint in Query 31 of *Newton's Opticks,* though Geoffroy never said what he thought affinity might be. Even these mistaken enthusiasms merely demonstrate how eagerly and effectively the message of the scientific revolution was received and spread. Eighteenth-century thinkers were unable to conceive a universe except in the terms formulated by the great minds of the scientific revolution, a mechanical universe of matter and motion, reasonable, rational, obeying fixed mathematical laws, to be ascertained by means of experiment.

Though this was eminently a rational, mechanical world, it was not a coldly mechanistic one. Paradoxically, the scientists who admired the ancient atomists for creating a mechanical *physical* world had no desire to emulate them in ascribing the origin of the world to chance or the fortuitous concourse of atoms. They were no atheists, nor were they as mechanistic in their view of man as was Hobbes, who, great philosopher that he was, never understood the new science developed in his old age. (Though he died in 1679, he was after all born in 1588; and though he read Galileo's *Dialogues* and Descartes' *Discourse on Method* he never understood the aims and methods of the creative scientists of the 1660's.) The men who made the scientific revolution—from Galileo to Newton —were all profoundly devout, and not merely conventionally so. They all felt that the contemplation of God's omnipotence and the study of His works were proper occupations for a natural philosopher. When, like Boyle, they insisted that it added to man's notion of God's supremacy to think that God had created the world by creating matter and motion and natural law, and that it did not require a series of miraculous interventions to keep it going, they were not just making a conventional bow toward proper religion but genuinely meant what they said. Newton thought, and said, that natural philosophy rightly led ultimately to the "First Cause, which is certainly not mechanical" because the ultimate cause of all things is God's will. His contemporaries agreed with him and knew that he was only following precedent. Only their enemies ever accused seventeenth-century scientists of being "atheists"; they were one and all truly devout men, many of them of the clergy, and they awarded due place to theology. They only insisted that, as Galileo had tried but failed to establish, God meant the book of nature to be read direct, not through the intermediary of Church doctrine and the literal words of the Bible.

It has sometimes been argued, indeed, that there was a direct connection between religious and scientific conviction, and that men who were radical in one were radical in another. This thesis— that Protestantism or (in the case of Protestant countries) Puritanism encouraged science—was widely held in the 1930's by such disparate thinkers as the American sociologist Robert Merton, the Belgian historian of science Jean Pelseneer, and the American historian Dorothy Stimson. It is an interesting and at first sight compelling thesis, strengthened by having its roots in a similar one in economic history strongly maintained by Max Weber and generally known to most historians through the work of R. H. Tawney. But although it was carefully studied and documented it cannot be sustained in the light of cold evidence. Tycho Brahe and Kepler were Protestants; the first was an anti-Copernican, the second a passionate pro-Copernican; Copernicus himself was a Catholic Churchman, Galileo a devout son of the Church; Huygens was Protestant but Descartes was Catholic; Gassendi, the reviver of Epicurean atomism (usually regarded as atheistical), was a minor Catholic cleric; and so on. In England John Wilkins, Seth Ward, and John Wallis certainly supported the Parliamentary side in the English Civil War, but they welcomed King Charles II at his restoration, when Wallis was confirmed in all his positions while Wilkins and Ward both became bishops. Robert Boyle had a tender conscience about oathtaking, but he showed no other Puritanical tendencies and was able to get on happily with both sides, secure in the knowledge that his family's outward loyalities were so evenly divided that they were all secure whichever side won. Newton was an exception in being an Arian (a kind of Unitarian), but Newton was always an exception; and the reasons for his religious convictions were as logical and scholarly as those for his scientific convictions. On the whole, true Puritanism does not seem to have encouraged science nor (in spite of Christopher Hill's massive barrage of names) is there any evidence that the political Puritans cared for science or were influenced by it. Excessive belief in the strictest religious dogma leaves little room for the frivolity of curiosity or the delights of an inquiring mind.

Somewhat akin to the thesis that Puritanism and science were intimately connected is the thesis that science and technology were intimately connected—either the view that the scientist derived his problems from the advancing technology of the day, science developing as the scholar met the craftsman in his study, or the less commonly held view that when what the scientist did could be applied, he must have worked to this end. In fact it is a far better thesis for some aspects of Renaissance science than it is for seventeenth-century science; mathematicians and astronomers in the fifteenth and sixteenth centuries genuinely had reacted to a practical need in developing navigational astronomy and mathematical cartography. But this had nothing to do with the main problems of current astronomy and contributed nothing to their solution. Copernicus might justify his efforts by suggesting that they would be useful for calendrical computation, but in fact the great Gregorian reform of the calendar in 1582 was effected by mathematicians who were emphatically not Copernicans.

Yet there was a feeling in the seventeenth century that there ought to be some connection between science and practical affairs. Bacon had insisted that the scientist could and should learn from the craftsman, pointing out that the crafts improved with time as the sciences, he thought, had not. He also prophesied that one day scientists would understand Nature well enough to command her, so that science might be used "for the relief of man's estate." These Baconian views profoundly influenced many amateurs and semiamateur scientists, the "virtuosi" who found science an interesting avocation and hoped to see it applied to agriculture, cider–making or inventions. These ideas also influenced the early Fellows of the Royal Society. They spent much effort in the collection of "Histories of Trades" (accounts of craft processes, from brewing to calico printing, glassmaking, coal mining, and so on), but it was never clear what was achieved through these efforts, or what might be achieved. The scientist's best aid to the craftsman was, as Boyle pointed out, the invention of new instruments and an increase in the demand for all scientific necessities and equipment. Even those who wrote most on the useful aspects of science never pretended to find any close connection between their most important theoretical work and technology, nor did they expect their ideas to be applied for direct and immediate use. They saw pure science as existing for scientists, though they were eager to increase the numbers of educated men who understood science and labored much to that end.

Indeed, as Bacon had foreseen, the use of experiment did help much to democratize science. Any reasonably adroit man (or woman) could perform a simple experiment, aided by a reasonably competent set of instructions. So Samuel Pepys and his young wife learned to manage their new microscope by reading the relevant sections of Henry Power's *Experimental Philosophy,* and "mighty pleased" they were at their accomplishments. Other amateurs went further than Pepys, and perhaps made minor discoveries, just as educated travelers would supply novel information about strange lands. Above all, the instructed layman could, with assistance, understand the discoveries of theoretical and mathematical science. It required an experimenter of genius to make truly new discoveries like those of Galileo with the telescope or Leeuwenhoek with the microscope. It required both superb experimental technique and theoretical understanding to plan and then to interpret pneumatic investigations or Newton's optical discoveries. Not all purely experimental science, even in the seventeenth century, was composed of random examination of a wide variety of objects; one has only to compare the use of the microscope by experimental biologists as an aid to planned investigation—as in the anatomical investigations of Malpighi or the entomological studies of Swammerdam—with the more random investigations of Leeuwenhoek, Power, and Hooke. Both kinds of experimental science were at this stage very fruitful, and both were readily comprehensible. All men might admire the Newtonian achievement set out so brilliantly in the *Principia,* but only mathematical physicists could advance upon it, or even comprehend it. Most men, even very many scientists, could only admire, then turn for enlightenment to *Opticks* or to Newton's optical papers published in the *Philosophical Transactions,* for the optical work was easier to grasp. The eighteenth-century *Newton for the Ladies* (in Italian, by Count Algarotti) was correctly given the title *Newton on Light and Colour* in its English translation. Newton's mechanical and cosmological work became accessible to the nonmathematician in the eighteenth century through the work of English, Dutch, and French scientists who presented Newton without mathematics for the reader with some knowledge of science. These were followed by a host of popularizers presenting to the layman some slight idea of the intellectual achievements of the scientific revolution which all could admire, if not really understand.

It must not be forgotten that the scientific revolution was, in the last analysis, an intellectual revolution—a revolution in what men thought about and the way in which they did this thinking. And however much it may be true that every man is a product of his times and influenced by the economic and social developments of those times, new ideas must ultimately emerge out of the intellectual climate rather than the purely social climate, for every man must build on the ideas of his predecessors. Alexandre Koyré illuminated the work of Galileo more by his *Galilean Studies* (*Études Galiléennes,* Paris, 1939)—in which he analyzed Galileo's ideas on moving bodies in the light of the ideas on the same subject by his predecessors and contemporaries—than all the discussion of the previous decade upon the social and economic forces at work in Galileo's lifetime. It is interesting to know that Galileo visited the arsenal at Venice, or that he learned from practical men that suction pumps could raise water only thirty feet; but neither of these facts contributes as much to our understanding of Galileo's mode of thought or of his achievements or his novelty as does Koyré's analysis of purely intellectual factors. Further, though *inventions* are things that may occur to more than one man at a particular time, and even broad general ideas may be simultaneously discovered by several men, the working out of such ideas once discovered is idiosyncratic and belongs to one man alone. Other men besides Galileo, Descartes, and Newton worked out *some* of their ideas and discoveries, but without Galileo, Descartes, and Newton the scientific revolution would have taken a very different shape. Inventions, once they leave the mind of the inventor, have an independent existence; ideas are forever shaped by the individual mind which conceived them and which expressed them in writing in a particular way. When all is said and done the scientific revolution was an intellectual revolution, and its roots and causes must be sought in intellectual developments.

Similarly, to appreciate the scientific revolution we must try to understand the modes of thought and expression in which the discoveries, ideas, and achievements were cast. Some of these will be found in the documents printed in this book. No selection, of course, can ever give the complete meaning of the whole. To appreciate fully the scientific revolution one must try to read as attentively as the men concerned did themselves both the successes and the failures, the novel attacks on new problems and the traditional restatement of old problems. Failing that, one can sample. Here, in mainly contemporary language, are some specimens of what is available.

I.

A New Age and an Old Tradition

EVEN the scientific revolution did not force men's minds to forsake the old completely, nor indeed did those who made it wish to do so. The scientists who proposed the clearest break with the past were very often traditionalists, while the men who wished to sever all connections with contemporary thought were often the least original. It is fitting that we should begin here with the astronomical revolution for two reasons: first because, as indicated in the Introduction, astronomy in the sixteenth century was the most widely understood and popular of sciences, and second because Copernicus thought of himself as reviving the best traditions of Greek astronomy. His *De Revolutionibus Orbium Coelestium* (*On the Revolutions of the Celestial Orbs*) could have been written by no one else in the first half of the sixteenth century because no astronomer beside Copernicus had the mathematical ability to write a great new work on mathematical astronomy. But others beside himself had thought that if Ptolemy's theory was unsatisfactory the solution might be found by considering other Greek suggestions for the construction of the universe and then applying Ptolemy's mathematical techniques, following established tradition.

Copernicus' *De Revolutionibus,* published after a lifetime's cogitation in 1543 (there is a tradition that Copernicus only saw a copy on his deathbed), is a difficult book, intended for the mathematically competent specialist. Few moderns have read more than the Dedication to Pope Paul, the Preface, and the general discussion in Book I. These have been often translated in modern times, are readily available, and for the thoughtful reader are of immense interest. But few contemporaries read even this much; nor did they need to do so for a general view of the Copernican system. They could read Copernicus' own early account (*The Commentariolus*) or the account of his disciple Rheticus (*Narratio Prima*), both now conveniently available in the Dover *Three Copernican Treatises* (edited by E. Rosen). They could read popular treatises which mentioned the main outlines of Copernicus' theory, and they could read incidental discussion in works dealing with other aspects of science.

Some examples of such sources are given in the selections below, which also attempt to give some idea of the spread of interest in the Copernican system before 1600, a period in which there may well have been more non-technical than technical discussions of the new and paradoxical theory. The first two selections were accessible to the nonlearned of the age, at least in England; the third was translated from Latin only in the late nineteenth century, but it was well known and quite widely read among the truly literate—those who had at least been to a grammar school. Each of these selections demonstrates the balance between old and new; the reverence for past giants and the search for new ones, often blended with the strange and even the mystic, is well displayed in each. The first selection is a sixteenth-century translation of Chapters 7 through 10 of Book I of De Revolutionibus by Thomas Digges (c. 1543–1595). Digges was a skilled astronomer and applied mathematician. In 1576 he undertook to issue a new edition of a very popular perpetual almanac first published by his father, Leonard Digges, in 1556. *A Prognostication Everlasting*, a cheap and handy little work, was originally, of course, based upon Ptolemaic astronomy. But intellectual conviction overcame filial piety, and Thomas Digges brought his father up to date by adding to the *Prognostication Everlasting* the *Perfect Description* printed below. It is the most complete vernacular account of the Copernican system before the time of Galileo and fully describes the main features of the system together with a refutation of the chief anti-Copernican arguments. In addition, it contains the first suggestion that the fixed stars are not attached to a crystalline, fixed celestial sphere but are scattered at varying distances. Digges quaintly combined the astronomical and theological "heavens" to suggest that the fixed stars extended as far as the divine heaven, to provide the jeweled lights among which the blessed souls of the Christian elect moved in eternal bliss. The text is presented complete, except for the omission of a few Latin tags and Digges' often reproduced diagram of his system.

The second selection, a translation of a popular French cosmological poem, *The Week, or the Creation of the World,* was published by Guillaume du Bartas (1544–1590) two years after Digges' defense of Copernicanism. Du Bartas, a French poet and soldier-adventurer, was thoroughly at ease in the old traditional cosmology; he saw no connection between the Copernican system and Paradise; rather he found Copernicanism was destructive of theology and manifestly contrary to common sense. Very likely du Bartas believed that a stone thrown upward from the ship *would* fall astern and that the atmosphere *would* be left behind by a revolving earth; in any case these were apparently clinching arguments, though astronomers and scientists knew better. The English translation of this poem by Joshua Sylvester, first published in 1605, was very widely read and familiarized many with the outlines of the Copernican system. Here du Bartas describes the traditional universe, from the earth in the center to the Zodiac and the constellations of fixed stars.

The third selection, *On the Magnet* (*De Magnete,* 1600), gives some idea of the use that might be made by scientists who accepted the Copernican system. William Gilbert (c. 1540–1603), physician to Queen Elizabeth, obsessed with the peculiar properties of loadstones (naturally occurring magnets) and the magnetic compass, a passionate experimenter, interested in improving navigation, was also a Renaissance natural magician who believed that mysterious forces of nature like magnetism could be studied and controlled. Having discovered that the earth itself was a magnet (which was, then, the reason why a magnetized compass needle oriented itself in a north-south line), he concluded that its magnetic properties gave the earth the possibility of motion (which is the significance of "soul" in the Aristotelian sense). Hence he used his experimental discoveries about magnetism to promote one portion of the Copernican theory—the diurnal motion of the earth on its axis. (it is not known whether Gilbert also accepted the annual revolution of the earth about the s   Gilbert's use of magnetism in astronomy was to be adopted by Kepler a few years later as a means of explaining what kept the moon in her orbit around the earth, so that it has some importance in the history of mechanics, as well as being an interesting example of a very characteristically Renaissance form of thought.

1. A Sixteenth-Century Exposition of Copernicanism

A prognostication everlasting of right good effect, fruitfully augmented by the author, containing plain, brief, pleasant, chosen rules to judge the weather by the Sun, Moon, Stars, Comets, Rainbow, Thunder, clouds, with other extraordinary tokens, not omitting the Aspects of Planets, with a brief judgment for ever, of Plenty, Lacke, Sickness, Dearth, Wars &c. opening also many natural causes worthy to be known. To these and other now at the last, are joined divers general, pleasant Tables, with many compendious Rules, easy to be had in memory, manifold ways profitable to all men of understanding, Published by Leonard Digges Gentleman. Lately corrected and augmented by Thomas Digges his son. Imprinted at London by Thomas Marsh. 1576.

SOURCE: Thomas Digges, *A Perfect Description of the Celestial Orbs* (London, 1576) annexed to Leonard Digges, *A Prognostication Everlasting.*

A Perfect Description of the Celestial Orbs,
According to the Most Ancient Doctrine of the Pythagoreans
Lately Revived by Copernicus
and by Geometrical Demonstrations
Proved to the Reader

Having of late (gentle Reader) corrected and reformed sundry faults that by negligence in printing have crept into my father's *General Prognostication*: Among other things I found a description or model of the world and situation of Celestial and Elementary Spheres according to the doctrine of Ptolemy, whereunto all Universities (led thereto chiefly by the authority of Aristotle) sithens have consented. But in this our age one rare wit (seeing the continual errors that from time to time more and more have been discovered, besides the infinite absurdities in their theories, which they have been forced to admit who would not confess any mobility in the ball of the Earth) hath by long study, painful practise, and rare invention delivered a new theory or model of the world, showing that the Earth resteth not in the centre of the whole world, but only in the centre of this our mortal world or globe of elements which, environed and enclosed in the Moon's orb, and together with the whole globe of mortality is carried yearly round about the Sun, which like a king in the midst of all reigneth and giveth laws of motion to the rest, spherically dispersing his glorious beams of light through all this sacred celestial temple. And the Earth itself to be one of the planets, having its peculiar and straying courses, turning every 24 hours round upon its own centre, whereby the Sun and great globe of fixed stars seem to sway about and turn, albeit indeed they remain fixed. So many ways is the sense of mortal men abused, but reason and deep discourse of wit having opened these things to Copernicus, and the same being with mathematical demonstrations most apparently by him delivered to the world, I thought it convenient together with the old theory also to publish this, to the end [that] such noble English minds as delight to reach above the baser sort of men might not be altogether defrauded of so noble a part of Philosophy. And to the end it might manifestly appear that Copernicus meant not, as some have fondly excused him, to deliver these grounds of the Earth's mobility only as feigned mathematical principles and not as philosophical [ones] truly averred,[1] I have also, from him, delivered both the philosophical reasons produced by Aristotle and others to maintain the Earth's stability, and also their solutions and insufficiency, wherein I cannot a little commend the modesty of that grave Philosopher Aristotle, who seeing (no doubt) the insufficiency of his own reasons in seeking to confute the Earth's motion, useth these words: This is the best explanation I can offer; howbeit his disciples have not with like sobriety maintained the same. Thus much for my own part in this case I will only say: there is no doubt but of a true ground truer effects may be produced than of principles that are false, and of true principles falsehood or absurdity cannot be inferred. If therefore the Earth be situated immovable in the centre of the world, why find we not Theories upon that ground to produce effects as true and certain as these of Copernicus? Why cast we not away those equant circles[2] and irregular motions, seeing our own Philosopher Aristotle himself, the light of our universities, hath taught us: The motion of simple bodies ought to be simple. But if it be found [on the] contrary impossible (the Earth's stability being granted) but that we must necessarily fall into these absurdities, and cannot by any means avoid them: Why shall we so much dote on the appearance of our senses, which many ways may be abused, and not suffer ourselves to be directed by the rule of reason, which the great *GOD* hath given us as a lamp to lighten the darkness of our understanding and the perfect guide to lead us to the golden branch of verity amid the forest of errors.

Behold a noble Question to be of the Philosophers and Mathematicians of our universities argued not with childish invectives, but with grave reasons Philosophical and irreprovable Demonstrations Mathematical. And let us not in matters of reason be led

1. *De Revolutionibus* was given by Copernicus to his disciple Rheticus to see through the press. He in turn delegated his responsibilities to a Lutheran clergyman Andreas Osiander who, apparently shocked by the novelties of the Copernican system, wrote a preface stating that the author intended his system as a mathematical hypothesis not as a true depiction of the physical universe. As this preface was unsigned it was often taken to have been written by Copernicus himself; evidently Copernicans were not deceived.

2. The equant was a point, not the center of its orbit, in relation to which the motion of a celestial body was uniform. It was one of the merits of the Copernican system that it dispensed with the equant, though the problem was not wholly clarified until Kepler enunciated his law.

away with authority and opinions of men. . . . The globe of elements enclosed in the orb of the Moon I call the globe of mortality because it is the peculiar empire of death. For above the Moon they fear not his force. . . .

In the midst of this globe of mortality hangeth this dark star or ball of earth and water balanced and sustained in the midst of the thin air only with that property which the wonderful workman hath given at the Creation to the centre of this globe with his magnetical force vehemently to draw and hale unto itself all such other elemental things as retain the like nature. This ball every 24 hours by natural, uniform and wonderful sly and smooth motion rolleth round, making with his period our natural day, whereby it seems to us that the huge, infinite, immovable globe should sway and turn about.

The Moon's orb that environeth and containeth this dark star and the other mortal, changeable, corruptible elements and elementary things is also turned round every 29 days 31 minutes 50 seconds, 8 thirds, 9 fourths, and 20 fifths, and this Period may most aptly be called the Month. The rest of the Planets' motions . . . [3] shall more largely be hereafter spoken of.

Herein good Reader, I have waded farther than the vulgar sort demonstratively and practically, and God sparing life, I mean, though not as a judge to decide, yet at the mathematical bar in this case to plead in such sort, as it shall manifestly appear to the world whether it be possible upon the Earth's stability to deliver any true or probable theory, and then refer the pronouncing of sentence to the grave senate of indifferent discreet mathematical readers.

Farewell and respect my travails as thou shalt see them tend to the advancement of truth and discovering the monstrous loathsome shape of error.

A PERFECT DESCRIPTION OF THE CELESTIAL ORBS

Although in this most excellent and difficult part of philosophy in all times [there] have been sundry opinions touching the situation and moving of the celestial bodies, yet in certain principles all philosophers of any account, of all ages, have agreed and consented. First, that the orb of the fixed stars is, of all others, the most high, the farthest distant and comprehendeth all the other

---

3. There is a reference here to a diagram on which is inscribed the period of revolution of each planet.

spheres of wandering stars. And of these straying bodies called *Planets* the old philosophers thought it a good ground in reason that the nighest to the centre should swiftliest move, because the circle was least and thereby the sooner overpassed, and the farther distant the more slowly. Therefore as the Moon, being swiftest in most high, the farthest distant, and comprehendeth all the other course, is found also by measure nighest, so have all agreed that the orb of Saturn, being in moving the slowest of all the Planets, is also the highest: Jupiter the next, and then Mars, but of Venus and Mercury there hath been great controversy, because they stray not every way from the Sun as the rest do. And therefore some have placed them above the Sun, as Plato in his *Timaeus:* others beneath, as Ptolemy, and the greater part of them that followed him. Alpetragius[4] maketh Venus above the Sun and Mercury beneath, and sundry reasons have been of all sides alleged in defence of their opinions. They that follow Plato (supposing that all stars should have obscure and dark bodies shining with borrowed light like the Moon) have alleged that if those Planets were lower than the Sun, then should they sometimes obscure some part of the body of the Sun, and also shine not with a light circular but segmentary, and that variable as in the Moon: which when they see by experience at no time to happen, they conclude with Plato. On the contrary part such as will maintain them beneath, frame a likelihood by reason of the large space between the orbs of Sun and Moon: for the greatest distance of the Moon is but 64½ semidiameters of the Earth and to the nighest of the Sun are 1160, so that there remaineth between the Moon and the Sun 1095 semidiameters of the Earth. And therefore that so huge a space should not remain empty, there they situate the orbs of Mercury and Venus. And by the distance of their Apsides, [5] whereby they search the thickness of their orbs, they find that they of all the rest best answer that situation, so as the lowest of Mercury's orb may reach down almost to the highest of the Moon's, and the top of Mercury to the inferior part of Venus' sphere, which with his Apsis should reach almost to the Sun. For between the Apsides of Mercury by their Theories they postulate 177 semidiameters of

4. Or al-Bitrugi, a Moslem living in Spain; his astronomical treatise was written about 1180.
5. The points on the planet's orbit when it was at its highest or lowest position in relation to the earth (or, for a Copernican, the sun).

the Earth and then the crassitude of Venus' orb, being 910 semi-diameters, doth very nigh supply and fill the residue. They therefore will not confess that these Planets have any obscurity in their bodies like the Moon, but that either with their own proper light, or else being thoroughly pierced with solar beams, they shine and show circular.[6] And having a straying course of latitude, they seldom pass between the Sun and us, or if they should, their bodies being so small could scarcely hide the hundredth part of the Sun, and so small a spot in so noble a light could hardly be discerned. And yet Averroes [7] in his Paraphrase on Ptolemy affirmeth that he saw a little spot in the Sun at such time as by calculation he had forecast a corporeal conjunction. But how weak this their reason is, it may soon appear if we consider how from the Earth to the lowest of the Moon's orb there is 38 semidiameters of the Earth, or by the truer computation according to Copernicus 52, and yet in all that so huge a space we know nothing but the air or fiery orb, if any such be. Again, the diameter of the circle whereby Venus should be carried nigh 45 degrees distant from the Sun, must needs be 6 times greater at the least than the distance of that circle's lowest part from the Earth; then if that whole circle comprehended within the orb of Venus should be turned about the Earth (as needs it must) if we will not attribute to the Earth any motion, we may easily consider what rule in the heavens so vast and huge an Epicycle,[8] containing a space so many times greater than the earth, air and orbs of the Moon and Mercury also, will make, especially being turned round about the Earth. Again, that reason of Ptolemy that the Sun must needs be placed in the midst of those Planets that wander from him at liberty and those that are, as it were, combined to him,[9] is proved senseless by the motion of the Moon, which we see no less to stray from the Sun than any of those other three superior Planets. But if

6. It was only after 1610 (when Galileo published his *Sidereal Messenger*) that it was clear that planets differed from stars and shone only by reflected light, like the moon.
7. A twelfth-century Spanish Moslem philosopher, widely read in Christian Europe, especially for his commentaries on Aristotle.
8. In Ptolemaic and later astronomical theory (including Copernican), the complex motion of the planets was represented by the mathematical device of locating the planet on the circumference of a rotating circle (the epicycle) whose center lay on a larger, also rotating circle (the deferent). The "orbs" on the other hand are solid crystalline spheres.
9. Venus and Mercury are never far from the sun.

they will have these two Planets' orbs within an orb of the Sun, what reason can they give why they should not depart from the Sun at large, as the other Planets do, considering the increase of swiftness in their motion must accompany the inferior situation, or else the whole order of theory should be disturbed. It is therefore evident that either there must be some other centre whereunto the order of these orbs should be referred, or else no reason in their order, nor apparent cause why we should rather to Saturn than to Jupiter or any of the rest attribute the higher or remoter orb. And therefore it seemeth worthy of consideration that Martianus Capella[10] wrote in his *Encyclopedia,* and certain other latins held, affirming that Venus and Mercury do run about the Sun in their peculiar spheres, and therefore could not stray farther from the Sun than the capacity of their orbs would give them leave, because they encompass not the Earth as the others do, but have their Apsides conversed after another manner; what other thing would they hereby signify, but that the orbs of these Planets should environ the Sun as their centre? So may the sphere of Mercury, being not of half the amplitude of Venus' orb, be situated well within the same. And if, in like sort, we situate the orbs of Saturn, Jupiter and Mars, referring them, as it were, to the same centre, so as their capacity be such as they contain and circulate the Earth also, haply we shall not err, as by evident demonstrations in the residue of Copernicus' *Revolutions* is demonstrated. For it is apparent that these Planets nigh the Sun are always least, and farthest distant and opposite are much greater in sight and nigher to us, whereby it cannot be but the centre of them is rather to the Sun than to the earth to be referred: as in the orbs of Venus and Mercury also. But if all these to the Sun as a centre be referred in this manner, then must there needs between the convex orb of Venus and the concave of Mars be left a huge space wherein the Earth and elementary frame enclosed with the lunar orb of duty must be situate. For from the Earth the Moon may not be far removed, being without controversy of all others nighest in place and nature to it; especially considering between the same orbs of Venus and Mars there is sufficient room. Therefore we need not to be ashamed to confess this whole globe of elements enclosed with the Moon's sphere together with the Earth as the centre of the

10. A late Latin writer, much read in the Middle Ages, who lived about A.D. 500.

same to be by this great orb, together with the other Planets, turned about the Sun, making by its revolution our year. And whatsoever seems to us to proceed by the moving of the Sun, the same to proceed indeed by the revolution of the Earth, the Sun still remaining fixed and immovable in the midst. And the distance of the Earth from the Sun to be such as, being compared with the other Planets, maketh evident alterations and diversity of aspects, but if it be referred to the orb of fixed stars, then hath it no sensible proportion, but [is] as a point or centre to a circumference, which I hold far more reasonable to be granted, than to fall into such an infinite multitude of absurd imaginations, as they were fain to admit that will needs wilfully maintain the Earth's stability in the centre of the world. But rather herein to direct ourselves by that wisdom we see in all God's natural works, where we may behold one thing rather endued with many virtues and effects, than any superfluous or unnecessary part admitted. And all these things, although they seem hard, strange and incredible, yet to any reasonable man that hath his understanding ripened with mathematical demonstration, Copernicus in his *Revolutions,* according to his promise, hath made them more evident and clear than the sunbeams. These grounds therefore admitted, which no man reasonably can impugn, that the greater orb requireth the longer time to run its period, the orderly and most beautiful frame of the Heavens doth ensue.

The first and highest of all is the immovable sphere of fixed stars containing itself and all the rest, and therefore fixed: as the universal place of rest, whereunto the motions and positions of all inferior spheres are to be compared. For albeit sundry astrologians, finding alteration in the declination and Longitude of stars, have thought that the same also should have its peculiar motion, yet Copernicus by the motions of the Earth salveth all, and utterly cutteth off the ninth and tenth spheres, which contrary to all sense the maintainers of the Earth's stability have been compelled to imagine.

The first of the movable orbs is that of Saturn, which being of all other next unto that infinite orb immovable, garnished with lights innumerable,[11] is also in his course most slow, and once only in thirty years passeth his period.

The second is Jupiter, who in 12 years performeth his circuit.

11. The heavens (which Digges imagined to extend into Heaven).

Mars in 2 years runneth his circular race.

Then followeth the great orb wherein the globe of mortality, enclosed in the Moon's orb as an epicycle, and holding the Earth as a centre by his own weight resting always permanent, in the midst of the air is carried round once in a year.

In the fifth place is Venus, making her revolution in 9 months.

In the sixth is Mercury, who passeth his circuit in 80 days.

In the midst of all is the Sun.

For in so stately a temple as this, who would desire to set his lamp in any other better or more convenient place than this, from whence uniformly it might distribute light to all, for not unfitly it is of some called the lamp or light of the world, of others the mind, of others the ruler of the world. . . .

Trismegistus[12] calleth it the visible god. Thus doth the Sun like a king sitting in his throne govern his courts of inferior powers: Neither is the Earth defrauded of the service of the Moon, but, as Aristotle saith, of all others the Moon with the Earth hath nighest alliance, so here are they matched accordingly.

In this form or frame may we behold such a wonderful symmetry of motions and situations, as in no other can be propounded. The times whereby we, the inhabitants of the Earth, are directed, are constituted by the revolutions of the Earth: the circulation of its centre causeth the year, the conversion of its circumference maketh the natural day, and the revolution of the Moon produceth the month. By the only view of this Theory, the cause and reason is apparent why in Jupiter the progressions and retrogradations are greater than in Saturn, and less than in Mars, why also in Venus they are more than in Mercury. And why such changes from Direct to Retrograde, Stationary,[13] etc. happeneth notwithstanding more rifely in Saturn than in Jupiter, and yet more rarely in Mars, why in Venus not so commonly as in Mercury. Also why Saturn, Jupiter and Mars are nigher the Earth in their Acronical than in their Cosmical or Heliacal rising.[14] Especially Mars, who rising at sunset, showeth in his ruddy fiery colour equal in quanti-

12. Hermes Trismegistus, a legendary Greco-Egyptian alchemist.
13. The outer planets (Mars, Jupiter, Saturn) seem when viewed from the earth to move at some points in their orbits in loops, so that they there appear to go backward ("retrograde" motion) and occasionally to stand still.
14. "Acronical" means rising at sunset; "Cosmical" at sunrise; "Heliacal" just before sunrise or after sunset, near the sun.

ty with Jupiter, and contrarywise setting little after the Sun, is scarcely to be discerned from a Star of the second light. All which alterations apparently follow upon the Earth's motion. And that none of these do happen in the fixed stars, that plainly argueth their huge distance and immeasurable Altitude, in respect whereof this great Orb wherein the Earth is carried is but a point, and utterly without sensible proportion being compared to that Heaven. For as it is demonstrated in perspective, every quantity hath a certain proportionable distance whereunto it may be discerned, and beyond the same it may not be seen, this distance therefore of that immovable Heaven is so exceedingly great, that the whole Great Orb vanisheth away, if it be referred to that heaven.

Herein we can never sufficiently admire this wonderful and incomprehensibly huge frame of God's work propounded to our senses, seeing first this ball of the Earth wherein we move, to the common sort seemeth great, and yet in respect of the Moon's orb is very small; but compared with the Great Orb wherein it is carried, it scarcely retaineth any sensible proportion, so marvelously greater is that orb of annual motion than this little dark star wherein we live. But that Great Orb being, as is before declared, but as a point in respect of the immensity of that immovable Heaven, we may easily consider what a little portion of God's frame our elementary corruptible world is, but never sufficiently be able to admire the immensity of the rest. Especially of that fixed orb, garnished with lights innumerable, and reaching up in spherical altitude without end. Of which lights celestial it is to be thought that we only behold such as are in the inferior parts of the same orb, and as they are higher, so seem they of less and lesser quantity, even till our sight, being not able farther to reach or conceive, the greatest part rest, by reason of their wonderful distance, invisible unto us. And this may well be thought by us to be the glorious court of the great God, whose unsearchable, invisible works we may partly by these His visible conjecture, to whose infinite power and majesty such an infinite place surmounting all others both in quantity and quality only is convenient. But because the world hath so long a time been carried with an opinion of the Earth's stability, that the contrary cannot but be now very impersuadable, I have thought good out of Copernicus also to give a taste of the philosophical reasons alleged for the Earth's stability, and their solutions, that such as are not able with geometrical eyes to behold the secret perfection of Copernicus' theory, may yet by these familiar, natural reasons be induced to search farther, and not rashly to condemn for fantastical, so ancient a doctrine revived, and by Copernicus so demonstratively approved.

WHAT REASONS MOVED ARISTOTLE AND OTHERS THAT THAT FOLLOWED HIM TO THINK THE EARTH TO REST IMMOVABLE AS A CENTRE TO THE WHOLE WORLD

The most effectual reasons that they produce to prove the Earth's stability in the middle or lowest part of the world, is that of Gravity and Levity. For of all others the Element of the earth, say they, is most heavy, and all ponderous things are carried unto it, striving as it were to sway even down to the inmost part thereof. For the Earth being round, into the which all weighty things on every side fall, making right angles on the superficies, must needs, if they were not stayed on the superficies, pass to the centre, seeing every right line that falleth perpendicularly upon the Horizon in that place where it toucheth the Earth must needs pass by the centre. And those things that are carried toward that medium [center], it is likely that there also they would rest. So much, therefore, the rather shall the Earth rest in the middle, and (receiving all things into it self that fall) by its own weight shall be most immovable. Again, they seek to prove it by reason of motion and its nature, for of one and the same simple body the motion must also be simple, saith Aristotle.

Of simple motions there are two kinds, right [straight] and circular: right are either up or down, so that every simple motion is either downward toward the centre or upward from the centre, or circular about the centre. Now unto earth and water, in respect of their weight, the motion downward is convenient to seek the centre. To air and fire, in regard to their lightness, upward and from the centre. So it is meet to attribute to these elements the right or straight motion, and to the heavens only it is proper circularly about this mean or centre to be turned round. Thus much Aristotle. If therefore, saith Ptolemy of Alexandria, the Earth should turn but only by the daily motion, things quite contrary to these should happen. For its motion should be most swift and violent, that in 24 hours should let pass the whole circuit of the Earth, and those things which by sudden turnings are stirred, are altogether unmeet to collect, but rather to disperse things united, unless they should by some firm fastening be kept together. And long ere this the Earth being dissolved in pieces should have been scattered through the heavens, which were a mockery to think of, and much more beasts and all other weights that are loose could not remain unshaken. And also things falling should not light on the places perpendicular under them, neither should they fall directly thereto, the same being in the meantime violently carried away. Clouds also and other things hanging in the air should always seem to us to be carried toward the West.

The Solution Of These Reasons
With Their Insufficiency

These and such other are the causes wherewith they prove the Earth to rest in the middle of the world and that out of all question: but he that will maintain the Earth's mobility may say that this motion is not violent but natural. And those things which are naturally moved have effects contrary to such as are violently carried. For such motions wherein force and violence are used must needs be dissolved and cannot be of long continuance, but those which are caused by nature still remain in their perfect state and are conserved and kept in their most excellent constitution. Without cause, therefore, Ptolemy did fear lest the Earth and all earthly things should be torn in pieces by this revolution of the Earth caused by the working of nature, whose operations are far different from those of Art or such as human intelligence may reach unto. But why should he not much more think and misdoubt the same of the world, whose motion must of necessity be so much more swift and vehement than this of the Earth, as the Heaven is greater than the Earth? Are therefore the Heavens made so huge in quantity that they might, with unspeakable vehemence of motion, be severed from the centre, lest, happily, resting it should fall, as some Philosophers have affirmed? Surely if this reason should take place, the magnitude of the Heavens should extend infinitely. For the more this motion should violently be carried higher, the greater should be the swiftness, by reason of the increasing of the circumference which must of necessity be passed over in 24 hours, and in like manner, by increase of the motion, the magnitude must also necessarily be augmented. Thus should the swiftness increase the magnitude, and the magnitude, the swiftness, infinitely. But according to the ground of nature, whatsoever is infinite can never be passed over. The Heavens therefore of necessity must stand and rest fixed. But, say they, without the Heavens there is no body, no place, no emptiness, no, not any thing at all whether the Heavens should or could extend farther. But this surely is very strange, that nothing should have such efficient power to restrain something, the same having a very essence and being. Yet if we would thus confess that the Heavens were indeed infinite upward, and only finite downward in respect of their spherical concavity, much more perhaps might that saying be verified, that without the Heavens is nothing, seeing everything in respect to the infiniteness thereof had sufficient place within the same. But then must they of necessity remain immovable. For the chiefest reason that has moved some to think the Heavens limited was motion, which they thought without controversy to be indeed in them. But whether the world have its bounds or be indeed infinite and without bounds, let us leave that to be discussed by philosophers; sure we are, that the Earth is not infinite but has a limited circumference, seeing therefore that all philosophers consent that limited bodies may have motion, and infinity cannot have any. Why do we yet stagger to confess motion in the Earth, it being most agreeable to its form and nature, whose bounds also and circumference we know, rather than to imagine that the whole world should sway and turn, whose end we know not, nor possibly can be known of any mortal man? And therefore the true motion indeed to be in the Earth, and the apparent only in the Heavens, and these appearances are no otherwise than if the Virgilian Aeneas should say, "We leave the harbor, and the land and city recede."

For a ship carried in a smooth sea doth pass away with such tranquility that all things on the shores and the seas to the sailors seem to move, and themselves only quietly to rest with all such things as are aboard them; so surely may it be in the Earth, whose motion, being natural and not forcible, is of all other the most uniform and imperceptible, whereby to us who sail therein the whole world may seem to roll about. But what shall we then say of clouds and other things hanging or resting in the air or tending upward, but that not only the Earth and sea, making one globe, but also no small part of the air is likewise circularly carried, and in like sort all such things as are derived from them, or have any manner of alliance with them: either for that the lower region of the air, being mixed with earthly and watery vapors, follows the same nature of the Earth or that it be gained and gotten from the Earth by reason of vicinity or contiguity. Which if any man marvel at, let him consider how the old philosophers did yield the same reason for the revolution of the highest region of the air, wherein we may sometimes behold comets carried circularly no otherwise than the celestial bodies seem to be, and yet hath that region of the air less convenience with the celestial orbs than this lower part with the Earth. But we affirm that part of the air in respect of its great distance to be destitute of this terrestrial motion, and that this part of the air that is next to the Earth doth appear most still and quiet by reason of its uniform natural accompanying of the Earth, and likewise things that hang therein, unless by winds or other violent accident they be tossed to and fro. For the wind in the air is nothing else but as the wave in the sea; and of things ascending and descending in respect of the world we must confess them to have a mixed motion of right and circular, albeit it seem to us right and straight; no otherwise than if in a ship under sail a man should softly let a plummet down from the top along by the mast, even to the deck: this plummet, passing always by the straight mast, seemeth also to fall in a right line, but being weighed by discourse of reason, its motion is found mixed of right and circular. For such things as naturally fall downward being of earthly nature, there is no doubt but as parts they retain the nature of the whole. No otherwise is it of these things that by fiery force are carried upward. For the earthly fire is chiefly nourished with earthly matter, and flame is defined to be nought else but a burning fume or smoke, and the property of fire is to extend the subject whereunto it entereth, the which it doth with so great violence as by no means or engines can it be constrained, but that with breach of bands it will perform its nature. This extensive motion is from the centre to the circumference, so that if any earthly part be fired it is carried violently upward. Therefore, whereas they say that of simple bodies the motion is altogether simple, of the circular is it chiefly verified, so long as the simple body remaineth in its natural place and perfect unity of composition, for in the same place there can be no other motion but circular, which remaining wholly in itself is most like to rest and immobility. But right or straight motions only happen to those things that stray or wander or are by any means thrust out of their natural place. But nothing can be more repugnant to the form and ordinance of the world, than that things *naturally* should be out of their natural place. This kind of motion therefore (that is, by right line) is only accidental to those things that are not in their right state or natural perfection, while parts are disjoined from their whole body and covet to return to the unity thereof again. Neither do these things which are carried upward or downward besides this circular moving make any simple, uniform or equal motion, for they cannot be tempered with their levity of [the] ponderosity of their body, but always as they fall (beginning slowly) they increase their motion, and the farther, the more swiftly, whereas, contrariwise, this our earthly fire (for other we cannot see) we may behold as it is carried upward to vanish and decay, as it were confessing the cause of violence to proceed only from its terrestrial matter. The circular motion always continues uniform and equal, by reason of its cause, which is indefinite and always continuing. But the other hasteneth to end and to attain that place where they leave [cease] longer to be heavy or light, and have attained that place, their motion ceaseth. . . . And whereas Aristotle hath distributed simple motion into these three kinds, *from* the centre *to* the centre, and *about* the centre, it must be only in reason and imagination, as we likewise in geometrical consideration sever a point, a line, and a superficies [surface], whereas indeed neither can stand without [the] other, nor any of them without a body.

Hereto we may adjoin that the condition of immobility is more noble and divine than that of change, alteration or instability, and therefore more agreeable to Heaven than to this Earth, where all things are subject to continual mutability. And, seeing by evident proof of geometrical mensuration we find that the planets are sometimes nigher to us and sometimes more remote, and that therefore even the maintainers of the Earth's stability are forced to confess that the Earth is not their orb's centre, this motion *about* the centre must be taken in more general sort, and that it may be understood that every orb has its peculiar *medium* and centre, in regard whereof this simple and uniform motion is to considered. Seeing therefore that these orbs have several centres, it may be doubted whether the centre of this earthly gravity be also the centre of the world. For gravity is nothing else but a certain proclivity or natural coveting of parts to be coupled with the whole, which, by divine providence of the Creator of all, is given and impressed into the parts, that they should restore themselves into their unity and integrity, concurring in spherical form; which kind of property or affection it is likely also that the Moon and other glorious bodies want not to knit and combine their parts together and to maintain them in their round shape—which bodies, notwithstanding, are by sundry motions conveyed sundry ways. Thus, as it is apparent by these natural reasons that the mobility of the Earth is more probable and likely than the stability, so if it be mathematically considered and every part of every theory examined with geometrical mensurations, the discreet student shall find that Copernicus not without great reason did propound this ground of the Earth's mobility.

2. Popular Anti-Copernicanism

The Fourth Day

Pure Spirit that rapt'st above the firmest sphere
In fiery coach thy faithful messenger,
Who, smiting Jordan with his plaited cloak,
Did erst divide the waters with the stroke,
O, take me up, that far from earth I may
From sphere to sphere see th'azure heavens today.
Be Thou my coachman, and now, cheek by jowl,
With Phoebus' chariot let my chariot roll,
Drive on my coach by Mars's flaming coach;
Saturn and Luna let my wheels approach;
That, having learned of their fire-breathing horses,
Their course, their light, their labor and their forces,
My muse may sing in sacred eloquence
To virtue's friends their virtuous eloquence.

SOURCE: Guillaume du Bartas, *La semaine, ou création du monde* (Paris, 1578); the text is from the English translation by Joshua Sylvester, *The Week, or the Creation of the World* (London, 1605).

And with the loadstone of my conquering verse
About the poles attract the most perverse.
  God's none of these faint, idle artisans,
Who, at the best, abandon their designs,
Working by halves; as rather, a great deal,
To do much quickly than to do it well;
But rather as a workman never weary,
And all-sufficient, He His works doth carry
To happy end, and to perfection
With sober speed brings what He hath begun.
  Having therefore the world's wide curtain spread
About the circuit of the fruitful bed
Where, to fill all with her unnumbered kin,
King Nature's self each moment lieth in,
To make the same forever admirable,
More stately-pleasant and more profitable,
He th'azure tester trimmed with golden marks
And richly spangled with bright, glistering sparks.
  I know those tapers, twinkling in the sky,
Do turn too swiftly from our hand and eye;
That men can never rightly reach to seeing
Their course and force and much, much less their being,
But if conjecture may extend above
To that great orb whose moving all doth move,
Th'imperfect light of the first day was it
Which for heaven's eyes did shining matter fit.
For God, selecting lightest of that light,
Garnished heaven's ceiling with those torches bright,
Or else divided it, and pressing close
The parts, did make the sun and stars of those.
  But if thy wit's thirst rather seek these things
In Grecian cisterns than in Hebrew springs,
I then conclude that, as of moistful matter
God made the people that frequent the water,
And of an earthly stuff the stubborn droves
That haunt the hills and dales and downs and groves,
So did He make of His almighty might
Th'heavens and stars of one same substance bright,
To th'end these lamps, dispersed in these skies,
Might with their orb, it with them, sympathize.

And as with us, under the oaken bark
The knurly knot with branching veins we mark
To be of substance all one with the tree,
Although far thicker and more rough it be,
So those gilt studs in th'upper story driven,
Are nothing but the thickest parts of heaven.
When I observe their light and heat are blent,
Mere accidents of th'upper element,
I think them fire, but not such fire as lasts
No longer than the fuel that it wastes;
For them I think all th'elements too little
To furnish them with only one day's victual.
And therefore smile I at those fable forges
Whose busy, idle style so stiffly urges
The heaven's bright cressets to be living creatures
Ranging for food, and hungry fodders eaters,
Still sucking up in their eternal motion
The earth for meat, and for their drink, the ocean.
Sure I perceive no motion in a star
But certain, natural and regular;
Whereas beasts' motions infinitely vary,
Confused, uncertain, diverse, voluntary.
I see not how so many golden posts
Should scud so swift about heaven's azure coasts,
But that the heavens must ope and shut sometimes,
Subject to passions which our earthly climes
Alter, and toss the sea, and th'air estrange,
From itself's temper with exceeding change.
I see not how, in those sound, blazing beams,
One should imagine any food-fit limbs,
Nor can I see how th'earth and sea should feed
So many stars, whose greatness doth exceed
So many times, if star-divines say troth,
The greatness of the earth and ocean both;
Since here our cattle in a month will eat
Seven times the bulk of their own bulk in meat.
These torches range not then at random o'er
The lightsome thickness of an unfirm floor,
As here below diversely moving them
The painted birds between two airs do swim;

But rather fixed unto turning spheres
Ay, will they nill they, follow their careers,
As cart nails fastened in a wheel without
Self motion turn with others' turns about.
  As the'ague-sick, upon his shivering pallet
Delays his health oft to delight his palate,
When willfully his tasteless taste delights
In things unsavory to sound appetites,
Even so some brain-sicks live there nowadays
That lose themselves still in contrary ways,
Preposterous wits that cannot row at ease
On the smooth channel of our common seas.
And such are those, in my conceit at least,
Those clerks that think—think how absurd a jest!—
That neither heavens nor stars do turn at all,
Nor dance about this great, round earthly ball,
But th'earth itself, this massy globe of ours,
Turns round about once every twice twelve hours,
And we resemble land-bred novices
New brought aboard to venture on the seas;
Who at first launching from the shore suppose
The ship stands still and that the firm earth goes.
  So twinkling tapers that heaven's arches fill,
Equally distant should continue still.
So never should an arrow shot upright
In the same place upon the shooter light;
But would do rather as at sea a stone
Aboard a ship upward uprightly thrown,
Which not within-board falls, but in the flood
Astern the ship if so the wind be good.
So should the fowls that take their nimble flight
From western marshes towards heaven's light,
And Zephyrus, that in the summer time
Delights to visit Eurus in his clime,
And bullets thundered from the cannon's throat,
Whose roaring drowns the heavenly thunder's note,
Should back recoil; sithence the quick career
That our round earth should daily gallop here
Must needs exceed a hundred fold for swift
Birds, bullets, winds; their winds, their force, their drift.

Armed with these reasons, 'twere superfluous
T'assail the reasons of Copernicus,
Who to salve [explain] better of the stars' appearance
Unto the earth a three-fold motion warrants,
Making the sun the center of this all,
Moon, earth and water in one only ball.
But since here nor time nor place doth suit
His paradox at length to prosecute,
I will proceed, grounding my next discourse
On th'heavens' motions, and their constant course.

Greatness I oft admire of mighty hills,
And pleasant beauty of the flowery fields,
And countless number of the ocean's sand,
And secret force of sacred adamant;
But much, much more, the more I mark their course,
Stars' glistering greatness, beauty, number, force.

Even as a peacock, pricked with love's desire,
To woo his mistress, strutting stately by her,
Spreads round the rich pride of his pompous veil,
His azure wings and golden starry tail,
With rattling pinions wheeling still about,
The more to let his beauteous beauty out,
The firmament, as feeling like above,
Displays its pomp; pranceth about its love,
Spreads its blue curtain mixed with golden marks,
Set with gilt spangles, sown with glittering sparks,
Sprinkled with eyes, specked with tapers bright,
Powdered with stars, streaming with glorious light,
T'inflame the earth the more, with lover's grace,
To take the sweet fruit of his kind embrace.

3. Renaissance Cosmology

To the Candid Reader,
Studious of the Magnetic Philosophy

Since in the discovery of secret things and in the investigation of hidden causes, stronger reasons are obtained from sure experiments and demonstrated arguments than from probable conjectures and the opinions of philosophical speculators of the common sort; therefore to the end that the noble substance of that great loadstone, our common mother (the earth), still quite unknown,[1] and also the extraordinary and exalted forces of this globe may the better be understood, we have decided first to begin with the common stony and ferruginous matter, and magnetic bodies, and the parts of the earth that we may handle and may perceive with the senses; then to proceed with plain magnetic experiments, and to penetrate to the inner parts of the earth. For after we had, in order to discover the true substance of the earth, seen and examined very many matters taken out of lofty mountains, or the depths of seas, or deepest caverns, or hidden mines, we gave much attention for a long time to the study of magnetic forces— wondrous forces they, surpassing the powers of all other bodies around us, though the virtues of all things dug out of the earth were to be brought together. Nor did we find this our labor vain or fruitless, for every day, in our experiments, novel, unheard-of properties came to light: and our Philosophy became so widened, as a result of diligent research, that we have attempted to set forth, according to magnetic principles, the inner constitution of the globe and its genuine substance, and in true demonstrations and in experiments that appeal plainly to the senses, as though we

SOURCE: William Gilbert, *De Magnete* (London, 1600); in the translation by P. Fleury Mottelay, *On the Magnet* (New York, 1893), pp. xlvii-li, 189-91, 313-14, 317-27.

1. It was Gilbert's greatest discovery, of which he was justly proud, that the Earth itself is a loadstone or magnet.

were pointing with the finger, to exhibit to mankind Earth, mother of all.

And even as geometry rises from certain slight and readily understood foundations to the highest and most difficult demonstrations, whereby the ingenious mind ascends above the æther: so does our magnetic doctrine and science in due order first show forth certain facts of less rare occurrence; from these proceed facts of a more extraordinary kind; at length, in a sort of series, are revealed things most secret and privy in the earth, and the causes are recognized of things that, in the ignorance of those of old or through the heedlessness of the moderns, were unnoticed or disregarded. But why should I, in so vast an ocean of books whereby the minds of the studious are bemuddled and vexed; of books of the more stupid sort whereby the common herd and fellows without a spark of talent are made intoxicated, crazy, puffed up; are led to write numerous books and to profess themselves philosophers, physicians, mathematicians, and astrologers, the while ignoring and condemning men of learning: why, I say, should I add aught further to this confused world of writings, or why should I submit this noble and (as comprising many things before unheard of) this new and inadmissible philosophy to the judgment of men who have taken oath to follow the opinions of others, to the most senseless corrupters of the arts, to lettered clowns, grammatists, sophists, spouters, and the wrong-headed rabble, to be denounced, torn to tatters and heaped with contumely. To you alone, true philosophers, ingenuous minds, who not only in books but in things themselves look for knowledge, have I dedicated these foundations of magnetic science— a new style of philosophizing. But if any see fit not to agree with the opinions here expressed and not to accept certain of my paradoxes; still let them note the great multitude of experiments and discoveries—these it is chiefly that cause all philosophy to flourish; and we have dug them up and demonstrated them with much pains and sleepless nights· and great money expense. Enjoy them, and, if you can, employ them for better purposes. I know how hard it is to impart the air of newness to what is old, trimness to what is gone out of fashion; to lighten what is dark; to make that grateful which excites disgust; to win belief for things doubtful; but far more difficult is it to win any standing for or to establish doctrines that are novel, unheard-of, and opposed to everybody's opinions. We care naught, for that, as we have held that philosophy is for the few.

We have set over against our discoveries and experiments large and smaller asterisks according to their importance and their subtility. Let whosoever would make the same experiments, handle the bodies carefully, skilfully and deftly, not heedlessly and bunglingly; when an experiment fails, let him not in his ignorance condemn our discoveries, for there is naught in these Books that has not been investigated and again and again done and repeated under our eyes. Many things in our reasonings and our hypotheses will perhaps seem hard to accept, being at variance with the general opinion; but I have no doubt that hereafter they will win authoritativeness from the demonstrations themselves. Hence the more advanced one is in the science of the loadstone, the more trust he has in the hypotheses, and the greater the progress he makes; nor will one reach anything like certitude in the magnetic philosophy, unless all or at all events most of its principles are known to him.

This natural philosophy is almost a new thing, unheard-of before; a very few writers have simply published some meagre accounts of certain magnetic forces. Therefore we do not at all quote the ancients and the Greeks as our supporters, for neither can paltry Greek argumentation demonstrate the truth more subtly nor Greek terms more effectively, nor can both elucidate it better. Our doctrine of the loadstone is contradictory of most of the principles and axioms of the Greeks. Nor have we brought into this work any graces of rhetoric, any verbal ornateness, but have aimed simply at treating knotty questions about which little is known in such a style and in such terms as are needed to make what is said clearly intelligible. Therefore we sometimes employ words new and unheard-of, not (as alchemists are wont to do) in order to veil things with a pedantic terminology and to make them dark and obscure, but in order that hidden things which have no name and that have never come into notice, may be plainly and fully published.

After the magnetic experiments and the account of the homogenic parts of the earth, we proceed to a consideration of the general nature of the whole earth; and here we decided to philosophize freely, as freely, as in the past, the Egyptians, Greeks, and Latins published their dogmas; for very many of their errors have been handed down from author to author till our own time; and as our sciolists[2] still take their stand on these foundations, they continue to stray about, so to speak, in perpetual darkness. To those men of early times and, as it were, first parents of philosophy, to Aristotle, Theophrastus, Ptolemaeus, Hippocrates, Galen, be due honor rendered ever, for from them has knowledge descended to those that have come after them: but our age has discovered and brought to light very many things which they too, were they among the living, would cheerfully adopt. Wherefore we have had no hesitation in setting forth in hypotheses that are provable, the things that we have through a long experience discovered. Farewell!

How Iron Acquires Verticity from the Loadstone and How this Verticity Is Lost or Altered

An oblong piece of iron, on being stroked with a loadstone, receives forces magnetic, not corporeal, not inhering in or consisting with any body. Plainly, a body. . . briskly rubbed on one end with a loadstone, and left for a long time in contact with the stone, receives no property of stone, gains nothing in weight; for if you weigh in the smallest and most accurate scales of a goldsmith a piece of iron before it is touched by the loadstone you will find that after the rubbing it has the same precise weight, neither less nor more. And if you wipe the magnetized iron with cloths, or if you rub it with sand or with a whetstone, it loses naught at all of its acquired properties. For the force is diffused through the entire body and through its inmost parts, and can in no wise be washed or wiped away. Test it, therefore, in fire, that fiercest tyrant of nature. Take a piece of iron the length of your hand and as thick as a goose-quill; pass it through a suitable round piece of cork and lay it on the surface of water, and note the end of the bar that looks north. Rub that end with the true smooth end of a loadstone; thus the magnetized iron is made to turn to the north. Take off the cork and put that magnetized end of the iron in the fire till it just begins to glow; on becoming cool again it will retain the virtues of the loadstone and will show verticity, though not so promptly as before, either because the action of the fire was not kept up long enough to do away all its force, or because the whole of the iron was not made hot, for the property is diffused through-

2. Conceited and pretentious claimants to learning.

out the whole. Take off the cork again, drop the whole of the iron into the fire, and quicken the fire with bellows so that it becomes all alive, and let the glowing iron remain for a little while. After it has grown cool again (but in cooling it must not remain in one position) put iron and cork once more in water, and you shall see that it has lost its acquired verticity. All this shows how difficult it is to do away with the polar property conferred by the loadstone. And were a small loadstone to remain for as long in the same fire, it too would lose its force. Iron, because it is not so easily destroyed or burnt as very many loadstones, retains its powers better, and after they are lost may get them back again from a loadstone; but a burnt loadstone cannot be restored.

Now this iron, stripped of its magnetic form, moves in a way different from any other iron, for it has lost the polar property; and though before contact with the loadstone it may have had a movement to the north, and after contact toward the south, now it turns to no fixed and determinate point; but afterward, very slowly,* after a long time, it turns unsteadily toward the poles, having received some measure of force from the earth. There is, I have said, a twofold cause of direction,—one native in the loadstone and in iron, and the other in the earth, derived from the energy that disposes things. For this reason it is that after iron has lost the faculty of distinguishing the poles and verticity, a tardy and feeble power of direction is acquired anew from the earth's verticity. From this we see how difficult, and how only by the action of intense heat and by protracted firing of the iron till it becomes soft, the magnetic force impressed in it is done away. When this firing has suppressed the acquired polar power, and the same is now quite conquered and as yet has not been called to life again, the iron is left a wanderer, and quite incapable of direction.

But we have to inquire further how it is that iron remains possessed of verticity. It is clear that the presence of a loadstone strongly affects and alters the nature of the iron, also that it draws the iron to itself with wonderful promptness. Nor is it the part rubbed only, but the whole of the iron, that is affected by the friction (applied at one end only), and therefrom the iron acquires a permanent though unequal power, as is thus proved.

Rub with a loadstone a piece of iron wire on one end so as to magnetize it and to make it turn to the north; then cut off part of it, and you shall see it move to the north as before, though weakly. For it is to be understood that the loadstone awakens in the whole mass of the iron a strong verticity (provided the iron rod be not too long), a pretty strong verticity in the shorter piece throughout its entire length, and, as long as the iron remains in contact with the loadstone, one somewhat stronger still. But when the iron is removed from contact it becomes much weaker, especially in the end not touched by the loadstone. And as a long rod, one end of which is thrust into a fire and made red, is very hot at that end, less hot in the parts adjoining and midway, and at the farther end may be held in the hand, that end being only warm,—so the magnetic force grows less from the excited end to the other; but it is there in an instant, and is not introduced in any interval of time nor successively, as when heat enters iron, for the moment the iron is touched by the loadstone it is excited throughout. For example, take an unmagnetized iron rod, 4 or 5 inches long: the instant you simply touch with a loadstone either end, the opposite end straightway, in the twinkling of the eye, repels or attracts a needle, however quickly brought to it.

Of the Globe of Earth as a Loadstone

Hitherto we have spoken of the loadstone and magnetic bodies, how they conspire together and act on each other, and how they conform themselves to the terrella and to the earth. Now we have to treat of the globe of earth itself separately. All the experiments that are made on the terrella,[3] to show how magnetic bodies conform themselves to it, may—at least the principal and most striking of them—be shown on the body of the earth; to the earth, too, all magnetized bodies are associate. And first, on the terrella the equinoctial circle, the meridians, parallels, the axis, the poles, are natural limits: similarly on the earth these exist as natural and not merely mathematical limits. As on the periphery of a terrella a loadstone or the magnetic needle takes direction to the pole, so on the earth there are revolutions special, manifest, and constant, from both sides of the equator: iron is endowed with verticity by being stretched toward the pole of the earth as toward the pole of a terrella; again, by being laid down and suffered to grow cool lying toward the earth's pole, after its prior verticity has been de-

3. A spherical loadstone.

stroyed by fire, its acquires new verticity conformed to the position earthward. And iron rods that have for a long time lain in the poleward direction acquire verticity simply by regarding the earth; just as the same rods, if they be pointed toward the pole of a loadstone, though not touching it, receive polar force. There is no magnetic body that draws nigh in any way to a loadstone which does not in like manner obey the earth. As a loadstone is more powerful at one end and at one side of the equator, so the same thing is shown with a small terrella on a large one. According to the difference in amount and mode of friction in magnetizing a piece of iron at a terrella, it will be powerful or weak in performing its functions. In movements toward the body of the earth, just as on a terrella, variation is produced by unlikeness and inequality of prominences and by imperfections of the surface; and all variation of the versorium or the mariner's compass all over the earth and everywhere at sea—a thing that has so bewildered men's minds—is found and recognized through the same causes. The dip of the magnetic needle (that wonderful turning of magnetic bodies to the body of the terrella by formal progression) is seen also in the earth most clearly. And that one experiment reveals plainly the grand magnetic nature of the earth, innate in all the parts thereof and diffused throughout. The magnetic energy, therefore, exists in the earth just as in the terrella, which is a part of the earth and homogenic in nature with it, but by art made spherical so it might correspond to the spherical body of the earth and be in agreement with the earth's globe for the capital experiments.

Of the Daily Magnetic Revolution of the Globes, as Against the Time-honored Opinion of a Primum mobile: A Probable Hypothesis

Among the ancients, Heraclides of Pontus, and Ecphantus, the Pythagoreans Nicetas of Syracuse and Aristarchus of Samos, and, as it seems, many others, held that the earth moves, that the stars set through the interposition of the earth, and that they rise through the earth's giving way: they do give the earth motion, and the earth being, like a wheel, supported on its axis, rotates upon it from west to east. The Pythagorean Philolaus would have the earth to be one of the stars, and to turn in an oblique circle toward the fire, just as the sun and moon have their paths: Philolaus was an illustrious mathematician and a very experienced investigator of nature. But when Philosophy had come to be handled by many, and had been given out to the public, then theories adapted to the capacity of the vulgar herd or supported with sophistical subtleties found entrance into the minds of the many, and, like a torrent, swept all before them, having gained favor with the multitude. Then were many fine discoveries of the ancients rejected and discredited—at the least were no longer studied and developed. First, therefore, Copernicus among moderns (a man most worthy of the praise of scholarship) undertook, with new hypotheses, to illustrate the *phaenomena* of bodies in motion; and these demonstrations of reasons, other authors, men most conversent with all manner of learning, either follow, or, the more surely to discover the alleged "symphony" [*phaenomena*] of motion, do observe. Thus the suppositions and purely imaginary spheres postulated by Ptolemy and others for finding the times and periods of movements, are not of necessity to be accepted in the physical lectures of philosophers.

It is then an ancient opinion, handed down from the olden time, but now developed by great thinkers, that the whole earth makes a diurnal rotation in the space of twenty-four hours. But since we see the sun, the moon, and the other planets, and the whole heavenly host, within the term of one day come and depart, then either the earth whirls in daily motion from west to east, or the whole heavens and all the rest of the universe of things necessarily speeds about from east to west. But in the first place, it is not probable that the highest heaven and all those visible splendors of the fixed stars are swept round in this rapid headlong career. Besides, what genius ever has found in one same (Ptolemaic) sphere those stars which we call fixed, or ever has given rational proof that there are any such adamantine spheres at all? No man hath shown this ever; nor is there any doubt that even as the planets are at various distances from earth, so, too, are those mighty and multitudinous luminaries ranged at various heights and at distances most remote from earth: they are not set in any sphaeric framework or firmament (as is supposed), nor in any vaulted structure. As for the intervals (between the spheres) imagined by some authors, they are matters of speculation, not of fact; those other intervals so far surpass them and are far more remote; and, situated as they are in the heavens, at various distances, in thinnest æther, or in that most subtile fifth essence, or in vacuity—how shall the stars keep their places in the mighty swirl of these enormous spheres composed of a substance of which no one knows aught? Astronomers have observed 1022 stars; besides these, innumerable other stars appear minute to our senses; as regards still others our sight grows dim, and they are hardly discernible save by the keenest eye; nor is there any man possessing the best power of vision that will not, while the moon is below the horizon and the atmosphere is clear, feel that there are many more indeterminable and vacillating by reason of their faint light, obscured because of the distance. Hence, that these are many and that they never can be taken in by the eye, we may well believe. What, then, is the inconceivable great space between us and these remotest fixed stars? And what is the vast immeasurable amplitude and height of the imaginary sphere in which they are supposed to be set? How far away from earth are those remotest of the stars: they are beyond the reach of eye, or man's devices, or man's thought. What an absurdity is this motion (of spheres).

It is evident, therefore, that all the heavenly bodies, being, as it were, set down in their destined places, in them are conglobed whatever elements bear to their own centres, and around them are assembled all their parts. But if they have a motion, it will be motion of each round its proper centre, like the earth's rotation; or it will be by a progression in an orbit, like that of the moon; in so multitudinous a scattered flock there will be no circular motion. And of the stars, those situate nigh the equator would seem to be borne around with greatest rapidity, while others nigher the pole have a rather less rapid movement; and others still, as though motionless, have but a small revolution. Yet no differences in the light, the mass, or the colors of the light are perceptible for us; for they are as brilliant, as clear, as resplendent, or as faint (sombre) toward the poles as nigh the equator and the zodiac; and in their seats do they remain and there are they placed, nor are they suspended from aught, nor fastened nor secured in any vault. Far more extravagant yet is the idea of the whirling of the supposititious primum mobile, which is still higher, deeper, more immeasurable; and yet this incomprehensible primum mobile would have to be of matter, of enormous altitude, and far surpassing all the creation below in mass, for else it could not make the whole universe down to the earth revolve from east to west, and we should have to accept a universal force, an unending despotism, in the governance of the stars, and a hateful tyranny. This primum mobile presents no visible body, is in no wise recognizable; it is a fiction believed in by some philosophers, and accepted by weaklings who wonder more at this terrestrial mass here than at those distant mighty bodies that baffle our comprehension.

But there cannot be diurnal motion of infinity or of an infinite body, nor, therefore, of this immeasurable primum mobile. The moon, neighbor of earth, makes her circuit in twenty-seven days; Mercury and Venus have a tardy movement; Mars completes his period in two years, Jupiter in twelve, Saturn in thirty. And the astronomers who ascribe motion to the fixed stars hold that it is completed, according to Ptolemy, in 36,000 years, or according to Copernicus's observations, in 25,816 years; thus in larger circles the motion and the completion of the course are ever more slow; and yet this primum mobile, surpassing all else in height and depth, immeasurable, has a diurnal revolution. Surely that is superstition, a philosophic fable, now believed only by simpletons and the unlearned; it is beneath derision; and yet in times past it was supported by calculation and comparison of movements, and was generally accepted by mathematicians, while the importunate rabble of philophasters egged them on.

The motions of the heavenly bodies (i.e., of the planets) seem all to be eastward, and according to the successions of the zodiacal signs; and mathematicians and philosophers of the vulgar sort do also believe that the fixed stars progress in the same way with a very slow movement: to these stars they must needs, through their ignorance of the truth, add a ninth sphere. But now this inadmissible primum mobile, this fiction, this something not comprehensible by any reasoning and evidenced by no visible star, but purely a product of imagination and mathematical hypothesis, accepted and believed by philosophers, and reared into the heavens and far beyond all the stars, —this must needs by a contrary incitation wheel from east to west, counter to the tendence of all the rest of the universe.

Whatever in nature moves naturally, the same is impelled by its own forces and by a consentient compact of other bodies. Such is the motion of the parts to a whole, of the globes and stars throughout the universe with each other accordant; such is the circular propulsion of the planets' bodies, each the other's career observing and inciting. But as regards this primum mobile with its contrary and most rapid career,—where are the bodies that incite it, that propel it? Where is the nature conspiring with it? And what made force lies beyond the primum mobile? For the agent force abides in bodies themselves, not in space. not in the interspaces

But he who supposes that all these bodies are idle and inactive, and that all the force of the universe pertains to those spheres, is as foolish as the one who, entering a man's residence, thinks it is the ceilings and the floors that govern the household, and not the thoughtful and provident good-man of the house. So, then, not by the firmament are they borne, not from the firmament have they movement or position; and far less are those multitudes of stars whirled round *en masse* by the primum mobile, and taken up at random and swept along in a reverse direction at highest velocity.

Ptolemy of Alexandria, it seems to me, was over-timid and scrupulous in apprehending a break-up of this nether world were earth to move in a circle. Why does he not apprehend universal ruin, dissolution, confusion, conflagration, and stupendous celestial and supercelestial calamities from a motion that surpasses all imagination, all dreams and fables and poetic licenses—a motion ineffable and inconceivable? So, then, we are borne round and round by the earth's daily rotation—a more congruous sort of motion; and as a boat glides over the water, so are we whirled round with the earth, the while we think we stand still and are at rest. This seems to some philosophers wonderful and incredible, because of the ingrained belief that the mighty mass of the earth makes an orbital movement in twenty-four hours: it were more incredible that the moon should in the space of twenty-four hours traverse her orbit or complete her course; more incredible that the sun and Mars should do so; still more that Jupiter and Saturn; more than wonderful would be the velocity of the fixed stars and firmament; and let them imagine as best they may the wonders that confront them in the ninth sphere. But it is absurd to imagine a primum mobile, and, when imagined, to give to it a motion that is completed in twenty-four hours, denying that motion to the earth within the same space of time. For a great circle of earth, as compared to the circuit of the primum mobile is less than a stadium as compared to the whole earth. And if the rotation of the earth seems headlong and not to be permitted by nature because of its rapidity, then worse than insane, both as regards itself and the whole universe, is the motion of the primum mobile, as being in harmony or proportion with no other motion. Ptolemy and the Peripatetics think that all nature must be thrown into confusion, and the whole structure and configuration of this our globe destroyed by the earth's so rapid rotation. The diameter of the earth is 1718 German miles; the greatest elongation of the new moon is 65, the least 55, semi-diameters of the earth; but probably its orbit is still larger. The sun at his greatest eccentricity is distant 1142 semi-diameters from earth; Mars, Jupiter, Saturn, as they are slow in movement, so are far more distant from the earth. The best mathematicians regard the distances of the firmament and the fixed stars as indeterminable; to say nothing of the ninth sphere, if the convexity of the primum mobile be fairly estimated in its proportion to the rest, it must travel over as much space in one hour as might be comprised within three thousand great circles of the earth, for on the convexity of the firmament it would travel over more than eighteen hundred such circles: but what structure of iron can be imagined so strong, so tough, that it would not be wrecked and shattered to pieces by such mad and unimaginable velocity? The Chaldees believed the heavens to be light. But in light there is no such firmness, neither in the fire-firmament of Plotinus, nor in the fluid or watery heavens of God-inspired Moses, nor in the supremely tenuous and transparent firmament that stands between our eye and the lights of the stars, but does not intercept the same. Hence we must reject the deep-seated error about this mad, furious velocity, and this forceful retardation of the rest of the heavens. Let the theologues reject and erase these old wives' stories of a so rapid revolution of the heavens which they have borrowed from certain shallow philosophers. The sun is not swept round by Mars' sphere (if sphere he have) and its motion, nor Mars by Jupiter's sphere, nor Jupiter by Saturn's: the sphere of the fixed stars, too, seems moderate enough, save that movements are attributed to the heavens that really are earth movements, and these produce a certain change in the phenomena. The higher do not tyrannize over the lower, for the heaven both of the philosopher and of the divine must be gentle, happy, tranquil, and not subject to changes; neither will the violence, fury, velocity, and rapidity of the primum mobile bear sway. That fury descends through all the celestial spheres and heavenly bodies, enters the elements of the philosophers, sweeps the fire along, whirls the air around, or at least the greater part thereof; leads in its train the universal ether, and causes it to whirl round as though it were a solid and firm body, whereas it is a most tenuous substance, that neither offers resistance nor is ductile; and leads captive the fires of the upper heavens. O wondrous steadfastness of the globe of earth, that alone is unconquered! And yet the earth is holden nor stayed in its place by any chains, by no heaviness of its own, by no contiguity of a denser or a more stable body, by no weights. The substance of the terrestrial globe withstands and resists universal nature.

Aristotle imagines a philosophy of motions simple or complex, holds that the heavens move with a simple circular motion, and his elements with motion in a right line; that the parts of the earth tend to the earth in right lines; that they impinge upon it at the superficies at right angles and seek its centre, and there always rest; and that hence the whole earth stands in its place, held together and compacted by its own weight. This coherence of parts and this consolidation of matter exists in the sun, the moon, the planets, the fixed stars,—in short, in all those spherical bodies whose parts cohere and seek their several centres; else would the heavens rush to destruction and their grand order disappear. But these heavenly bodies have a circular motion, and hence the earth, too, may have its motion, for this motion is not, as some suppose, adverse to cohesion nor to production. For, inasmuch as this motion is intrinsic in the earth and natural, and as there is nothing without that may convulse it or with contrary motions impede it, it revolves untroubled by any ill or peril; it moves on under no external compulsion; there is nought to make resistance, nothing to give way before it, but the path is open. For since it revolves in a space void of bodies, the incorporeal æther, all atmosphere, all emanations of land and water, all clouds and suspended meteors, rotate with the globe: the space above the earth's exhalations is a vacuum; in passing through vacuum even the lightest bodies and those of least coherence are neither hindered nor broken up. Hence the entire terrestrial globe, with all its appurtenances, revolves placidly and meets no resistance. Causelessly, therefore, and superstitiously, do certain faint hearts apprehend collisions, in the spirit of Lucius Lactantius,[4] who, like the most unlearned of the

4. Lactantius, a fourth-century Latin Church Father, utterly rejected all Greek learning as being unnecessary, distracting, probably contrary to Scripture, and certainly contrary to common sense. For example he rejected the notion of Antipodes where men would clearly be upside down.

vulgar, or like an uncultured bumpkin, treats with ridicule the mention of antipodes and of a round globe of earth.

From these arguments, therefore, we infer, not with mere probability, but with certainty, the diurnal rotations of the earth; for nature ever acts with fewer rather than with many means; and because it is more accordant to reason that the one small body, the earth, should make a daily revolution than that the whole universe should be whirled around it. I pass by the earth's other movements, for here we treat only of the diurnal rotation, whereby it turns to the sun and produces the natural day (of twenty-four hours). And, indeed, nature would seem to have given a motion quite in harmony with the shape of the earth, for the earth being a globe, it is far easier and far more fitting that it should revolve on its natural poles, than that the whole universe, whose bounds we know not nor can know, should be whirled round; easier and more fitting than that there should be fashioned a sphere of the primum mobile—a thing not received by the ancients, and which even Aristotle never thought of or admitted as existing beyond the sphere of the fixed stars; finally, which the holy Scriptures do not recognize, as neither do they recognize a revolution of the whole firmament.

II.

The Astronomical Revolution

THROUGHOUT most of the sixteenth century, acceptance of Copernicanism resulted from those intangible intellectual preferences which make men choose one side or another of any important issue. There were no compelling reasons for accepting the Copernican system in 1543, and no new reasons were to be announced until 1610. Yet between 1572 and 1610 revolutionary ideas about the nature and structure of the heavens were to be enunciated, based upon astronomical observation; and Galileo's telescopic discoveries of 1609 further worked to destroy the old, traditional astronomy.

Tycho Brahe

The most truly revolutionary of the three scientists represented in this section was, probably, Tycho Brahe (1546-1601), a professional, observational astronomer who rejected the Copernican system on the basis of empirical "fact" (the behavior of stones dropped from towers and guns fired toward both east and west) combined with the evidence of Scripture (he was a devout Lutheran). His observations upon the new star or nova of 1572 convinced him that it was indeed a new object appearing in the supposedly immutable celestial regions. Firm in the conviction that his observations were correctly made and assessed, he cheerfully rejected the age-old view that only the terrestrial regions experienced change, generation, and corruption, while the heavens were eternal and unchanging. Five years later, having carefully observed the great comet of 1577, Tycho concluded that it, too, was a celestial object, and further that it moved through and across the paths of the planets. Whereupon he equally cheerfully abandoned another very old view of nature, the view that the heavens were filled with a series of solid, crystalline spheres. These spheres (or orbs) were thought to hold the planets at their fixed distances from one another and from the earth. and by their motion to cause the planets to move in their respective orbits. When Tycho discarded the "crystalline spheres," he introduced a problem into astronomy of which he was seemingly unaware: if there were no crystalline spheres, what kept the planets in their places? (Gilbert, as we have seen, inclined to believe that it was magnetism.) Tycho's own system, described below—Document 4—from his treatise on the comet (*On the Most Recent Phenomena of the Aetherial World, 1588*), is happily unconcerned with any such problems of celestial mechanics. Yet the Tychonic system, precisely because it was based on the newest astronomical evidence without insisting upon the apparent absurdities and real difficulties of the Copernican system, was widely accepted after Tycho's death in 1601. Tycho's even more important contribution to radical astronomy, however, was the mass of precise, accurate observations, especially observations of planetary positions, which he accumulated with unflagging industry and unfailing care and which he bequeathed to his young assistant Johannes Kepler.

Johannes Kepler

Johannes Kepler (1571-1630) was the first astronomer to have been educated in the Copernican system. Though Michael Maestlin, Kepler's professor of astronomy at Tübingen, never wrote a book in support of Copernicus, it is clear from his pupil's letters that he lectured on the Copernican system and encouraged his students to support it in debate. Kepler's reasons for accepting Copernicanism were peculiar, for he was a strong adherent of a vaguely Platonic mathematical mysticism that repelled many of his contemporaries, and he saw, even more strongly than Copernicus, that the new cosmology offered more opportunity for mathematical harmony than the old. But his respect for Tycho was such that, searching with the aid of Tycho's data for a better theory of the motion of Mars, he worked ceaselessly to discover mathematical expressions worthy of Tycho's accuracy. In the Introduction to the *Astronomia Nova* (*New Astronomy*, 1609) Kepler describes his underlying point of view and refers to what we now call Kepler's first and second laws. The first states that the planets move in elliptical orbits, and the second describes their rate of motion about this elliptical path. Kepler did not think of them as "laws," but as two among many (some more powerful) arguments in support of the Copernican system. He stressed three main points: the importance of the mathematical approach to astronomy; his new arguments for the reconcileableness of Copernicanism and Scripture; and his new, quasi-magnetic, doctrine of gravity. All these are discussed at length in the Introduction to the book, for Kepler loved to prepare his readers.

In the selection below (Document 5), Kepler's specific references to the analytical table of contents and to various chapters of the book itself are omitted, while his more lengthy arguments are abridged where this does not curtail or falsify the argument. From what remains the reader should be able to gain a fairly good notion of Kepler's scientific personality and notion of the proper way of reasoning in science, in a form not previously available in English. Here will be found Kepler's profound Copernicanism, tempered only by his respect for the work and views of Tycho Brahe, without whose observational data Kepler would never have been able to work. Though Kepler makes difficult reading (his Latin is often very difficult indeed) it is well worth the attempt, if only because without Kepler later celestial mechanics would have been impossible. It is ironic that acceptance of his work was hampered by the mathematical mysticism with which he surrounded it.

Galileo

To Kepler the fact of being able to establish these mathematical laws or relationships was proof enough of the truth of the Copernican system. Galileo (1564-1642) saw things otherwise. True, Galileo was a Copernican at least by the time he was thirty, when he was professor of mathematics at the University of Padua, but it was not mathematical mysticism that made him so. Galileo might be convinced, as indeed he was, that God designed the universe in mathematical fashion, but he had no use for mysticism. He could not follow Kepler in speculating upon the geometrical solids that might, like wooden eggs, nest one inside another; he was concerned with motion and its mathematical laws, laws of physics which could be observed in the world close at hand, not merely in the stars. For Galileo rejected completely any notion of a distinction between celestial and terrestrial physics; for him one of the beauties of the Copernican system was that it made possible a uniform cosmos, in which all objects followed the same laws. As early as 1604 he demonstrated, to his own satisfaction, the law of falling bodies. Then, in 1609, he heard of the new optical instrument being demonstrated in Holland, guessed its composition, made one, and turned it upon the heavens. What he saw is described in the *Sidereal Messenger* (*Sidereus Nuncius*, 1610) and redescribed in the *Dialogue on the Two Chief Systems of the World* (*Dialogo sopra i due Massini Sistemi del Mondo*), part of which appears below as Document 6. The discovery that the moon is, in appearance, very like the earth; that planets are not just moving stars, but shine by reflected light; that the earth is not unique in possessing a moon, since Jupiter has four; that Venus has phases like the moon—all these seemed to show that the Copernican system was more probable, that it was not unlikely that the earth was a planet, and the sun not. It did nothing to offer any real evidence, let alone proof, of the reality of the earth's motion. Galileo in 1616 became convinced that he possessed such proof, because he thought he could convincingly explain the tides on the basis of the earth's motion. He did not immediately publish his ideas; ever since 1610 he had experienced opposition from within the Church, although the highest ecclesiastical figures in Rome were friendly and sympathetic. Eventually, convinced from conversations with the pope that he might lawfully defend the Copernican system as long as he did not insist that God had necessarily created the world in the Copernican fashion, he began writing the *Dialogue on the Two Chief Systems of the World* (1632). It is a brilliant, witty, semipopular, rather long-winded presentation of his views through conversations between Salviati (a convinced Galilean and Copernican), Sagredo (an intelligent amateur), and Simplicio, a slightly slow-witted but very earnest adherent of the Aristotelian world view. The result of publication was a storm; it was in vain for Galileo to plead, as he tried to do, that he had complied with the letter of the pope's injunctions, when he had clearly violated their spirit. He was tried by the Holy Office (the Inquisition) and condemned to abjuration and perpetual detention (which in fact soon meant house arrest in his own villa outside Florence). In the *Dialogue* Galileo mustered every argument he could in support of Copernicanism, from dynamics to telescopic observation. This is the source from which most scientists later in the century learned of Galileo's discoveries in mechanics as well as his view of nature.

Here in the necessarily long selection below (the *Dialogue* is a large and leisurely work) are some examples of Galileo's method of argument, as well as of his positive achievement. The Introduction shows Galileo's clever but perhaps sly ability to put his tongue in his cheek and circumvent authority; at the same time it is the work of a proud and patriotic man. Next is a section of conventional arguments and a brief description of the Copernican system. The following section, on the behavior of moving bodies, shows Galileo at his most effective. In an interesting exchange on scientific method, the Aristotelian Simplicio relies far more on the evidence of the senses than does the more sophisticated and Platonist Salviati. Then in quick exchanges of conversation Galileo demolishes the Aristotelian theory of bodies in motion and establishes his own view—a limited inertial view—in its place. This passage is particularly revealing both because it contains an important discussion in mechanics and because it shows how Galileo incorporated his new discoveries in mechanics into his cosmological system, so that for the first time the Copernican system received support from a new physics. The conclusion once again shows Galileo at his more urbane. Yet it contains Galileo's greatest psychological blunder, the placing of the argument suggested by the pope—

that God cannot be thought to be constrained by human logic, so that strong arguments from science cannot cause a change in theology—in the mouth of the so often defeated Simplicio. It was treating a delcate matter too lightly, and it lost him the pope's sympathy, a loss for which he was to pay dearly.

The translation, by Thomas Salusbury, was part of what was apparently planned as a vindication of Galileo. For Salusbury's *Mathematical Collections and Translations* (London, 1661-65) was to include not only Galileo's *Dialogue, The Discourses on Two New Sciences, Floating Bodies,* and the *Letter to the Grand Duchess Christina* (a plea for religious liberty for science) but also supporting treatises on the relation of religion and science (like the excerpt from Kepler, included in Document 5), and to crown all, a life of Galileo. This last is known mainly from contemporary references, for it was to appear at the end of the whold, and this volume was totally destroyed in the printer's shop by the Great Fire of London in 1666. One copy was said in the nineteenth century to have survived, but even this has now disappeared. The sections reprinted here are from the first volume, published in 1661 and fortunately fairly widely distributed before the Great Fire.

4. The Tychonic System

Of the Discovery of the Place of Space
Between the Celestial Revolutions of the Planets
Where the Comet May Fitly Run Its Course
and of the Construction of an Hypothesis
by Which Its Apparent Motion Is Approximately Represented

Thus from what has gone before, it was made obvious and beyond any controversy that our Phaenomenon [the comet of 1577] had nothing in common with the elementary world, but was shown to have a motion in the Aether far up above the Moon, its tail perpetually maintaining an Olympian relationship to certain stars. It remains now, and seems to be especially fitting, that we should assign to it also some particular place in the very wide space of the same Aether, in order that we may establish between which orbs of the Secundum Mobile it will direct its path. Indeed the Aetherial World comprises incredible vastness, so that if we assume that this elementary world [measures] from the centre of the Earth to the nearest limits of the Moon about 52 Earth-radii (each of which contains 860 of our common or German miles) this will be contained 235 times in the rest of the space of the Secundum Mobile, that is to say as far as the extreme distance of Saturn from the Earth. In this enormously vast interval, seven planets perform incessantly their wonderful and almost divine periodic motions; so that I can say nothing about that immense distance of the Eighth Sphere, which is beyond doubt greater by far than that of Saturn at his furthest point. On the other hand, according to the Copernican hypothesis, that space between Saturn and the Fixed Stars will be many times greater than the distance of the Sun from the Earth (which however is such that it includes the semidiameter of the elementary world about twenty

SOURCE: Tycho Brahe *De Mundi aetherei recentoribus phaenomenis* (Uraniborg, 1588); in the translation by A. Rupert Hall and Marie Boas Hall, *On the Most Recent Phenomena of the Aetherial World*, in *Occasional Notes of the Royal Astronomical Society, III, no. 21* (1959), 257-63.

times). For otherwise the annual revolution of the Earth in the great orb, according to his speculation, will not turn out to be insensible with respect to the Eighth Sphere, as it ought. Because the region of the Celestial World is of so great and such incredible magnitude as aforesaid, and since in what has gone before it was at least generally demonstrated that this comet continued within the limits of the space of the Aether, it seems that the complete explanation of the whole matter is not given unless we are also informed within narrower limits in what part of the widest Aether, and next to which orbs of the planets, [the comet] traces its path, and by what course it accomplished this. So that this may be more correctly and intelligibly understood, I will set out my reflections of more than four years ago about the disposition of the celestial revolutions, or synthesis of the whole system of the world. These were referred to before, but postponed to this point in the Astronomical Work, where they are required.

I considered that the old and Ptolemaic arrangement of the celestial orbs was not elegant enough, and that the assumption of so many epicycles, by which the appearances of the planets towards the Sun and the retrogradations and stations of the same, with some part of the apparent inequality, are accounted for, is superfluous; indeed, that these hypotheses sinned against the very first principles of the Art, while they allow, improperly, uniform circular motions not about [the orbit's] own centre, as it ought to be, but about another point, that is an eccentric centre which for this reason they commonly call an equant. At the same time I considered that newly introduced innovation of the great Copernicus, in these ideas resembling Aristarchus of Samos (as Archimedes shows in his *Sand-Reckoner*),[1] by which he very elegantly obviates those things which occur superfluously and incongruously in the Ptolemaic system, and does not at all offend against mathematical principles. Nevertheless the body of the Earth, large, sluggish and inapt for motion is not to be disturbed

1. Archimedes' *Sand-Reckoner* is an exercise in mathematical ingenuity, which discusses how to handle very large numbers. For this purpose Archimedes purported to "count" the number of grains of sand in a universe, picking the largest universe known to him, that of his elder contemporary, Aristarchos of Samos, who is the only astronomer before Copernicus to have postulated a true heliostatic, heliocentric universe. This, the clearest and best description of the Aristarchan system, was not known to Copernicus himself.

by movement (especially three movements) any more than the Aetherial Lights are to be shifted, so that such ideas are opposed to physical principles and also to the authority of Holy Writ which many times confirms the stability of the Earth (as we shall discuss more fully elsewhere). Consequently I shall not speak now of the vast space between the orb of Saturn and the Eighth Sphere left utterly empty of stars by this reasoning, and of the other difficulties involved in this speculation. As (I say) I thought that both these hypotheses admitted no small absurdities, I began to ponder more deeply within myself, whether by any reasoning it was possible to discover an hypothesis, which in every respect would agree with both Mathematics and Physics, and avoid theological censure, and at the same time wholly accord with the celestial appearances. And at length almost against hope there occurred to me that arrangement of the celestial revolutions by which their order becomes most conveniently disposed, so that none of these incongruities can arise; this I will now communicate to students of celestial philosophy in a brief description.

I am of the opinion, beyond all possible doubt, that the Earth, which we inhabit, occupies the centre of the universe, according to the accepted opinions of the ancient astronomers and natural philosophers, as witnessed above by Holy Writ, and is not whirled about with an annual motion, as Copernicus wished. Yet, to speak truth, I do not agree that the centre of motion of all the orbs of the Secundum Mobile is near the Earth, as Ptolemy and the ancients believed. I judge that the celestial revolutions are so arranged that not only the lamps of the world, useful for discriminating time, but also the most remote Eighth Sphere, containing within itself all others, look to the Earth as the centre of their revolutions. I shall assert that the other circles guide the five planets about the Sun itself, as their Leader and King, and that in their courses they always observe him as the centre of their revolutions, so that the centres of the orbs which they describe around him are also revolved yearly by his motion. For I have found out that this happens not only with Venus and Mercury, on account of their small elongations from the Sun, but also with the three other superior planets. The apparent inequality of motion in these three remoter planets, including the Earth, the whole elementary world and at the same time the confines of the Moon in the vastness of their revolutions about the Sun, which the ancients accounted for by means of epicycles, and Copernicus by the annual motion of the Earth, is in this way most aptly represented through a coalescence of the centres of their spheres with the Sun in an annual revolution. For thus as suitable an opportunity is offered for the appearance of the stations and retrogradations of these planets and of their approach to and recession from the Earth and for their apparent variations in magnitude and other similar events, as either by the pretext of epicycles or by the assumption of the motion of the Earth. From all these things, when the former treatment by epicycles is understood, are deduced the lesser circuits of Venus and Mercury about the Sun itself, but not around the Earth, and the refutation of the ancient views about the disposition of epicycles above and below the Sun. Thus a manifest cause is provided why the simple motion of the Sun is necessarily involved in the motions of all five planets, in a particular and certain manner. And thus the Sun regulates the whole Harmony of the Planetary Dance in order that all the celestial appearances may subject themselves to his rule as if he were Apollo (and this was the name assigned to him by the ancients) in the midst of the Muses.

So much, indeed, for the rest of the more particular differences of the apparent inequality [of motion], which the ancients conceived to be represented by eccentrics and equants, and Copernicus by an epicycle on the circumference of the eccentric, having the same angular velocity. These [differences] can also easily be represented in our hypothesis, either by a circle of a sufficient size in an eccentric orb about the Sun, or by a double circle in some concentric orb. Thus [in our system] no less than in the Copernican, all circular motions take place with respect to their own centre, since we have rejected Ptolemaic disorder. The manner of this we shall explain more particularly and fully in the work on the restoration of astronomy which (God willing) we have decided to elaborate. There we specifically discuss this hypothesis of celestial motions and shall demonstrate both that all the appearances of the planets agree perfectly among themselves and that these more correctly correspond [with our hypothesis] than with all others hitherto employed. So that this our new invention for the disposition of the celestial orbs may be better understood, we shall now exhibit its picture [fig. 1].

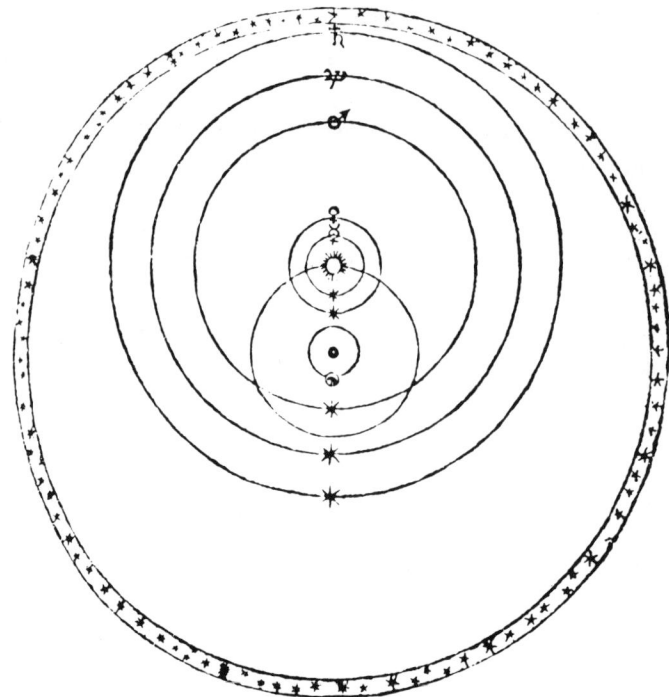

1. A new hypothesis of a world system newly invented by
   the author.
   From Tycho Brahe, *On the Most Recent Phenomena of
   the Aetherial World*

I have, in truth, constructed a fuller explanation of the new
disposition of the celestial orbs, in which are important corollaries
of all the present cogitations. I shall add this near the end of the
work, and there it will be shown first of all from the motions of
comets, and then clearly proved, that the machine of Heaven is
not a hard and impervious body stuffed full of various real
spheres, as up to now has been believed by most people. It will be
proved that it extends everywhere, most fluid and simple, and
nowhere presents obstacles as was formerly held, the circuits of
the planets being wholly free and without the labour and whirling
round of any real spheres at all, being divinely governed under a
given law. Whence also it will be established that no absurdity in
the arrangement of the celestial orbs follows from the fact that
Mars in opposition is nearer to the Earth than the Sun itself. For
in this way there is not admitted any real and incongruous penetration of the orbs (since these are really not in the sky, but are

postulated solely for the sake of teaching and understanding the business) nor can the bodies of any of the planets ever run into one another nor for any reason disturb the harmony of the motions which each of them observes. So that the imaginary orbs of Mercury, Venus, and Mars are mixed with that of the Sun, and cross it, as will be more clearly and extensively declared in that place near (as I said) the colophon of the whole book, especially in our astronomical volume where we deal explicitly with these things.

Now, however, we shall borrow at least that part from this same new scheme of the aetherial revolutions, which satisfies for the moment the difficulty in alloting a place for this comet, and providing an hypothesis to facilitate the ordering of its appearances.

The basis for these celestial revolutions having been laid down, I say that everything most aptly agrees with the apparent motion of this comet, if we understand that it also, as if it were an adventitious and extraordinary planet, has, no less than the other planets, revealed that the centre of its revolution is in the Sun. It has traced about this centre that portion of its own sphere by which it goes beyond not only the Sphere of Mercury, but even that of Venus; for it can depart from the Sun a sixth part of the heavens, while Venus is elongated not much more than an eighth part. Indeed, the comet proceeded in this orb in such a way that if it is assumed to be at the lowest parts of its orb and nearest to the Earth when it was joined to the mean motion of the Sun, it may be allowed to have proceeded thence in the order of the Signs [Eastward] toward the apogee of its orb, otherwise than occurs with Venus and Mercury, the constant centre of this revolution agreeing with the simple motion of the Sun. To perceive all these things rightly, we must now submit to the eyes a suitable arrangement of the construction of the orbs [fig. 2].

By A is understood the globe of the Earth located in the centre of the universe, closest to which revolves the Moon in the orb BEFD, in which all the region of the elements is contained. That the comet can in no way be discovered between these bounds of the Lunar orb was, however, sufficiently demonstrated by us in the Sixth Chapter. Above this let CHIG be the annual orb of the Sun revolving about the Earth, in which the Sun is represented near C, upon which are located the centres of all the orbs of all the rest of the 5 planets, according to our renovation of the celestial

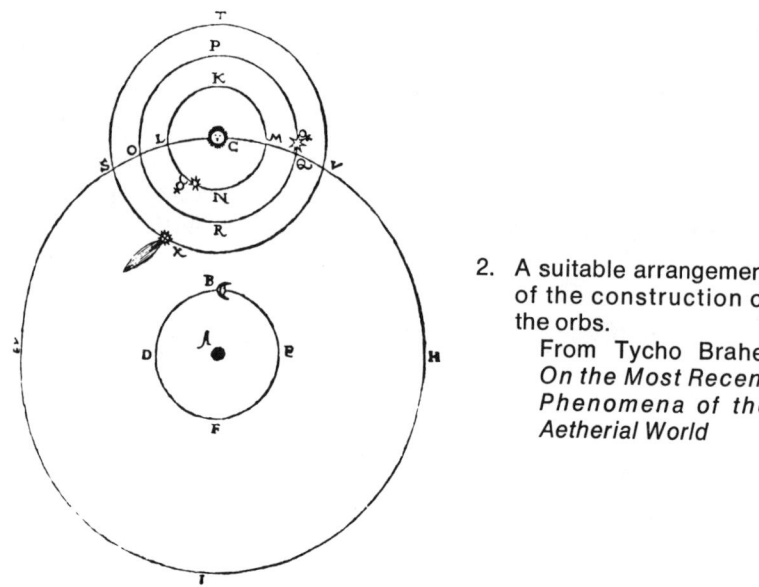

2. A suitable arrangement
   of the construction of
   the orbs.
     From Tycho Brahe,
   *On the Most Recent
   Phenomena of the
   Aetherial World*

hypothesis. And since the star of Mercury revolves closest to the Sun in the orb LKMN and a little above this the star of Venus revolves in the orb OPQR, it happens fitly that the comet revolves in yet a little greater orb described about the Sun. [The comet's orb] includes the orbs of Mercury and Venus only; it does not [include] the Lunar orb together with the Earth (as the star of Mars on its revolution does) because it cannot digress from the Sun by more than 60 degrees. And this same orb which we impute to the comet may be understood by the circle STVX, with the comet itself near X, in which situation it was when seen at the time of our first observation. It has a motion in this orb in the order of the Signs, contrary to the revolutions of Venus and Mercury, so that it goes round from X through S to T. And the centre of the same orb observes its simple motion allied to the Sun perpetually. And this disposition of the revolution of the comet between the celestial orbs being accepted, I assert that it is possible to satisfy its apparent motion, as it is perceived by us dwelling on A, the Earth.

However, this is to be observed, that the comet, led in this same circular path about the Sun, does not always show the same speed. At the beginning, when located in the lower part of its orb, which is nearer the Earth, it is moved more slowly; thereafter indeed it will increase its motion more and more, and this in such a manner that whereas about the IXth and Xth of November it completed barely ten twelfths of a degree in its circle in one day, by the XXth it completed a whole degree in each day. Indeed, at the beginning of December it increased its motion little by little from one degree and five twelfths until in the first days after the XXth of December it was brought up to a degree and a half, beyond which limit it did not extend its haste, but gradually returned to a slackening. All the same, its variation was slight, so that up to the XXVIth day of January, when it was last seen by us, no more than five twenty-fourths of a degree had been lost from the one and a half degrees of its motion. For there was about the end of January once again a motion of one degree and five twelfths during the natural day, in the same measure as for the whole of December and January; that is, it did not alter its daily progress, except at the most five twenty-fourths. So little in so long a time did its revolution about the Sun deviate from perfect equality. Indeed, in November it made from day to day generally more, by a little quicker variation; and all these things may be seen more fully from the fourth row of its table, which we shall subjoin to the end of the following chapter [2]

Indeed, I show, as may be more convenient, that the comet in its own orb through the whole of its duration completes an equal arc in an equal interval of time. For thus a simple uniformity of revolution is more rightly preserved, namely with that same regularity by which the planets themselves constantly observe a perpetual equality in their circuits. And this permits the inequality of the comet which occurs in its own circumgyrations to be bounded and corrected; by twisting the centre of its orb about the Sun circularly and in a due proportion in the opposite direction, or by adding to the circular circumference of the same, the motion may be now inhibited, now released. Nevertheless, because this business acquires through such complexities more obscurity and involvement than light and clarity, I was unwilling to bring up the third intricate arrangement of the various motions for an equal degree of consideration, especially as it would be very inappropriate to make such quickly vanishing bodies as comets liable to follow artificially compounded and much involved curves of motion. And

2. The table does indeed show such a variation as Tycho describes.

so I choose to retain those daily paths of the comet in its orb about the Sun which experience itself so abundantly supplied, notwithstanding that at the beginning they were a very little slower, and soon after this returned more quickly in each successive passage; especially when through the greatest and longest time of their visibility, they conformed to a nearly constant equality. For in December and January, for two whole months, the motion did not vary from equality by more than five twenty-fourths (as I indicated before) which truly is very little and almost of no moment; in November alone, and at most for half the month, it admitted a sensible alteration; so that only about one fifth part of the whole duration was subject to inequality, for the four remaining ones were almost exempt from it.

Nor is it the case that anyone may think that our hypotheses are to be overthrown because of the short duration or great inequality of the motion. For it is probable that comets, just as they do not have bodies as perfect and perfectly made for perpetual duration as do the other stars which are as old as the beginning of the world, so also they do not observe so absolute and constant a course of equality in their revolutions—it is as though they mimic to a certain extent the uniform regularity of the planets, but do not follow it altogether. This will be clearly shown by comets of subsequent years, which will no less certainly be located in the Aethereal region of the world. Therefore either the revolution of this our comet about the Sun will not be at all points exquisitely circular, but somewhat oblong, in the manner of the figure commonly called ovoid; or else it proceeds in a perfectly circular course, but with a motion slower at the beginning, and then gradually augmented. However this may be, the comet in fact revolves around the Sun just the same, even though with a certain inequality, which yet is not confused or irregular.

5. Mathematical Copernicanism

Introduction to This Work

It is extremely difficult nowadays to write mathematical books, especially astronomical ones. For unless use is made of an exact precision in the propositions, explanations, demonstrations, and conclusions the book will not be mathematical; if such use is made, then the reading of it becomes very difficult, especially in Latin, which lacks articles and that charm characteristic of Greek. . . . This is why there are nowadays few suitable readers; most generally despise and reject such works. How many mathematicians are there who will undertake the labor of reading through the *Conics* of Apollonios of Perga? And yet his material is of a kind which lends itself to explanation by figures and lines much better than Astronomy does.

I myself, who pass for a mathematician, tire my brain in reading my own work, when I try to evoke from the figures the demonstrations which I myself formerly expressed in the figures and the text. And if I should mitigate the obscurity of the material by introducing roundabout expressions, then I should seem, mathematically speaking, to fall into the opposite fault of prolixity.

And prolixity of expression lends itself to obscurity no less than concise brevity does. This latter escapes the eyes of the mind, while the former distracts them; the latter lacks light, while the former suffers from a superabundance of it; the latter does not affect the sight, the former clearly overwhelms it.

[Kepler next describes the table of contents, which he has designed as a careful guide to the whole book.]

I reveal [these things], however, chiefly for the satisfaction of those who profess physics and who are incensed against me and still more against Copernicus and so with remote antiquity, be-

SOURCE: Translated by A. Rupert Hall from Johannes Kepler, *Astronomia Nova* (Prague 1609), with the exception of the section on astronomy and Scriptures which is taken from Thomas Salusbury, *Mathematical Collections,* I (London, 1661), 461-67.

cause the fundamentals of science have been shattered by the motion of the earth; as I say, I faithfully disclose the design of the main chapters relating to this question and set before the readers' eyes all the principles of the demonstrations buttressing my conclusions, which they dislike so much. . . .

Let the reader understand that there are two sects of astronomers: the one distinguished by having Ptolemy as their leader and by the allegiance of many of the ancients; the other assigned to the moderns, though the more ancient.[1] The former sect treats each of the planets separately and assigns the causes of the motion of each to its individual sphere,[2] while the latter compares the planets together, deducing from one and the same common cause whatever is found to be common to their motions. This latter sect is further subdivided; for Copernicus like Aristarchos of old finds in the motion of the earth, our dwelling place, the cause of the planets' apparent stations and retrogradations, and with this I agree; but Tycho Brahe finds it in the sun, in whose vicinity is the connection (as he says) or focus (which is not to be understood as physical, but as quantitative) of the eccentric circles of all the five planets, and this focus or center revolves about the motionless earth with the body of the sun. . . .

Now my design in this work is chiefly to correct astronomical theory (especially in respect of the motion of Mars)[3] in each of its three forms, so that the results of computations from tables may be found to agree with the celestial appearances, which hitherto could not be done with sufficient certainty. For example, in August, 1608 Mars was a little less than four degrees beyond the place assigned to it in the *Prutenic Tables*.[4] In August and September, 1593 this error amounted to a little less than five degrees, which now by my computations is quite removed.

1. Because it goes back to Aristarchos, who lived five centuries before Ptolemy.
2. That is, each planet is conceived as fixed to a solid sphere, though its motion is mathematically defined by a series of circles—eccentrics, epicycles, deferents, and equant circles.
3. Almost all Kepler's immensely laborious calculations were made from observational data (supplied by Tycho Brache) of the planet Mars. He extended his theories to the other planets by analogy.
4. Compiled by Erasmus Reinhold in 1551 on the basis of data found in Copernicus' *De Revolutionibus*. They superseded the thirteenth-century *Alphonsine Tables* and were in turn about to be superseded by Kepler's *Rudolphine Tables* (1627).

Now while setting this out [in the book] and continuing successfully, I digress into Aristotle's metaphysics, or rather celestial physics, and I inquire into the causes of the natural motions [of the planets]. From the discussion of these things certain pretty plain conclusions emerge in time, by which Copernicus' theory of the world is, with a few changes, upheld, while the other two theories are condemned as false. . . .

The first step toward the investigation of the physical causes of the motions [of the planets] was that I should demonstrate that the focus of their eccentrics lies in no other place than the very center of the body of the sun, contrary to the belief of both Copernicus and of Tycho.

If this correction of mine be introduced into the Ptolemaic theory, it will instruct the Ptolemeians to investigate not the motion of the center of the epicycle (around which the epicycle moves uniformly) but the motion of some other point, which is as far from that center in proportion to the diameter as the center of the solar sphere is, according to Ptolemy, from the earth, and [lies] in the same straight line or in one parallel to it.

The Tychonians might object against me that I am a rash innovator, for they stick to the received opinion of the ancients and place the focus of the eccentrics not in the sun but close by; yet they have framed a calculus thence which corresponds to the heavens. And in translating the Tychonic numbers into the Ptolemaic form, Ptolemy might say to me, that while he upheld and set out the observations, he imagined to himself no other eccentric than that described by the center of the epicycle, about which the epicycle moves freely. Thus I must certainly beware, lest by a new way of practice I should furnish that which they have already set out in the ancient way.

So that this objection may be met, I demonstrate in the first part of this work that the same things may be clearly derived or obtained by this new way, as by their ancient method.

In the second part of the work I tackle the question itself, and I have expressed the places of Mars in apparent opposition to the sun not less but much more accurately than they expressed Mars' positions in mean opposition to the sun by the old method.[5]

5. Kepler had set himself the task of working out his computations in terms of the Ptolemaic, Tychonic, and Copernican systems; he was naturally pleased to be able to present a more accurate picture of the Ptolemaic

Throughout the whole of the second part I have, when considering the geometrical demonstration from the observations, left in suspense the question of whether they or I are more correct, since we both agree about certain observations. ... I have partly showed in the first part of the book (Chapter 6, especially) that the physical causes are consonant with my method and incompatible with the ancient method.

At last (in Part IV, Chapter 52) I have demonstrated most definitely by certain observations no less infallible than the former ones, with which my theory agrees beautifully and the ancient one not at all, that the eccentric of Mars is so placed that its line of apsides[6] meets the very center of the sun's body, not some other point close to the center; and thus all the eccentrics meet in the sun itself. ...

I have demonstrated in the third part [of the book], that whether the aforesaid theory is valid (that which uses the mean motion of the sun), or my new one (which uses the apparent motion), in either case some of the causes of the first inequality are mingled with the second inequality which is common to all the planets. So I have shown the Ptolemaic astronomers that their epicycles do not have as a center that point about which their motions are uniform. And I have shown the Copernicans that the circle in which the earth moves about the sun does not have as a center that point about which its motion is regular and uniform. So I have shown the Tychonians that the circle in which the aforesaid center or focus of the eccentrics goes round does not have as a center that point about which its motion is regular and uniform. For if I admit to the Tychonians that the focus of the eccentrics differs from the center of the sun, it is necessary for them to say that the orbit of that focus (which in size and duration is obviously equal to the sun's orbit) is eccentric and goes toward Capricorn when the eccentric orbit of the sun approaches Cancer. And the same happens to the Ptolemaic epicycle.

However, if I set the center or focus of the eccentrics [of Tycho] in the very center of the sun's body, then this common orbit of both that focus and the sun is eccentric to the earth and

system than the followers of Ptolemy could do, though worried lest the Ptolemaians might take comfort from this and refuse to see the superior accuracy and convenience of the Copernican system.

6. The line joining the aphelion and perihelion, i.e., the positions at which the planet is at its nearest or farthest point from the sun.

goes toward Cancer, but by only one-half of the eccentricity of that point about which the sun's motion is regular and uniform.

The same arguments apply to the Copernican and Ptolemaic systems.

When these matters have been demonstrated by an infallible method, the first step toward ascertaining the physical causes has been confirmed and a new one toward them prepared, very clearly if the views of Copernicus and Tycho are adopted, obscurely but at least plausibly in the Ptolemaic system.

For either the earth is moved, or the sun; and it is certainly demonstrated that the body which moves is moved nonuniformly; that is to say, slowly when it is remote from the body at rest, quickly when it approaches the latter closely.

Accordingly, there now immediately appears a means of distinguishing in physics between the three opinions—by conjecture, it is true, but with no less certainty than is found in the conjecture of physicians concerning physiology, or those in any other part of physics.

First of all the Ptolemaic system is exploded. For who will believe that there are as many theories for the sun (exactly similar to one another, true, and equal) as there are planets? It seems that a single theory of the sun is enough to enable Tycho to perform the same functions, and it is a very generally accepted maxim in physics that nature uses as few means as possible.

That Copernicus' system is superior to Tycho's as regards celestial physics is proved on many grounds. . . . For Tycho no less than for Ptolemy, each planet is moved not only by its own proper motion but also by the true motion of the sun, the two being combined together in a mixed motion, forming helices; whence it follows that there are no solid spheres, as Tycho has very solidly demonstrated. Copernicus, on the other hand, has quite freed the five planets from this irrelevant motion, as one caused by a deception arising from the circumstances in which it is observed. Hence, in Tycho's system as in Ptolemy's, motions are multiplied to no purpose.

Secondly, if there are no solid spheres, the lot of the intelligences and moving spirits becomes exceedingly hard when they are ordered to look after so many things as they must if they are to drive the planets with double, mixed motions. For at the very least they must endeavor at one and the same time to look after the principles, centers, and periods of both motions. But if the earth moves, many things can be accomplished, as I show, by powers which are not animate, but physical, and especially magnetical. . . .

It is further demonstrated that the law governing the speed of the earth's motion (assuming it moves) is quantitatively related to its approach to and recession from the sun.[7] And it is the same with the remainder of the planets; their motion is increased or decreased by their approach to or recession from the sun. The demonstration of these things is to this extent geometrical.

From this very reliable demonstration may be drawn a physical hypothesis—that the source of the motions of the five planets is in the sun itself.[8] It is further very probable that the source of the Earth's motion is in the same place as the source of the motion of the remaining five planets: that is, in the sun also. That the earth moves is, accordingly, probable, since there is a similar apparent cause of its motion.

On the contrary, for this and many other reasons, it is highly probable that the sun is fixed in its place in the center of the world, for in it is the source of the motion of at least five planets. . . . And the source of all motion is more likely to remain at rest than to move.

[Kepler argues next that in Tycho's view the earth must be the source of the motion of the sun and indirectly the source of the motion of the five planets carried around by the sun; this, he says, is absurd. Furthermore, the earth's 365-day period fits between the periods of Venus (225 days) and Mars (687 days), making all things harmonious.

He then turns to objections against the motion of the earth, especially those deriving from the fall of heavy bodies toward the earth, or toward the center of the universe in the old system. Kepler maintains that bodies do not fall to a *point* (the center) nor

7. This argument is connected with what we know as Kepler's second law (clearly stated in the body of the book), which he applied here to the motion of the planet in an eccentric circle; the law states that a line drawn from the planet to the sun sweeps out equal areas in equal times. Thus is the apparently uneven motion of the planets—faster when nearer the sun, slower when more remote—accounted for in simple fashion.

8. Kepler realized (as Tycho had not done) that if there are neither solid spheres nor "Intelligences," the physicist-astronomer was required to seek for the force which kept the planets in their orbits. Kepler found it in a quasi-magnetical force, derived ultimately from Gilbert; he regarded his discussion of this force as one of the highlights of his book.

are they repelled inward from the boundary of the universe. The traditional notions of heaviness—the cause of fall—are false.]

The true theory of heaviness (or gravity) rests upon the following axioms:

All corporeal substance, insofar as it is corporeal, is capable of being at rest in every place where it is alone, outside the orbit of the power of related bodies.

Heaviness is a mutual corporeal disposition between related bodies toward union or conjunction (which, in the order of things, the magnetic faculty is also), so that it is much rather the case that the earth attracts a stone than that the stone seeks the earth.

Heavy bodies (even if we locate the earth at the center of the world) are not borne to the center of the world as such, but to the center of related spherical bodies, to wit, the earth. So that wherever the earth is located, or wherever the point to which it is displaced by its animate faculty, it will always draw heavy bodies to itself.

If the earth were not round, heavy bodies would not be drawn from all points straight to the center of the earth, but would be drawn to different points from different directions.

If two stones were placed near to each other at some place in the universe outside the sphere of activity of some third related body, those stones would, like two magnetic bodies, meet at some intermediate point, each moving toward the other through a distance proportional to the other's bulk.

If the earth and the moon were not each retained in its orbit by an animate force or some other equivalent force, the earth would rise up toward the moon through one-fifty-fourth part of the distance [between them] and the moon would descend toward the earth about fifty-three fifty-fourths of the distance, and there they would meet; that is, this will follow on the assumption that the matter of each is of the same density.

If the earth should cease to attract its oceans, the water in all its seas would fly up and flow round the body of the moon.

The sphere of attractive power in the moon extends as far as the earth and, imperceptibly in land-locked seas, but measurably in the wide oceans . . . draws up the waters under the torrid zone. . . .

And it follows, if the attractive power of the moon extends as far as the earth, the attractive power of the earth must be even more likely to extend to the moon and far beyond, so that nothing in any way composed of terrestrial matter can, by being borne aloft, ever escape from the mighty embrace of this attractive power . . . .

Since no projectile ascends one-hundredth-thousandth part of the diameter of the earth above its surface, nor do the clouds or smoke containing a very slight amount of earthy matter rise as much as one-two-thousandth part, it follows that the resistance of these clouds, smoke, or projectiles, and their natural inclination to repose, can do nothing to obstruct this embrace [of the earth], in proportion to which any resistance belonging to these things is negligible. And so what is thrown upward returns to its [former] place without experiencing any interference from the motion of the earth, for this cannot be neutralized, but draws along with itself whatever flies through the air, constraining them by its magnetic power no less than if those bodies were in actual contact with it. . . .

AN ABSTRACT OF THE LEARNED TREATISE
OF JOHANNES KEPLERUS
THE EMPEROURS MATHEMATICIAN
ENTITLED HIS INTRODUCTION UPON MARS

It must be confessed, that there are very many who are devoted to Holinesse, that dissent from the judgement of Copernicus, fearing to give the lie to the Holy Ghost speaking in the Scriptures, if they should say, that the Earth moveth, and the Sun stands still. But let such consider, that since we judge of very many, and those the most principal things by the sense of seeing, it is impossible that we should alienate our speech from this sense of our eyes. Therefore many things daily occur, of which we speak according to the sense of sight, when as certainly we know that the things themselves are otherwise. . . .

But now the Sacred Scriptures, speaking to men of vulgar matters (in which they were not intended to instruct men) after the manner of men, so that they might be understood by men, do use such expressions as are granted by all, thereby to insinuate other things more mysterious and divine.

What wonder is it then, if the Scripture speaks according to man's apprehension, at such time when the truth of things doth dissent from the conception that all men, whether learned or unlearned have of them? Who knows not that it is a poetical allusion,

Psalm 19, where, whilst under the similitude of the Sun, the course of the Gospel, as also the peregrination of our Lord Christ in this world, undertaken for our sakes, is described, the Sun is said to come forth of his Tabernacle of the horizon, as a bridegroom out of his chamber, rejoicing as a giant to run a race? Which Vergil thus imitates; "Aurora leaving the golden bed of Tithonus." For the first poets were amongst the Jews.

The Psalmist knew that the Sun went not forth of the horizon, as out of its tabernacle, and yet it seemeth to the eye so to do: Nor did he believe, that the Sun moved, for that it appeared to his sight to do so. And yet he saith both, for that both were so to his seeming. Neither is it to be adjudged false in either sense: for the perception of the eyes hath its verity, fit for the more secret purpose of the Psalmist in shadowing forth the current passage of the Gospel, as also the peregrination of the Son of God. . . .

But the 104 Psalm is thought by some [the anti-Copernicans] to contain a discourse altogether physical, in regard it only concerns natural philosophy. Now God is there said, "To have laid the foundations of the Earth, that it should not be removed for ever." But here also the Psalmist is far from speculation of physical causes: For he doth wholly acquiesce in the greatness of God, who did all these things, and sings a hymn to God the Maker of them, in which he runneth over the world in order, as it appeared to his eyes. And if you well consider this Psalm, it is a paraphrase upon the six days work of the Creation . . . . He therefore here observeth that the Earth after so many ages hath not faltered, tired or decayed; when as notwithstanding no man hath yet discovered upon what it is founded. He goeth not about to teach men what they do not know, but putteth them in mind of what they neglect, to wit, the greatness and power of God in creating so huge a mass so firm and steadfast. If an astronomer should teach that the Earth is placed among the planets, he overthroweth not what the Psalmist here saith, not doth he contradict common experience; for it is true notwithstanding that the Earth, the Structure of God its Architect, doth not decay (as our buildings are wont to do) by age, or consume by worms, nor sway and lean to this or that side; that the seats and nests of living creatures are not molested; that the mountains and shores stand immovable against the violence of the winds and waves, as they were at the beginning. . . .

And I do also beseech my reader, not forgetting the divine

goodness conferred on mankind; the consideration of which the Psalmist doth chiefly urge, that when he returneth from the Temple, and enters into the school of astronomy, he would with me praise and admire the Wisdom and Greatness of the Creator, which I discover to him by a more narrow explication of the world's form, the disquisition of causes, and detection of the errors of sight: And so he will not only extol the bounty of God in the preservation of living creatures of all kinds, and establishment of the Earth; but even in its motion also, which is so strange, so admirable, he will acknowledge the wisdom of the Creator. But he who is so stupid as not to comprehend the science of astronomy, or so weak and scrupulous as to think it an offence of piety, to adhere to Copernicus, him I advise, that leaving the study of astronomy, and censuring the opinions of the philosophers at pleasure, he betake himself to his own concerns, and that desisting from further pursuit of these intricate studies, he keep at home and manure his own ground; and with these eyes wherewith alone he seeth, being elevated towards this to be admired heaven, let him pour forth his whole heart in thanks and praises to God the Creator; and assure himself that he shall therein perform as much worship to God, as the astronomer, on whom God hath bestowed this gift, that though he seeth more clearly with the eye of the understanding; yet whatever he hath attained to, he is both able and willing to extol his God above it. And thus much concerning the authority of the Sacred Scripture. . . .[9]

But enough of the truth of the Copernican hypothesis. Let us return to the design I had at the beginning of this Introduction.

I began to say that in this work I treat the whole of astronomy according to physical causes, not according to fictitious hypotheses. I have striven toward this end in two steps [of argument]: by discovering, first, that the eccentrics of the planets focus upon the sun, and second, that the theory of the earth required an equant circle and the bisection of its eccentricity.[10]

Here now is the third step: for it was demonstrated . . . that the eccentricity of the equant of Mars is also to be precisely

9. This is the end of the text from Salusbury's translation.
10. An equant circle in Ptolemaic astronomy is the center of uniform motion of the planet. To bisect the eccentricity is to divide the line joining the equant and the center of the eccentric circle. All these complications were necessary if perfectly circular motion was to be employed, before the discovery of Kepler's second law.

bisected; long ago Copernicus and Tycho were in doubt about this.

Whence by induction from all the planets . . . it is demonstrated that, since there are no solid spheres (as Tycho showed from the paths of comets), the body of the sun is the source of the power which drives all the planets. I have also reasoned in such a way that it becomes clear that the sun, remaining in its place but rotating on its axis, emits through the breadth of the universe an immaterial *species*[11] of its body analogous to the immaterial *species* of its light. This *species* of the rotation of the body of the sun is itself also rotated like a very rapid vortex throughout the breadth of the world, and it bears along with itself in its revolution the bodies of the planets with a more or less intense violence, according to the law of the greater or less density of its emission.

Having set forth this common power by which the planets—each in its circle—are borne around the sun, it follows logically in my reasoning that individual moving agents should be attributed to each planet and located in their globes, since I have already rejected the solid spheres on the authority of Tycho.

It is incredible how much labor the moving powers constituted by this way of argument caused me in the fourth part [of the book], giving false distances of the planets from the sun, irreconcilable with the observations, when I tried to work out the equations of the eccentric. This was not because these moving powers were wrongly invoked, but because I had forced them to tramp around in circles like donkeys in a mill, being bewitched by common opinion. Restrained by these fetters, they could not do their work.

Nor was my burdensome task finished before I had taken the fourth step in the building of my physical hypotheses; after most laborious demonstrations and analyses of many observations, I learned that the path of a planet in the sky is not a circle, but is rather an oval course, perfectly elliptical.[12]

Geometry then came to my aid and taught me that such a path

11. A technical term in scholastic philosophy for an emission or emanation—a concept which was ultimately replaced by the combined concepts of properties and forces.

12. This is Kepler's first law, which he found extraordinarily laborious to calculate. This was in part because the ellipse did not appear to him to be such a simple and natural shape as it does to us; as he said at the beginning of the Preface, conic sections (ellipse, parabola, hyperbola) were little studied in his day.

would be traced if we assign to each of the moving agents of the planets the task of librating[13] its body in a straight line toward the sun. Not only this, but equations of the eccentric that are accurate and in conformity with the observations are effected by such a libration.

Thus at length the coping stone was placed on the building, and it was demonstrated geometrically that a libration of this sort could be performed by a corporeal magnetic faculty. So these very moving agents of the planets are shown to be most probably nothing but attributes of the planetary bodies themselves, like the attributes of a magnet such as seeking for the Pole and attracting iron; thus the whole theory of the celestial motions may be based upon merely corporeal, that is magnetic, faculties, except for the gyration of the body of the sun remaining in its space, for which a vital faculty seems to be needed . . . .

6. The New Copernicanism

The Author's Introduction

Judicious Reader,

There was published some years since in Rome a salutiferous Edict, that, for the obviating of the dangerous scandals of the present Age, imposed a seasonable silence upon the Pythagorean Opinion of the Mobility of the Earth.[1] There want not such as unadvisedly affirm, that that Decree was not the production of a sober scrutiny, but of an ill informed passion; and one may hear

SOURCE: Galileo, *Dialogo sopra i due Massimi Sistemi del Mondo* (Florence, 1632); in the translation of Thomas Salusbury, *Dialogue on the Two Chief Systems of the World*, in *Mathematical Collections*, I (London, 1661), Introduction, pp. 54-60, 126-134, 423-24.

13. Moving in apparent oscillation, or rocking.

1. The Copernican system was often called "Pythagorean," both because it was genuinely believed at this time that the Pythagoreans had devised a solar system equivalent to the Copernican and because to emphasize its antiquity gave the system greater respectability. The decree mentioned by Galileo was that of 1616, when the Copernican theory was condemned by the Holy Office and *De Revolutionibus* was placed on the Index "until corrected" (which in fact no one ever bothered to do).

some mutter that Consultors altogether ignorant of Astronomical Observations ought not to clip the Wings of Speculative Wits with rash prohibitions. My zeal cannot keep silence when I hear these inconsiderate complaints. I thought fit, as being thoroughly acquainted with that prudent Determination, to appear openly upon the Theatre of the World as a Witness of the naked Truth. I was at that time in Rome, and had not only the audiences, but applauds of the most Eminent Prelates of that Court; nor was that Decree published without previous notice given me thereof. Therefore it is my resolution in the present case to give Foreign Nations to see, that this point is as well understood in Italy, and particularly in Rome, as Transalpine diligence can imagine it to be: and collecting together all the proper speculations that concern the Copernican System, to let them know, that the notice of all preceded the censure of the Roman Court; and that there proceed from this Climate not only Doctrines for the health of the Soul, but also ingenious Discoveries for the recreating of the mind.

To this end I have personated the Copernican in this Discourse; proceeding upon an Hypothesis purely Mathematical; striving by all artificial ways to represent it superior, not to that of the immobility of the Earth absolutely, but according as it is mentioned by some, that retain no more, but the name of Peripateticks, and are content, without going farther, to adore Shadows, not philosophizing with requisite caution, but with the sole remembrance of four Principles, but badly understood.

We shall treat of three principal heads. First I will endeavour to show that all Experiments that can be made upon the Earth are insufficient means to conclude its Mobility, but are indifferently applicable to the Earth moveable or immoveable: and I hope that on this occasion we shall discover many observable passages unknown to the Ancients. Secondly we will examine the Celestial Phenomena that make for the Copernican Hypothesis, as if it were to prove absolutely victorious; adding by the way certain new observations, which yet serve only for the Astronomical Facility, not for Natural Necessity. In the third place I will propose an ingenuous fancy. I remember that I have said many years since, that the unknown problem of the Tides might receive some light, admitting the Earth's Motion. This position of mine passing from one to another had found charitable Fathers that adopted it for the issue of their own wit. Now, because no stranger may ever appear that defending himself with our arms, shall charge us with want of caution in so principal an Accident, I have thought good to lay down those probabilities that would render it credible, admitting that the Earth did move. I hope, that by these considerations the World will come to know, that if other Nations have navigated more than we, we have not studied less than they; and that our returning to assert the Earth's stability, and to take the contrary only for a Mathematical Capriccio, proceeds not from inadvertency of what others have thought thereof, but (had we no other inducements) from those reasons that Piety, Religion, the knowledge of the Divine Omnipotency, and a consciousness of the incapacity of man's understanding dictate unto us.

Withall I conceived it very proper to express these conceits by way of Dialogue, which, as not being bound up to the rigid observance of Mathematical Laws, gives place also to digressions that are sometimes no less curious than the principal argument.

I chanced to be several years since, at several times, in the stupendous City of Venice, where I conversed with Signore Francesco Sagredo of Noble Extraction, and piercing wit. There came thither from Florence at the same time Signore Filippo Salviati, whose least glory was the Eminence of his Blood, and Magnificence of his Estate: a sublime wit that fed not more hungrily upon any pleasure than on elevated speculations. In the company of these two I often discoursed of these matters before a certain Peripatetick Philosopher who seemed to have no greater obstacle in understanding of the Truth, than the Fame he had acquired by Aristotelical Interpretations.

Now, seeing that inexorable Death hath deprived Venice and Florence of those two great Lights in the very Meridian of their years, I did resolve, as far as my poor ability would permit, to perpetuate their lives to their honour in these leaves, bringing them in as interlocutors in the present controversy. Nor shall the honest Peripatetick want his place, to whom for his excessive affection toward the commentaries of Simplicius, I thought fit, without mentioning his own name, to leave that of the Author he so much respected. Let those two great Souls, ever venerable to my heart, please accept this public monument of my never dying Love; and let the remembrance of their Eloquence assist me in delivering to posterity the considerations that I have promised.

There casually happened (as was usual) several discourses at times between these Gentlemen, the which had rather inflamed than satisfied in their wits the thirst they had to be learning; whereupon they took a different resolution to meet together for certain days, in which all other business set aside, they might betake themselves more methodically to contemplate the Wonders of God in Heaven, and in the Earth: the place appointed for their meeting being in the Palace of the Noble Sagredo; after the due, but very short complements, Signore Salviati began in this manner . . . .

The Probability of the Earth's Motion

SALV. Let our contemplation begin therefore with this consideration, that whatsoever motion may be ascribed to the Earth, it is necessary that it be to us, (as inhabitants upon it, and consequently partakers of the same) altogether imperceptible, and as if it were not at all, so long as we have regard only to terrestrial things; but yet it is on the contrary, as necessary that the same motion do seem common to all other bodies, and visible objects, that being separated from the Earth participate not of the same. So that the true method to find whether any kind of motion may be ascribed to the Earth, and that found, to know what it is, is to consider and observe if in bodies separated from the Earth, one may discover any appearance of motion, which equally suiteth to all the rest; for a motion that is only seen, e.g. in the Moon, and that hath nothing to do with Venus or Jupiter, or any other Stars, cannot any way belong to the Earth, or to any other save the Moon alone. Now there is a most general and grand motion above all others, and it is that by which the Sun, the Moon, the other Planets, and the Fixed Stars, and in a word, the whole Universe, the Earth only excepted, appeareth in our thinking to move from the East towards the West, in the space of twenty-four hours; and this, as to this first appearance, hath no obstacle to hinder it, that it may not belong to the Earth alone, as well as to all the world besides, the Earth excepted; for the same aspects will appear in the one position, as in the other. Hence it is that Aristotle and Ptolemy, as having hit upon this consideration, in going about to prove the Earth to be immoveable, argue not against any other than this diurnal motion; save only that Aristotle hinteth something in obscure terms against another motion ascribed to it by an Ancient, of which we shall speak in its place. . . .

SAGR. I find my fancy disturbed with certain conjectures so confusedly sprung from your later discourses; that, if I would be enabled to apply myself with attention to what followeth, I must of necessity attempt whether I can better methodize them, and gather thence their true construction, if haply any can be made of them; and peradventure, the proceeding by interrogations may help me the more easily to express myself. Therefore, I demand first to Simplicius, whether he believeth, that divers motions may naturally agree to one and the same moveable body, or else that it be requisite its natural and proper motion be only one.

SIMPL. To one single moveable, there can naturally agree but one sole motion, and no more; the rest all happen accidentally and by participation; like as to him that walketh upon the deck of a ship, his proper motion is that of his walk, his motion by participation that which carrieth him to his port, whither he would never with his walking have arrived, if the ship with its motion had not wafted him thither.

SAGR. Tell me secondly. That motion, which is communicated to any moveable by participation, whilst it moveth by itself, with another motion different from the participated, is it necessary, that it do residue in some certain subject by itself, or else can it subsist in nature alone, without other support?

SIMPL. Aristotle giveth you an answer to all these questions, and tells you, that as of one sole moveable the motion is but one; so of one sole motion the moveable is but one; and consequently, that without the inherence in its subject, no motion can either subsist, or be imagined.

SAGR. I would have you tell me in the third place, whether you believe that the Moon and other Planets and celestial bodies, have their proper motions, and what they are.

SIMPL. They have so, and they be those according to which they run through the Zodiac, the Moon in a month, the Sun in a year, Mars in two, the Starry Sphere in those so many thousand. And these are their proper, or natural motions.

SAGR. But that motion wherewith I see the fixed Stars, and with them all the Planets go unitedly from East to West, and return round to the East again in twenty-four hours, how doth it agree with them?

SIMPL. It suiteth with them by participation.

SAGR. This then resides not in them, and not residing in them, nor being able to subsist without some subject in which it is resident, it must of force be the proper and natural motion of some other sphere.

SIMPL. For this purpose Astronomers, and Philosophers have found another high sphere, above all the rest, without Stars, to which natural agreeth the diurnal motion; and this they call the Primum mobile; the which carrieth along with it all the inferior spheres, contributing and imparting its motion to them.

SAGR. But when, without introducing other spheres unknown and hugely vast, without other motions or communicated raptures, with leaving to each sphere its sole and simple motion, without intermixing contrary motions, but making all turn one way, as it is necessary that they do, depending all upon one sole principle, all things proceed orderly, and correspond with most perfect harmony, why do we reject this Phenomenon, and give our assent to those prodigious and laborious conditions?

SIMPL. The difficulty lieth in finding out this so natural and expeditious way.

SAGR. In my judgement this is found. Make the Earth the Primum mobile, that is, make it turn round its own axis in twenty-four hours, and towards the same point with all the other spheres; and without participating this same motion to any other Planet or Star, all shall have their risings, settings, and in a word, all their other appearances.

SIMPL. The business is, to be able to make the Earth move without a thousand inconveniences.

SALV. All the inconveniences shall be removed as fast as you propound them: and the things spoken hitherto are only the primary and more general inducements which give us to believe that the diurnal conversion may not altogether without probability be applied to the Earth, rather than to all the rest of the universe: the which inducements I impose not upon you as inviolable axioms, but as hints, which carry with them somewhat of likelihood. And in regard I know very well, that one sole experiment, or concludent demonstration, produced on the contrary part, sufficeth to batter to the ground these and a thousand other probable arguments; therefore it is not fit to stay here, but proceed forwards and hear what Simplicius answereth, and what greater probabilities, or stronger arguments he allegeth on the contrary.

SIMPL. I will first say something in general upon all these considerations together, and then I will descend to some particulars. It seems that you universally bottom all you say upon the greater simplicity and facility of producing the same effects, whilst you hold, that as to the causing of them, the motion of the earth alone, serveth as well as that of all the rest of the world, the Earth deducted: but as to the operations, you esteem that much easier than this. To which I reply, that I am also of the same opinion, so long as I regard my own not only finite, but feeble power; but having a respect to the strength of the mover, which is infinite, it's no less easy to move the universe, than the Earth, yea than a straw. And if his power be infinite, why should he not rather exercise a greater part thereof than a less? Therefore, I hold that your discourse in general is not convincing.

SALV. If I had at any time said, that the universe moved not for want of power in the Mover, I should have erred, and your reproof whould have been seasonable; and I grant you, that to an infinite power, it is as easy to move a hundred thousand, as one. But that which I did say, concerns not the Mover, but only hath respect to the moveables; and in them, not only to their resistance, which doubtless is lesser in the Earth, than in the universe; but to the many other particulars, but even now considered. As to what you say in the next place, that of an infinite power it is better to exercise a great part than a small: I answer, that of infinite one part is not greater than another, since both are infinite; nor can it be said, that of the infinite number, a hundred thousand is a greater part than two, though that be fifty thousand times greater than this; and if to the moving of the universe there be required a finite power, though very great in comparison of that which sufficeth to move the Earth only; yet is there not implied therein a greater part of the infinite power, nor is that part less infinite which remaineth unimployed. So that to apply unto a particular effect, a little more, or a little less power, importeth nothing, besides that the operation of such virtue, hath not for its bound or end the diurnal motion only; but there are several other motions in the world, which we know of, and many others there may be, that are to us unknown. Therefore if we respect the moveables, and granting it as out of question, that it is a shorter and easier way to move the Earth, than the universe; and moreover, having an eye to the so many other abbreviations, and facilities that only this way are to be obtained, an infallible maxim of Aristotle, which he teacheth us, that "it is vain to assign many causes when few will do," rendereth it more probable that the diurnal motion belongs to the Earth alone, than to the universe, the Earth subducted.

SIMPL. In reciting that axiom, you have omitted a small clause, which importeth as much as all the rest, especially in our case, that is to say, the words "equally well." It is requisite therefore to examine whether this hypothesis doth *equally well* satisfy in all particulars, as the other.

SALV. The knowledge whether both these positions do equally well satisfy, may be comprehended from the particular examination of the appearances which they are to satisfy; for hitherto we have discoursed, and will continue to argue *ex hypothesi,* namely, supposing, that as to the satisfaction of the appearances, both the assumptions are equally accommodated. As to the clause which you say was omitted by me, I have more reason to suspect that it was superfluously inserted by you. For the expression "equally well," is a relative that necessarily requireth two terms at least, for a thing cannot have relation to itself, nor do we say, e.g. rest to be "equally good," as rest. And because, when we say,"that is done in vain by many means, which may be done with fewer," we mean, that that which is to be done, ought to be the same thing, not two different ones; and because the same thing cannot be said to be done as well as itself; therefore, the addition of the phrase "equally well" is superfluous. . . .

The Behaviour of Moving Bodies

SIMPL. If you should refer me to any other means than to experience, I verily believe our Disputations would not come to an end in haste; for this seemeth to me a thing so remote from all human reason, as that it leaveth not the least place for credulity or probability.[2]

SALV. And yet it hath left place in me for both.

SIMPL. How is this? You have not made a hundred, no nor one

2. Salviati has just argued that a stone dropped from the top of the mast of a ship falls in exactly the same way, and hits the same point of the deck, whether the ship is moving or standing still; hence the behavior of falling bodies gives no disproof of the earth's motion. In the section that follows, Galileo indulges in some particularly ingenious argument, as well as publishing for the first time some of his conclusions on the behavior of moving bodies.

proof thereof, and do you so confidently affirm it for true? I for my part will return to my incredulity, and to the confidence I had that the experiment hath been tried by the principal authors who made use thereof, and that the event succeeded as they affirm.

SALV. I am assured that the effect will ensue as I tell you; for so it is necessary that it should: and I farther add, that you know yourself that it cannot fall out otherwise, however you feign or seem to feign that you know it not. Yet I am so good at taming of wits, that I will make you confess the same whether you will or no. But Sagredus stands very mute, and yet, if I mistake not, I saw him make an offer to speak somewhat.

SAGR. I had an intent to say something, but to tell you true, I know not what it was, for the curiosity that you have moved in me, by promising that you would force Simplicius to discover the knowledge which he would conceal from us, hath made me to depose all other thoughts: therefore I pray you to make good your vaunt.

SALV. Provided that Simplicius do consent to reply to what I shall ask him, I will not fail to do it.

SIMPL. I will answer what I know, assured that I shall not be much put to it, for that of those things which I hold to be false, I think nothing can be known, in regard that science respecteth truths and not falsehoods.

SALV. I desire not that you should say or reply, that you know anything, save that which you most assuredly know. Therefore tell me; if you had here a flat superficies as polished as a looking-glass, and of a substance as hard as steel, and that it were not parallel to the horizon, but somewhat inclining, and that upon it you did put a ball perfectly spherical, and of a substance grave [heavy] and hard, as suppose of brass; what think you it would do being let go? Do not you believe (as for my part I do) that it would lie still?

SIMPL If that superficies were inclining?

SALV. Yes: for so I have already supposed.

SIMPL. I cannot conceive how it should lie still: nay, I am confident that it would move towards the declivity with much propenseness.

SALV. Take good heed what you say, Simplicius, for I am confident that it would lie still in whatever place you should lay it.

SIMPL. So long as you make use of such suppositions, Salviatus, I shall cease to wonder if you infer most absurd conclusions.

SALV. Are you assured then, that it would freely move towards the declivity?

SIMPL. Who doubts it?

SALV. And this very verily believe, not because I told you so, (for I endeavoured to persuade you think the contrary) but of yourself, and upon your natural judgment.

SIMPL. Now I see what you would be at; you spoke not this as really believing the same; but to try me, and to wrest matter out of my own mouth wherewith to condemn me.

SALV. You are in the right. And how long would that ball move, and with what velocity? But take notice that I instanced in a ball exactly round, and a plane exquisitely polished, that all external and accidental impediments might be taken away. And so would I have you remove all obstructions caused by the air's resistance to division, and all other casual obstacles, if any other there can be.

SIMPL. I very well understand your meaning, and as to your demand, I answer, that the ball would continue to move in infinitum, if the inclination of the plane should so long last, and continually with an accelerating motion; for such is the nature of ponderous moveables, that they acquire force in moving: and the greater the declivity was, the greater the velocity would be.

SALV. But if one should require that the ball should move upwards on that same superficies, do you believe that it would so do?

SIMPL. Not spontaneously; but being drawn, or violently thrown, it may.

SALV. And in case it were thrust forward by the impression of some violent impetus from without, what and how great would its motion be?

SIMPL. The motion would go continually decreasing and retarding, as being contrary to nature; and would be longer or shorter, according to the greater or less impulse, and according to the greater or less acclivity.

SALV. It seems, then, that hitherto you have explained to me the accidents of a moveable upon two different planes; and that in the inclining plane, the grave [heavy] moveable doth spontaneously descend and goeth continually accelerating, and that to retain it in rest, force must be used therein: but that on the ascending plane, there is required a force to thrust it forward, and also to stay it in rest, and that the motion impressed goeth continually diminishing, till that in the end it cometh to nothing. You say yet farther, that in both the one and the other case, there do arise differences from the planes having a greater or less declivity or acclivity; so that the greater inclination is attended with the greater velocity; and contrariwise, upon the ascending plane, the same moveable thrown with the same force, moveth a greater distance, by how much the elevation is less; Now tell me, what would befall the same moveable upon a superficies that had neither acclivity nor declivity?

SIMPL. Here you must give me a little time to consider of an answer. There being no declivity, there can be no natural inclination to motion: and there being no acclivity, there can be no resistance to being moved; so that there would arise an indifference between propension and resistance of motion; therefore, methinks it ought naturally to stand still. But I had forgot myself: it was but even now that Sagredus gave me to understand that it would so do.

SALV. So I think, provided one did lay it down gently: but if it had an impetus given it towards any part, what would follow?

SIMP. There would follow, that it should move towards that part.

SALV. But with what kind of motion? with the continually accelerated, as in declining planes; or with the successively retarded, as in those ascending?

SIMPL. I cannot tell how to discover any cause of acceleration, or retardation, there being no declivity or acclivity.

SALV. Well: but if there be no cause of retardation, much less ought there to be any cause of rest. How long therefore would you have the moveable to move?

SIMPL. As long as that superficies, neither inclined nor declined shall last.

SALV. Therefore if such a space were interminate, the motion upon the same would likewise have no termination, that is, would be perpetual.

SIMPL. I think so, if so be the moveable be of a matter durable.

SALV. That hath been already supposed, when it was said, that all external and accidental impediments were removed, and the brittleness of the moveable in this our case, is one of those impediments accidental. Tell me now, what do you think is the cause that that same ball moveth spontaneously upon the inclining plane, and not without violence upon the erected?

SIMPL. Because the inclination of grave bodies is to move towards the centre of the Earth, and only by violence upwards towards the circumference; and the inclining superficies is that which acquireth vicinity to the centre, and the ascending one, remoteness.

SALV. Therefore a superficies, which should be neither declining nor ascending, ought in all its parts to be equally distant from the centre. But is there any such superficies in the world?

SIMPL. There is no want thereof: Such is our terrestrial globe, if it were more even, and not as it is rough and mountainous; but you have that of the water, at such time as it is calm and still.

SALV. Then a ship which moveth in a calm at sea, is one of those moveables, which run along one of those superficies that are neither declining nor ascending, and therefore disposed, in case all obstacles external and accidental were removed, to move with the impulse once imparted incessantly and uniformly.

SIMPL. It should seem to be so.

SALV. And that stone which is on the round top, doth not it move, as being together with the ship carried about by the circumference of a circle about the centre; and therefore consequently by a motion in it indelible, if all external obstacles be removed? And is not this motion as swift as that of the ship?

SIMPL. Hitherto all is well. But what followeth?

SALV. Then in good time recant, I pray you, that your last conclusion, if you are satisfied with the truth of all the premises.

SIMPL. By my last conclusion, you mean, That that same stone moving with a motion indelibly impressed upon it, is not to leave, nay rather is to follow the ship, and in the end to light in the selfsame place, where it falleth when the ship lieth still; and so I also grant it would do, in case there were no outward impediments that might disturb the stone's motion, after its being let go, the which impediments are two, the one is the moveable's inability to break through the air with its mere impetus only, it being deprived of that of the strength of oars, of which it had been partaker, as part of the ship, at the time that it was upon the mast; the other is the new motion of descent, which also must needs be an hinderance of that other progressive motion.

SALV. As to the impediment of the air, I do not deny it you; and if the thing falling were a light matter, as a feather, or a lock of wool, the retardation would be very great, but in a heavy stone is very exceeding small. And you yourself but even now did say, that the force of the most impetuous wind sufficeth not to stir a great stone from its place; now do but consider what the calmer air is able to do, being encountered by a stone no more swift than the whole ship. Nevertheless, as I said before, I do allow you this small effect, that may depend upon such an impediment; like as I know, that you will grant to me, that if the air should move with the same velocity that the ship and stone hath, then the impediment would be nothing at all. As to the other of the additional motion downwards; in the first place it is manifest, that these two, I mean the circular, about the centre, and the straight, towards the centre, are not contraries, or destructive to one another, or incompatible. Because that as to the moveable, it hath no repugance at all to such motions, for you yourself have already confessed the repugnance to be against the motion which removeth from the centre, and the inclination to be towards the motion which approacheth to the centre; Whence it doth of necessity follow, that the moveable hath neither repugnance, nor propension to the motion which neither approacheth, nor goeth from the centre, nor consequently is there any cause for the diminishing in it the faculty impressed. And forasmuch as the moving cause is not one alone, which it hath attained by the new operation of retardation; but that they are two, distinct from each other, of which, the gravity attends only to the drawing of the moveable towards the centre, and the virtue impressed to the conducting it about the centre, there remaineth no occasion of impediment.

SIMPL. Your argumentation, to give you your due, is very probable; but in reality it is enveloped with certain intricacies, that are not easy to be extricated. You have all along built upon a supposition, which the Peripatetick Schools will not easily grant you, as being directly contrary to Aristotle, and it is to take for known and manifest, That the project [projectile] separated from the projicient, continueth the motion by virtue impressed on it by the said projicient, which virtue impressed is a thing as much detested in Peripatetick Philosophy, as the passage of any accident from one subject into another. Which doctrine doth hold, as I believe it is well known unto you, that the project is carried by the medium, which in our case happeneth to be the air. And therefore if that stone let fall from the round top, ought to follow the motion of the ship, that effect should be ascribed to the air, and not to the virtue impressed. But you presuppose that the air doth not follow the motion of the ship, but is tranquil. Moreover, he that letteth it fall, is not to throw it, or to give it impetus with his arm, but ought barely to open his hand and let it go; and by this means, the stone, neither through the virtue impressed by the projicient, nor through the help of the air, shall be able to follow the ship's motion, and therefore shall be left behind.

SALV. I think then that you would say, that if the stone be not thrown by the arm of that person, it is no longer a projection.

SIMPL. It cannot be properly called a motion of projection.

SALV. So then that which Aristotle speaks of the motion, the moveable, and the mover of the projects, hath nothing to do with the business in hand; and if it concern not our purpose, why do you allege the same?

SIMP. I produce it on the occasion of that impressed virtue, named and introduced by you, which having no being in the world, can be of no force; for the non existent has no operations, and therefore not only of projected, but of all other preternatural motions, the moving cause ought to be ascribed to the medium, of which there hath been no due consideration had; and therefore all that hath been said hitherto is to no purpose.

SALV. Go to now, in good time. But tell me, seeing that your instance is wholly grounded upon the nullity of the virtue impressed, if J shall demonstrate to you, that the *medium* hath nothing to do in the continuation of projects, after they are separated from the projicient, will you admit of the impressed virtue, or will you make another attempt to overthrow it?

SIMP. The operation of the medium being removed, I see not how one can have recourse to anything else save the faculty impressed by the mover.

SALV. It would be well, for the removing, as much as is possible, the occasions of multiplying contentions, that you would explain with as much distinctness as may be, what is that operation of the medium in continuing the motion of the project.

SIMP. The projicient hath the stone in his hand, and with force and violence throws his arm, with which jactation the stone doth not move so much as the circumambient air; so that when the stone at its being forsaken by the hand, findeth itself in the air, which at the same time moveth with impetuosity, it is thereby borne away; for, if the air did not operate, the stone would fall at the foot of the projicient or thrower.

SALV. And were you so credulous, as to suffer yourself to be persuaded to believe these fopperies, so long as you had your senses about you to confute them, and to understand the truth thereof? Therefore tell me, that great stone, and that cannon bullet, which but only laid upon a table, did continue immoveable against the most impetuous winds, according as you a little before did affirm, if it had been a ball of cork or other light stuff, think you that the wind would have removed it from its place?

SIMP. Yes, and I am assured that it would have blown it quite away, and with so much more velocity, by now much the matter was lighter, for upon this reason we see the clouds to be transported with a velocity equal to that of the wind that drives them.

SALV. And what is the wind?

SIMP. The wind is defined to be nothing else but air moved.

SALV. Then the moved air doth carry light things more swiftly, and to a greater distance, than it doth heavy.

SIMP. Yes certainly.

SALV. But if you were to throw with your arm a stone, and a lock of cotton wool, which would move swiftest and farthest?

SIMP. The stone by much, nay the wool would fall at my feet.

SALV. But, if that which moveth the projected substance, after it is delivered from the hand, be no other than the air moved by the arm, and the moved air do more easily bear away light than grave matters, how cometh it that the project of wool flieth not farther, and swifter than that of stone? Certainly it argueth that the stone hath some other impulse besides the motion of the air. Furthermore, if two strings of equal length did hang at yonder beam, and at the end of one there was fastened a bullet of lead, and a ball of cotton wool at the other, and both were carried to an equal distance from the perpendicular, and then let go; it is not to be doubted, but that both the one and the other would move towards the perpendicular, and that being carried by their own impetus, they would go a certain space beyond it, and afterwards return thither again. But which of these two pendent globes do you think, would continue longest in motion, before that it would come to rest in its perpendicularity?

SIMP. The ball of lead would swing to and again many times, and that of wool but two or three at the most.

SALV. So that that impetus and that mobility whatsoever is the cause thereof, would conserve itself longer in grave substances, than light; I proceed now to another particular, and demand of you, why the air doth not carry away that lemon which is upon that same table?

SIMP. Because that the air itself is not moved.

SALV. It is requisite then, that the projicient confer motion on the air, with which it afterward moveth the project. But if such a motion cannot be impressed (i.e. imparted), it being impossible to make an accident pass out of one subject into another, how can it pass from the arm into the air? Will you say that the air is not a subject different from the arm?

SIMP. To this it is answered that the air, in regard it is neither heavy nor light in its own region, is disposed with facility to receive every impulse, and also to retain the same.

SALV. But if those penduli even now named, did prove unto us, that the moveable, the less it had of gravity, the less apt it was to conserve its motion, how can it be that the air which in the air hath no gravity at all, doth of itself alone retain the motion acquired? I believe, and know that you by this time are of the same opinion, that the arm doth not sooner return to rest, than doth the circumambient air. Let's go into the chamber, and with a towel let us agitate the air as much as we can, and then holding the cloth still, let a little candle be brought, that was lighted in the next room, or in the same place let a leaf of beaten gold be left at liberty to fly any way, and you shall by the calm vagation of them be assured that the air is immediately reduced to tranquility. I could allege many other experiments to the same purpose, but if one of these should not suffice, I should think your folly altogether incurable.

SAGR. When an arrow is shot against the wind, how incredible a thing is it, that that same small filament of air, impelled by the bow string, should in despite of fate go along with the arrow? But I would willingly know another particular of Aristotle, to which I intreat Simplicius would vouchsafe me an answer. Supposing that with the same bow there were shot two arrows, one just after the usual manner, and the other side ways, placing it long ways upon the bow string, and then letting it fly, I would know which of them would go farthest. Favour me, I pray you with an answer, though the question may seem to you rather ridiculous than otherwise; and excuse me, for that I, who am, as you see, rather blockish, than not, can reach no higher with my speculative faculty.

SIMPL. I have never seen an arrow shot in that manner, yet nevertheless I believe, that it would not fly side-long, the twentieth part of the space that it goeth end-ways.

SAGR. And for that I am of the same opinion, hence it is, that I have a doubt risen in me, whether Aristotle doth not contradict experience. For as to experience, if I lay two arrows upon this table, in a time when a strong wind bloweth, one towards the course of the wind, and the other side-long, the wind will quickly carry away this latter, and leave the other where it was; and the same to my seeming, ought to happen, if the doctrine of Aristotle were true, of those two shot out of a bow: forasmuch as the arrow shot sideways is driven by a great quantity of air, moved by the bow string, to wit by as much as the said string is long, whereas the other arrow receiveth no greater a quantity of air, than the small circle of the string's thickness. And I cannot imagine what may be the reason of such a difference, but would fain know the same.

SIMP. The cause seemeth to me sufficiently manifest; and it is because the arrow shot endways, hath but a little quantity of air to penetrate, and the other is to make its way through a quantity as great as its whole length.

SALV. Then it seems the arrows shot, are to penetrate the air? but if the air goeth along with them, yea is that which carrieth them, what penetration can they make therein? Do you not see that, in this case, the arrow would of necessity move with greater velocity than the air? and this greater velocity, what doth confer it on the arrow? Will you say the air giveth them a velocity greater than its own? Know then, Simplicius, that the business proceeds quite contrary to that which Aristotle saith, and that the medium conferreth the motion on the project, is as false, as it is true, that it is the only thing which procureth its obstruction; and having known this, you shall understand without finding any thing whereof to make question, that if the air be really moved, it doth much better carry the dart along with it long-ways, then endways, for that the air which impelleth it in that posture, is much, and in this very little. But shooting with the bow, forasmuch as the air stands still, the transverse arrow, being to force its passage through much air, comes to be much impeded, and the other that was knocked easily overcometh the obstruction of the small quantity of air, which opposeth itself thereto. . . .

The Conclusion

SALV. Now because it is time to put an end to our discourses, it remaineth, that I intreat you, that if, at more leisure going over the things again that have been alleged, you meet with any doubts, or scruples not well resolved, you will excuse my oversight, as well for the novelty of the notion, as for the weakness of my wit, as also for the grandeur of the subject, as also finally, because I do not, nor have pretended to that assent from others, which I myself do not give to this conceit, which I could very easily grant to be a chimera and a mere paradox; and you Sagredus, although in the discourses past you have many times, with great applause, declared, that you were pleased with some of my conjectures, yet do I believe, that that was in part more occasioned by the novelty than by the certainty of them, but much more by your courtesy, which did think and desire, by its assent, to procure me that content which we naturally use to take in the approbation and applause of our own matters: and as your civility hath obliged me to you; so am I also pleased with the ingenuity of Simplicius. Nay, his constancy in maintaining the doctrine of his Master, with so much strength and undauntedness, hath made me much to love him. And as I am to give you thanks, Sagredus, for your courteous affection; so of Simplicius, I ask pardon, if I have sometimes moved him with my too bold and resolute speaking: and let him be assured that I have not done the same out of any inducement of sinister affection, but only to give him occasion to set before us more lofty fancies that might make me the more knowing.

SIMP. There is no reason why you should make all these excuses, that are needless, and especially to me, that being accustomed to be at conferences and public disputes, have a hundred times seen the disputants not only to grow hot and angry with one another, but likewise to break forth into injurious words, and sometimes to come very near to blows. As for the past discourses, and particularly in this last, of the reason of the ebbing and flowing of the sea, I do not, to speak the truth, very well apprehend the same, but by that slight idea, what ever it be, that I have formed thereof to myself, I confess that your conceit seemeth to me far more ingenuous [ingenious] than any of all those I have ever heard besides, but yet nevertheless I esteem it not true and concluding:

but keeping always before the eyes of my mind a solid doctrine that I have learned from a most learned and ingenuous person, and with which it is necessary to sit down; I know that both [of] you being asked, Whether God, by his infinite Power and Wisdom, might confer upon the element of water the reciprocal motion which we observe in the same in any other way, than by making the containing vessel to move; I know, I say, that you will answer, that He might, and knew how to have done the same many ways and those unimaginable to our shallow understanding: upon which I forthwith conclude, that this being granted, it would be an extravagant boldness for any one to go about to limit and confine the Divine power and Wisdom to some one particular conjecture of his own.

SALV. This of yours is admirable, and truly angelical doctrine, to which very exactly that other accords, in like manner divine, which whilst it giveth us leave to dispute, touching the constitution of the world, addeth withall (perhaps to the end, that the exercise of the minds of men might neither be discouraged nor made bold) that we cannot find out the works made by His hands. Let therefore the disquisition permitted and ordained us by God, assist us in the knowing, and so much more admiring His greatness, by how much less we find ourselves too dull to penetrate the profound abysses of His infinite wisdom.

SAGR. And this may serve for a final close of our four days disputations, after which, if it seem good to Salviatus, to take some time to rest himself, our curiosity must, of necessity, grant him the same, yet upon condition that when it is less incommodious to him, he will return and satisfy my desire in particular concerning the problems that remain to be discussed and that I have set down to be propounded at one or two other conferences, according to our agreement: and above all, I shall very impatiently wait to hear the elements of the new science of our academick about the natural and violent local motions.[3] And in the mean time, we may, according to our custom, spend an hour in taking the air in the gondola that waiteth for us.

3. "Our academick" is Galileo. The discussions will be found in his *Two New Sciences* (1638), English translation by A. de Salvio and H. Crew now available in Dover paperback. There are two modern English editions of the *Dialogue* (one based on the translation used here); all Galileo's other major works are also available in modern translations.

III.

The New Philosophy

THE INNOVATIONS of the scientific revolution were by no means confined to astronomy, or indeed to empirical discovery and theoretical novelty. Equally revolutionary was the development of a new awareness of man's potentiality, of his ability to understand the world around him, and of the possible results of that understanding. Perhaps most important of all the new currents of thought was the realization that the new natural philosophy of the sixteenth and seventeenth centuries was truly new, and that in consequence it deserved examination. The scientists of the seventeenth century commonly spoke of "the new learning," the "new philosophy," the "new experimental philosophy," the "new mechanical philosophy," seeing, like the humanists but more truly, the novelty of the work upon which they were engaged.

Francis Bacon (1561-1626) was born into the ferment of Elizabethan England and was himself a good example of a versatile Elizabethan intellect. Like all intelligent men of his age, he was struck by the sterility of ancient, traditional learning. Uniquely, he became obsessed with the potentialities of learning rightly undertaken, and particularly with the study of nature, which he thought far too little cultivated in his own day. He saw himself as a prophet of what might be achieved in the future; it must be remembered that he died two years before Harvey wrote and six years before Galileo's *Dialogue* was published. The selection below (Document 7), from the *Advancement of Learning* was published as early as 1605. In it Bacon devoted much but not exclusive attention to science. His *Novum Organum* (*New Logic*) of 1620 on the other hand, was almost exclusively devoted to scientific method. Bacon saw most clearly that scientific theory would be truer to the real world if it were based upon experiment; he also thought there was a simple way to achieve this. He advocated extensive "natural histories" (inquiries into nature), which were principally collections of fact, and these he saw as a necessary preliminary to scientific theory (natural philosophy), perhaps an oversimplification, but a useful one. Further, he thought that if nature were properly understood, this knowledge could be employed for useful ends—for the "benefit and use" of mankind. Bacon thus not only saw the possibilities of applied science (basing his optimism upon the successes achieved by applying mathematics in the sixteenth century to navigation and cartography), but he believed that a better world—a Utopia, such as he described in the *New Atlantis*—would arise when all men, discarding the trammels of the past, sought knowledge of nature through observations and experiment. Above all he stressed the importance of experimental science.

The vision of René Descartes (1596-1650) was very different; yet he too wished to reform learning, discard the old, traditional modes of thought, and explore nature. Born thirty-five years after Bacon, on the threshold of the new century, and educated in a strict mathematical tradition in a Jesuit school rather than in the more empirical atmosphere of an English university, Descartes saw the key to knowledge in the mind, not the senses, in reason, not experiment. Gifted with great mathematical and philosophical ability, Descartes made important contributions in mathematics (analytical geometry), optics (theory of vision and of the rainbow), natural and pure philosophy. His *Discourse on Method* (*Discoure de la méthode,* 1637) is a classic philosophical text; it should not be forgotten that Descartes saw it as a preface to essays on mathematics, optics and theory of matter. His *Principles of Philosophy* (*Principia Philosophiae* 1644) was a revelation to his contempo raries: here Descartes began with principles of reasoning and then, using those principles, established matter and motion as the fundamental principles of physical nature. Using these physical principles, and having established the laws governing their behavior, he (mentally) constructed a universe. It was an awe-inspiring achievement, it dazzled his readers, and Cartesianism ultimately became as rigid and doctrinaire as Aristotelianism. Before that happened, however, Descartes inspired several generations of scientists.

The selection below (Document 8) is the Preface which Descartes wrote specially for the French translation of the *Principles* in 1647; it clearly and simply states his principles of rational thought and his conviction that if men will only put their minds to serious, careful thought they can attain the same certainty in philosophical and scientific questions as in the geometrical ones. This method of logical, deductive thought was, after Descartes, widely known as "mathematical" reasoning.

By way of contrast, the other selection (Document 9) is taken from the works of a practicing English scientist, who accepted many of Descartes' scientific principles but who believed in the power of Baconian empiricism. Henry Power (1623-1668), a medical man living in the quiet of the northern English provinces, was nonetheless in touch with the new currents of European discovery. His *Experimental Philosophy* (published in 1664, but mostly written ten or twelve years earlier) is partly devoted to microscopy, partly to pneumatic experiments. He was an able scientist, though no great original thinker; the selection demonstrates his happy conviction in the scientific progress of his age. His style was much influenced by his older contemporary, Thomas Browne, with whom he studied briefly. Browne is now remembered less as a medical man than as a stylist, and Power's more florid sentences reflect Browne's earlier love of ornate prose. Most striking is Power's great faith in the ability of mankind to achieve new discoveries and new truths, in which he was profoundly influenced by both Bacon and Descartes. His concluding paragraphs rather touchingly reveal the optimism of the age which saw the founding of the Royal Society, an age in which all nature's mysteries seemed open to revelation by means of the spirit and method of scientific inquiry. His relatively short life, and perhaps his isolation from London, prevented him from achieving more, though his *Experimental Philosophy* shows him to have been a competent practitioner as well as an exponent of the new learning. He just escapes being an amateur like his patrons the Towneleys, who also had much to do with spreading the ideas of the new experimental and mechanical philosophy among men educated previously only in Peripatetic, scholastic learning filled with semantic logic, or in classical literature. These men were necessary and useful intermediaries and perhaps had as much to do as more original minds with changing and modernizing the climate of opinion.

7. Baconianism

The Errors of Learning

The sciences themselves, which have had better intelligence and confederacy with the imagination of man than with his reason, are three in number; astrology, natural magic, and alchemy: of which sciences, nevertheless, the ends or pretences are noble. For astrology pretendeth to discover that correspondence or concatenation which is between the superior globe and the inferior: natural magic pretendeth to call and reduce natural philosophy from variety of speculations to the magnitude of works: and alchemy pretendeth to make separation of all the unlike parts of bodies which in mixtures of nature are incorporate. But the derivations and prosecutions to these ends, both in the theories and in the practices, are full of error and vanity; which the great professors themselves have sought to veil over and conceal by enigmatical writings and referring themselves to auricular traditions and such other devices, to save the credit of impostures: and yet surely to alchemy this right is due, that it may be compared to the husband-man whereof AEsop makes the fable; that, when he died, told his sons that he had left unto them gold buried under ground in his vineyard; and they digged over all the ground, and gold they found

SOURCE: Francis Bacon, *The Advancement of Learning* (London, 1605), pp. 33-34, 76-78, 98-111, in vol. I of *The Works of Francis Bacon* (10 vols., London, 1803).

none; but by reason of their stirring and digging the mould about the roots of their vines, they had a great vintage the year following; so assuredly, the search and stir to make gold hath brought to light a great number of good and fruitful inventions and experiments, as well for the disclosing of nature as for the use of man's life. . . .

The Classification of Learning

The parts of human learning have reference to the three parts of man's understanding, which is the seat of learning: history to his memory, poesy to his imagination, and philosophy to his reason. Divine learning receiveth the same distribution; for the spirit of man is the same, though the revelation of oracle and sense be diverse: so as theology consisteth also of the history of the church; of parables, which is divine poesy; and of holy doctrine or precept: for as for that part which seemeth supernumerary, which is prophecy, it is but Divine History; which hath that prerogative over human, as the narration may be before the fact as well as after.

History is natural, civil, ecclesiastical and literary; whereof the first three I allow as extant, the fourth I note as deficient. For no man hath propounded to himself the general state of learning to be described and represented from age to age, as many have done the works of nature, and the state civil and ecclesiastical; without which the history of the world seemeth to me to be as the statue of Polyphemus with his eye out; that part being wanting which doth most show the spirit and life of the person: and yet I am not ingnorant that in diverse and particular sciences, as of the jurisconsults, the mathematicians, the rhetoricians, the philosophers, there are set down some small memorials of the schools, authors and books; and so likewise some barren relations touching the invention of arts or usages.

But a just story of learning, containing the antiquities and originals of knowledge and their sects, their inventions, their traditions, their diverse administrations and managings, their flourishings, their oppositions, decays, depressions, oblivions, removes, with the causes and occasions of them, and all other events concerning learning, throughout the ages of the world, I may truly affirm to be wanting.

The use and end of which work I do not so much design for

curiosity or satisfaction of those that are the lovers of learning, but chiefly for a more serious and grave purpose; which is this, in a few words, that it will make learned men wise in the use and administration of learning. For it is not St. Augustine's nor St. Ambrose's works that will make so wise a divine, as ecclesiastical history, thouroughly read and observed; and the same reason is of learning.

History of nature is of three sorts; of nature in course, of nature erring or varying, and of nature altered or wrought; that is, history of creatures, history of marvels, and history of arts. The first of these, no doubt, is extant, and that in good perfection; the two latter are handled so weakly and unprofitably, as I am moved to note them as deficient. For I find no sufficient or competent collection of the works of nature which have a digression and deflection from the ordinary course of generations, productions and motions; whether they be singularities of place and region, or the strange events of time and chance, or the effects of yet unknown properties, or the instances of exception to general kinds. It is true, I find a number of books of fabulous experiments and secrets and frivolous impostures for pleasure and strangeness; but a substantial and severe collection of the heteroclites or irregulars of nature, well examined and described, I find not: especially not with due rejection of fables and popular errors: for as things now are, if an untruth in nature be once on foot, what by reason of the neglect of examination, and countenance of antiquity, and what by reason of the use of the opinion in similitudes and ornaments of speech, it is never called down.

The use of this work, honoured with a precedent in Aristotle, is nothing less than to give contentment to the appetite of curious and vain wits, as the manner of Mirabilaries is to do; but for two reasons, both of great weight; the one to correct the partiality of axioms and opinions, which are commonly framed only upon common and familiar examples; the other because from the wonders of nature is the nearest intelligence and passage towards the wonders of art: for it is no more but by following, and as it were hounding nature in her wanderings, to be able to lead her afterwards to the same place again. Neither am I of opinion, in this history of marvels, that superstitious narrations of sorceries, witchcrafts, dreams, divinations and the like, where there is an assurance and clear evidence of the fact, be altogether excluded. For it is not

yet known in what cases and how far effects attributed to superstition do participate of natural causes: and therefore howsoever
the practice of such things is to be condemned, yet from the
speculation and consideration of them light may be taken, not
only for the discerning of the offences, but for the further disclosing of nature. Neither ought a man to make scruple of entering
into these things for inquisition of truth, as your majesty [King
James I] hath showed in your own example; who with the two
clear eyes of religion and natural philosophy have looked deeply
and wisely into these shadows, and yet proved yourself to be of
the nature of the sun, which passeth through pollutions, and itself
remains as pure as before. But this I hold fit, that these narrations,
which have mixture with superstition, be sorted by themselves,
and not be mingled with the narrations which are merely and
sincerely natural. But as for the narrations touching the prodigies
and miracles of religions, they are either not true, or not natural;
and therefore impertinent for the story of nature. . . .

The Nature and Role of Science

Leaving therefore divine philosopny or natural theology (not
divinity or inspired theology, which we reserve for the last of all,
as the haven and sabbath of all man's contemplations) we will now
proceed to natural philosophy.

If then it be true that Democritus said, "That the truth of
nature lieth hid in certain deep mines and caves"; and if it be true
likewise that the alchemists do so much inculcate, that Vulcan is a
second nature, and imitateth that dexterously and compendiously,
which nature worketh by ambages [circuitously] and length of
time, it were good to divide natural philosophy into the mine and
the furnace: and to make two professions or occupations of
natural philosophers, some to be pioneers and some smiths; some
to dig, and some to refine and hammer: and surely I do best allow
of a division of that kind, though in more familiar and scholastical
terms; namely, that these be the two parts of natural philosophy,—the inquisition of causes, and the production of effects;
speculative, and operative; natural science, and natural prudence.

For as in civil matters there is a wisdom of discourse, and a
wisdom of direction; so is it in natural. And here I will make a
request, that for the latter, or at least for a part thereof, I may
revive and reintegrate the misapplied and abused name of natural

magic; which, in the true sense, is but natural wisdom, or natural prudence; taken according to the ancient acception; purged from vanity and superstition.

Now although it be true, and I know it well, that there is an intercourse between causes and effects, so as both these knowledges, speculative and operative, have a great connection between themselves; yet because all true and fruitful natural philosophy hath a double scale or ladder, ascendent and descendent: ascending from experiments to the invention of new experiments; therefore I judge it most requisite that these two parts be severally considered and handled.

Natural science or theory is divided into physic and metaphysic: wherein I desire it may be conceived that I use the word metaphysic in a differing sense from that that is received: and in like manner, I doubt not but it will easily appear to men of judgement, that in this and other particulars, wheresoever my conception and notion may differ from the ancient, yet I am studious to keep the ancient terms. For hoping well to deliver myself from mistaking, by the order and perspicuous expressing of that I do propound; I am otherwise zealous and affectionate to recede as little from antiquity, either in terms of opinions, as may stand with truth and the proficience of knowledge.

And herein I cannot a little marvel at the philosopher Aristotle, that did proceed in such a spirit of difference and contradiction towards all antiquity: undertaking not only to frame new words of science at pleasure, but to confound and extinguish all ancient wisdom: insomuch as he never nameth or mentioneth an ancient author or opinion, but to confute and reprove; wherein for glory, and drawing followers and disciples, he took the right course. . . .

But to me, on the other side, that do desire as much as lieth in my pen to ground a sociable intercourse between antiquity and proficience, it seemeth best to keep way with antiquity; and therefore to retain the ancient terms, though I sometimes alter the uses and definitions, according to the moderate proceeding in civil government. . . .

To return therefore to the use and acceptation of the term Metaphysic, as I do now understand the word; it appeareth, by that which hath been already said, that I intend philosophia prima, Summary Philosophy and Metaphysic which heretofore have been confounded as one, to be two distinct things. For, the one I have made as a parent or common ancestor to all knowledge; and the other I have now brought in as a branch or descendant of natural science. It appeareth likewise that I have assigned to summary philosophy the common principles and axioms which are promiscuous and indifferent to several sciences: I have assigned unto it likewise the inquiry touching the operation of the relative and adventive characters of essences, as quantity, similitude, diversity, possibility, and the rest: with this distinction and provision; that they be handled as they have efficacy in nature, and not logically. It appeareth likewise, that Natural Theology, which heretofore hath been handled confusedly with Metaphysic, I have inclosed and bounded by itself.

It is therefore now a question what is left remaining for Metaphysic; wherein I may without prejudice preserve thus much of the conceit of antiquity, that Physic should contemplate that which is inherent in matter, and therefore transitory; and Metaphysic; that which is abstracted and fixed.

And again, that Physic should handle that which supposeth in nature only a being and moving; and Metaphysic should handle that which supposeth further in nature a reason, understanding, and platform. But the difference, perspicuously expressed, is most familiar and sensible. For as we divided natural philosophy in general into the inquiry of causes, and productions of effects: so that part which concerneth the inquiry of causes, we do subdivide according to the received and sound division of causes; the one part which is Physic, inquireth and handleth the material and efficient causes; and the other, which is Metaphysic, handleth the formal and final causes.

Physic, taking it according to the derivation, and not according to our idiom for medicine, is situate in a middle term or distance between Natural History and Metaphysic. For natural history describeth the variety of things; physic, the causes, but variable or respective causes; and metaphysic, the fixed and constant causes. . . .

Fire is the cause of induration [hardening], but respective to clay; fire is the cause of colliquation [melting], but respective to wax; but fire is no constant cause either of induration or colliquation; so then the physical causes are but efficient and the matter. Physic hath three parts; whereof two respect nature united or collected, the third contemplateth nature diffused or distributed. Nature is collected either into one entire total, or else into the same principles or seeds. So as the first doctrine is touching the contexture or configuration of things, as on the world, on the universality of things. The second is the doctrine concerning the prinicples or originals of times. The third is the doctrine concerning all variety and particularity of things; whether it be of the differing substances, or their differing qualities and natures; whereof there needeth no enumeration, this part being but as a gloss, or paraphrase, the attendeth upon the text of natural history. Of these three I cannot report any as deficient. In what truth or perfection they are handled, I make not now any judgement; but they are parts of knowledge not deserted by the labour of man.

For Metaphysic, we have assigned unto it the inquiry of formal and final causes; which assignation, as to the former of them, may seem to be nugatory and void; because of the received and inveterate opinion, that the inquisition of man is not competent to find out essential forms or true differences; of which opinion we will take this hold, that the invention of forms is of all other parts of knowledge the worthiest to be sought, if it be possible to be found. As for the possibility, they are ill discoverers that think there is no land, when they can see nothing but sea. But is is manifest that Plato, in his opinion of Ideas, as one that had a wit of elevation situate as upon a cliff, did descry, that forms were the true object of knowledge; but lost the real fruit of his opinion, by considering of forms as absolutely abstracted from matter, and not confined and determined by matter; and so turning his opinion upon theology, wherewith all his natural philosophy is infected. But if any man shall keep a continual watchful and severe eye upon action, operation and the use of knowledge, he may advise and take notice what are the forms, the disclosures whereof are fruitful and important to the state of man. For as to the forms of substances. . .as they are now by compounding and transplanting multiplied, are so perplexed, as they are not to be inquired; no more than it were either possible or to purpose to seek in gross the forms of those sounds which make words, which by composition and transposition of letters are infinite.

But on the other side, to inquire the form of those sounds of voices which make simple letters, is easily comprehensible; and being known, induceth and manifesteth the forms of all words, which consist and are compounded of them. In the same manner to inquire the form of a lion, of an oak, of gold; nay, of water, of air, is a vain pursuit; but to inquire the forms of sense, of voluntary motion, of vegetation, of colours, of gravity and levity, of density, of tenuity, of heat, of cold, and all other natures and qualities, which, like an alphabet, are not many, and of which the essences, upheld by matter, of all creatures do consist; to inquire, I say, the true forms of these, is that part of metaphysic which we now define of.

Not but that Physic doth make inquiry, and take consideration of the same natures; but how? Only as to the material and efficient causes of them, and not as to the forms. For example; if the cause of whiteness in snow or froth be inquired, and it be rendered thus, that the subtle intermixture of air and water is the cause, it is well rendered; but, nevertheless, is this the form of whiteness? No; but it is the efficient, which is ever but a vehicle of form. This part of metaphysic I do not find laboured and performed: whereat I marvel not: because I hold it not possible to be invented by that course of invention which hath been used; in regard that men, which is the root of all error, have made too untimely a departure and too remote a recess from particulars.

But the use of this part of Metaphysic, which I report as deficient, is of the rest the most excellent in two respects: the one, because it is the duty and virtue of all knowledge to abridge the infinity of individual experience, as much as the conception of truth will permit, and to remedy the complaint of life is short, the art long, which is performed by uniting the notions and conceptions of sciences: for knowledges are as pyramids, whereof history is the basis. So of natural philosophy, the basis is natural history, the stage next the basis is physic; the stage next the vertical point is metaphysic. . . .

But to those who refer all things to the glory of God, they are as the three acclamations, Sancte, sancte, sancte! holy in the description or dilation of his works; holy in the connection or concatenation of them; and holy in the union of them in a perpetual and uniform law. And therefore the speculation was excellent in Parmenides and Plato, although but a speculation in them, that all things by scale did ascend to unity. So then always that knowledge is worthiest which is charged with least multiplicity; which appeareth to be metaphysic; as that which considereth the simple forms or differences of things, which are few in number, and the degrees and co-ordinations whereof make all this variety. . . .

The second part of metaphysic is the inquiry of final causes, which I am moved to report not as omitted, but as misplaced; and yet if it were but a fault in order, I would not speak of it: for order is matter of illustration, but pertaineth not to the substance of sciences. But this misplacing hath caused a deficience, or at least, a great improficience in the sciences themselves. For the handling of final causes, mixed with the rest in physical inquiries, hath intercepted the severe and diligent inquiry of all real and physical causes, and given men the occasion to stay upon these satisfactory and specious causes, to the great arrest and prejudice of further discovery. . . .

Nevertheless there remaineth yet another part of Natural Philosophy, which is commonly made a principal part and holdeth rank with Physic special and Metaphysic, which is Mathematic; but I think it more agreeable to the nature of things, and to the light of order, to place it as a branch of Metaphysic; for the subject of it being quantity (not quantity indefinite, which is but a relative, and belongeth to philosophia prima, as hath been said, but quantity determined or proportionable), it appeareth to be one of the essential forms of things; as that that is causative in nature of a number of effects; insomuch as we see, in the schools both of Democritus and of Pythagoras, that the one did ascribe figure to the first seeds of things and the other did suppose numbers to be the principles and originals of things; and it is true also that of all other forms, as we understand forms, it is the most abstracted and separable from matter, and therefore most proper to Metaphysic which hath likewise been the cause why it hath been better laboured and inquired than any of the other forms, which are more immersed in matter.

For it being the nature of the mind of man, to the extreme prejudice of knowledge, to delight in the spacious liberty of generalities, as in a champain region, and not in the inclosures of particularity; the Mathematics of all other knowledge were the goodliest fields to satisty that appetite. But for the placing of this science, it is not much material: only we have endeavoured, in these our partitions, to observe a kind of perspective, that one part may cast light upon another.

The Mathematics are either pure or mixed. To the pure mathematics are those sciences belonging which handle quantity determinate, merely severed from any axioms of natural philosophy: and these are two, Geometry, and Arithmetic; the one handling quantity continued, and the other dissevered.

Mixed hath for subject some axioms or parts of natural philosophy, and considereth quantity determined, as it is auxiliary and incident unto them. For many parts of nature can neither be invented with sufficient subtlety, nor demonstrated with sufficient perspicuity, nor accommodated unto us with sufficient dexterity, without the aid and intervening of the mathematics; of which sort are perspective, music, astronomy, cosmography, architecture, enginery and divers others.

In the Mathematics I can report no deficience, except it be that men do not sufficiently understand the excellent use of the Pure Mathematics, in that they do remedy and cure many defects in the wit and faculties intellectual. For if the wit be too dull, they sharpen it; if too wandering they fix it; if too inherent in the sense, they abstract it. So that as tennis is a game of no use in itself, but of great use in respect it maketh a quick eye and a body ready to put itself into all postures; so in the mathematics, that use which is collateral and intervenient is no less worthy than that which is principal and intended.

And as for the Mixed Mathematics, I may only make this prediction, that there cannot fail to be more kinds of them, as nature grows further disclosed. Thus much of natural science, or the part of nature speculative.

For Natural Prudence, or the part operative of Natural Philosophy, we will divide it into three parts, experimental, philosophical, and magical; which three parts active have a correspondence and analogy with the three parts speculative, natural history, physic, and metaphysic; for many operations have been invented, sometimes by a casual incidence and occurrence, sometimes by a purposed experiment, and of those which have been found by an intentional experiment, some have been found out by varying or extending the same experiment, some by transferring and compounding divers experiments the one into the other, which kind of invention an empiric may manage. . . .

For it is a thing more probable, that he that knoweth well the natures of weight, of colour, of pliant and fragile, in respect of the hammer, of volatile and fixed in respect of the fire and the rest, may superinduce upon some metal the nature and form of gold by such mechanics as belongeth to the production of the natures afore rehearsed, than that some grains of the medicine projected should in a few moments of time turn a sea of quicksilver or other material into gold: so it is more probable that he that knoweth the nature of arefaction [rarefaction], the nature of assimilation of nourishment to the thing nourished, the manner of increase and clearing of spirits, the manner of the depredations which spirits make upon the humours and solid parts, shall by ambages of diets, bathings, annointing, medicines, motions, and the like, prolong life, or restore some degree of youth or vivacity, than that it can be done with the use of a few drops or scruples of a liquor or receipt. To conclude, therefore, the true Natural Magic, which is that great liberty and latitude of operation which dependeth upon the knowledge of forms, I may report deficient, as the relative thereof is.

To which part, if we be serious, and incline not to vanities and plausible discourse, besides the deriving and deducing the operations themselves from metaphysic there are pertinent two points of much purpose, the one by way of preparation, the other by way of caution: the first is that there be made a kalendar, resembling an inventory of the estate of man, containing all the inventions, being the works or fruits of nature or art, which are now extant, and whereof man is already possessed; out of which doth naturally result a note, what things are yet held impossible, or not invented; which kalendar will be the more artificial and serviceable, if to every reputed impossibility you add what thing is extant which cometh the nearest in degree to that impossibility; to the end that by these optatives and potentials man's inquiry may be the more awake in deducing direction of works from the speculation of causes: and secondly, that those experiments be not only esteemed which have an immediate and present use, but those principally which are of most universal consequence for invention of other experiments, and those which give most light to the invention of causes; for the invention of the mariner's needle, which giveth the direction, is of no less benefit for navigation than the invention of the sails which give the motion.

8. Cartesianism

The Author's Letter to the Translator,
Which May Serve Here by Way of a Preface

Sir,

The version of my *Principles* which you have labored to make is
so clear and so polished that I hope it will be read by more people
in its French than in its Latin form and better understood. I only
fear lest the title put off some who have not been nurtured in
learning, or rather who have a poor opinion of philosophy, because what philosophy they have been taught did not please them.
This makes me believe that it would be well to add a preface
informing such persons what this book is about, what my purpose
in writing it was, and what use they can derive from it. But although it is my responsibility to write this preface—for I ought to
know these things better than anyone else—I cannot think of anything better to put here than a summary of the chief points which
it seems to me ought to be treated, and I leave it to your discretion to publish as much as you think proper.

I should like first of all to explain the nature of philosophy,
beginning with the most obvious points, such as, that the word
"philosophy" signifies the study of wisdom, and that by wisdom is
to be understood not only discretion in business but a perfect
understanding of everything which a man can know, whether for
regulating his life, preserving his health, or discovering all the arts
and crafts. This kind of understanding must necessarily be deduced from first causes; hence, in order to set out to acquire it
(what is properly called "to philosophize"), it is necessary to begin
with the search for these first causes, that is, for first principles.
Moreover, these principles must conform to two conditions: first,
they must be so clear and evident that the human mind cannot
doubt their truth when it attentively considers the matter; second,

SOURCE: Translated by the editor from René Descartes, *Principes de la
  Philosophie* (Paris, 1647).

they must be principles upon which the understanding of other things depends, so that they may be known without these other things, but not vice versa. And after this it is necessary to try to deduce from these principles the understanding of the things which depend upon them in such a manner that everything which follows in the course of these deductions is manifestly plain. Truly only God is perfectly wise, that is to say, only He has entire understanding of the truth of all things; but men may be called more or less wise according as they have more or less understanding of the most important truths. And I believe that there is nothing in all this with which any learned man will fail to agree.

Next I would ask the reader to consider the usefulness of this philosophy and would show him that since it is applicable to everything within the scope of the human mind, it is right to believe that it is philosophy alone which distinguishes us from the most savage and barbarous beings, and that the measure of civilization and polish of each nation is to be judged by the degree to which its members cultivate philosophy; and hence the greatest good which any state can experience is to contain true philosophers. And further, that it is not only useful for each individual to live alongside those who apply themselves to this study, but it is incomparably better for each to apply himself to it, in the same way as it is certainly better for everyone to use his own eyes to go about and to enjoy the beauty of colors and of light, than to have his eyes closed and follow where another leads. Yet even this last is better than to keep them closed and find one's own way. And to live without philosophizing is, rightly speaking, to have one's eyes shut and never try to open them; the pleasure of seeing those things which our sight discovers is not comparable with the satisfaction which arises from the understanding found through philosophy. Finally, that study is more essential to the proper conduct of our lives and morals than is the use of our eyes to guide our footsteps. Brute beasts who have only their bodies to look after spend all their time in looking for nourishment, but men, because their minds make up their greater part, ought to employ their efforts principally in the search for wisdom, which is the mind's true nourishment; and I am sure that there are many men who would not fail to do this if they had any hope of success and if they knew how fitted they were for it. No man, however ignoble, remains so firmly attached to the objects of the senses as to fail sometimes to turn away toward hope for some other, greater good, in spite of the fact that often he does not know of what this good consists. Even those most favored by fortune, who have health, honors, and wealth in abundance, are no more exempt from this desire than other men; on the contrary, I am convinced that it is these very men who most ardently long for another, more dominant good than that they possess. Now this dominant good acquired by natural reason, without the revelation of faith, is nothing else but the comprehension of truth from its first causes, that is to say wisdom, whose study is philosophy. And since all these things are entirely true, it will not be difficult to persuade such men, provided they are properly guided.

But since men are often prevented from believing these things by experience, which demonstrates that those who profess to be philosophers are often less wise and less reasonable than those who have never applied themselves to this study, I should at this point briefly explain the content of this science at the present time and set out stages in the wisdom we have acquired. The first consists of those ideas which are so clear in themselves that they can be grasped without reflection. The second comprises all that can be learned through the evidence of the senses. The third comprehends what we learn from conversation with others. To which we can add, for the fourth, the reading not of all books, but of those written by persons capable of instructing us properly, for it is a kind of conversation which we have with their authors when we read such books. And it seems to me that all the wisdom which we are accustomed to acquire comes only through these four means; for I do not take account here of divine revelation, since that does not lead us on by stages but raises us at one stroke into an infallible faith.

Now there have at all times been great men who have sought to find a fifth way of attaining wisdom, incomparably loftier and more certain than the other four. This is to seek for the first causes and true principles from which it is possible to deduce the reasons of all that we can know; and it is particularly those who have sought this way whom we call philosophers. Nevertheless I do not know whether there have been any up to now who have succeeded at this design. The first and chief men of this sort whose writings survive are Plato and Aristotle, between whom the only difference is that the first, following in the footsteps of his master,

Socrates, ingenuously confessed that he could find nothing certain and so contented himself with writing what seemed to him likely, divising to that effect several principles by which he tried to explain other things; while Aristotle was less open and although he had been Plato's disciple for twenty years and knew no principles that were not his master's, entirely altered the manner of retailing them and put them forward as true and certain, although there is no evidence that Plato ever thought of them as being so. Now these men had high intelligence and great wisdom acquired by the four stages mentioned above, all of which gave them great authority, and to such an extent that those who came after them set themselves rather to follow their opinions than to look for something better.

And the chief dispute among their disciples was to know whether all things should be regarded as doubtful or whether there were some which were certain. Which carried both sides into extravagant errors, for some of those who took the side of doubt even carried this into daily life, so that they failed to use discretion in their conduct; and those on the side of certainty, assuming that it ought to derive from the senses, relied totally on them, to such a point that Epicurus dared to assert that the sun is no larger than it appears to the eye. This is a fault always noticeable in disputes, that as truth is halfway between the two opinions being upheld, each side draws farther away from it the more disputatious its adherents become. But the error of the doubters was not followed for long and that of the other side was somewhat corrected when it was recognized that the senses deceive us in many cases. All the same, I do not know that it has been entirely removed by demonstrating that certainty lies not in the senses, but in the understanding alone, when it perceives things plainly; and that while knowledge is only to be acquired in the four chief stages of wisdom, it is wrong to doubt things which concern every day conduct when they seem true, although it is also wrong for us to regard them as so certain as to be unable to change our opinion when obliged to do so by the evidence of reason. Failing to learn this truth, or even, having learned it, failing to use it, most of those in past centuries who wished to be philosophers have followed Aristotle, and in such a fashion as to have spoiled the sense of his writings by attributing to him various opinions which he would never recognize to be his if he were to reappear on earth. And those who did not follow him (among whom have been some of the best minds) could not escape being imbued with his opinions in youth (because they are the only ones taught in the schools); this so preoccupied them that they have been unable to arrive at an understanding of the true principles.

And although I respect them all and would not wish to make myself odious by reproving them, I can give proof for what I say, which I think none can fail to acknowledge: that is, that they have all taken as *principles* things which they could not perfectly understand. For example, I know of none who has not assumed weight as a characteristic of terrestrial bodies; but although experience very clearly teaches us that bodies called "heavy" fall toward the center of the earth, we do not at all know in spite of this, the nature of what is called "weight", that is to say, the cause or principle which makes these bodies fall in this way, and we must learn it elsewhere. The same may be said of the vacuum, and of atoms, of heat and cold, of dry and moist, and of salt, sulfur, and mercury, and of all similar things which some have taken as their principles. Now conclusions derived from principles which are not certain cannot be certain themselves, even though they may be clearly derived from these principles, from which it follows that all the reasoning which men have based on such principles has been unable to give them certain knowledge of anything or, consequently, enabled them to advance even one step toward the search for wisdom. And if they have discovered anything true, it has only been by one of the four methods adduced above.

At the same time I do not at all wish to diminish the honor which each of them claims; I am only forced to say (to console those who have not studied these things at all) that it is the same as in traveling: those who turn their backs on their goals go farther and farther away from them the longer and faster they journey; so that even if they are later put upon the right road, they cannot arrive as soon as if they had not traveled at the beginning. So, when a man holds false principles, the more he cultivates them and carefully applies them to discovering various consequences, thereby thinking to philosophize properly, the further he strays from the understanding of the truth and of wisdom. From which it necessarily follows that those who have learned least of all that has hitherto been known as philosophy are the most suited to learning true philosophy.

After having made the reader understand these things, I should advance arguments tending to prove that the true principles for reaching the highest degree of wisdom, the highest pitch of human good, are those which I have set out in my book. Two arguments alone suffice for this purpose. The first is that they are very clear, and the second that all other things can be deduced from them; and only these two are necessary. Now I can easily prove that they are very clear: first, by my method of arriving at them, which was by rejecting everything of which I could entertain the slightest doubt, for it is certain that those things which cannot be rejected when carefully considered are the clearest and most certain things which the human mind can grasp. So, reflecting that anyone wishing to doubt everything can yet not at the same time doubt that he exists while he is in the act of doubting, and that what is engaged in reasoning thus—that is, in not doubting of its own existence while nevertheless doubting of the existence of everything else— is not what we call our body, but what we call our soul or our mind, I took the being or the existence of this mind for the first principle from which I have very clearly derived the following.

The first is that there is a God who is the creator of everything in the world and who, being the source of all truth, has not created our understanding of such a nature that it could be mistaken in its judgment of whatever it very clearly and distinctly perceives to be true. These are all the principles I make use of concerning immaterial or metaphysical matters and from these I very clearly derive the following principle about corporeal or physical substances: that there are bodies having extension (length, breadth, and depth), of various shapes, and moving in various ways.

These, in short, are all the principles I require and from these I derive the truth about other things. The other fact which proves that these principles are clear and distinct is that they have been known since the beginning of time and even accepted as true and indisputable by all men—except only the existence of God, which has been questioned by some because they have relied too much upon the evidence of the senses and God cannot be touched or seen. But although all the truths which I number among my principles have been known always by everybody, at the same time there has, as far as I know, been nobody up to the present who has recognized that they are the principles of philosophy, that is to say that they are the principles from which can be derived the understanding of all other things in the world. This is why it is necessary for me here to prove that there are such principles, and I do not think I can do better than to make it appear from experience, that is, by convincing readers that they should read this book. For though I have not dealt with everything—which would be impossible—I think I have explained everything which I do discuss in such a way that anyone who reads attentively will be convinced that there is no need to look for any other principles than those which I have given in order to attain the highest degree of understanding of which the human mind is capable. Especially if, having perused my book, readers will take the trouble to consider how many different questions are there examined, whereas in reading the works of others they will see how few probable reasons these others have been able to give for answering the same questions by means of principles different from mine. And that readers may more easily understand this, I could tell them that those imbued with my views have much less difficulty in understanding the writings of others and appreciating their true worth than those who are not so imbued; just the opposite of what I formerly remarked of those who begin with ancient philosophy, for the further they have pursued that, the less they are commonly fitted to comprehend the true philosophy.

I should also add a word of advice about how to read this book. I should like the reader to run through the whole things first just as if it were a novel, without straining his attention or stopping at any difficulties encountered on the way, so as to get a general notion of the material treated; after that, if the reader finds that it deserves more examination, and has some curiosity about the causes of the things treated, he may read the book a second time in order to follow the consequences of my statements. But he must not be put off if he cannot everywhere understand them completely or if he does not understand them all; he must only mark with a pen the places he finds difficult and continue to read uninterruptedly to the end. Then if he takes up the book for a third time, I venture to think that he will find the solution to the greater part of the difficulties noted before and that if any do still remain, these will be solved on rereading.

Carefully examining the natural ability of various minds, I have determined that there are almost none which are so heavy and backward as to be incapable of appreciating sound opinions or even of comprehending the most rarefied sciences if they are properly introduced. And this can be proved rationally, for if the principles are clear and if nothing is derived from them except by very convincing arguments, anyone will have enough understanding to comprehend the things so derived. Moreover, aside from the hindrance produced by the prejudice from which none of us is quite free (although those who have studied bad learning are most affected by it), it almost always happens that ordinary minds neglect hard study because they think themselves incapable of it, while others, more eager, are too hasty, whence it happens that they often accept principles which are not certain and from these draw imperfect conclusions. This is why I should like to assure the diffident that there is nothing in my writings which is not completely comprehensible, if only they will take the trouble to examine it; and at the same time to warn the rest that even the most brilliant minds require plenty of time and care if they wish to be aware of everything which I intend them to understand.

Finally, so that the reader may appreciate my aim in publishing these things, I should very much like here to explain the road which, so it seems to me, must be traversed by someone who wishes to be well informed. First, a man who has no more than the common and imperfect sort of knowledge which can be acquired in the four stages of wisdom mentioned above ought first of all to try to formulate a system of ethics by which he can govern his daily life; for this brooks no delay, and we ought all to try to live an upright life. After that he ought to try to study logic; not the logic of the schools, for that is, properly speaking, only a dialectic which teaches a man how to explain to others what he knows, or even to produce some unreflecting words about what he does not know, and thus it destroys good sense more than it improves it; but rather the logic which teaches a man how to order his thoughts in the right way for discovering truths of which he is ignorant. And since that depends a great deal upon practice, it is well to spend a good deal of time practicing the methods of reasoning upon easy and simple problems, like those of mathe-

matics. When a man has some facility in discovering the truth in
these problems, he ought to begin in earnest to apply himself to
true philosophy. Of this the first part is metaphysics, which consists of the principles of learning, among which is the explanation
of the principal attributes of God, of the immateriality of our
souls, and of all the clear and simple notions which we hold. The
second part is physics; this, after the discovery of the true principles of material things, consists of the general examination of the
composition of the whole universe, followed by a particular examination of the nature of the earth and of all the bodies which
surround it, like air, water, fire, the loadstone, and other minerals.
Following this it is necessary to examine in particular the nature
of plants, of animals, and above all, of man, so as to be able later
to comprehend the other sciences useful to him. Thus all philosophy is like a tree—the roots are metaphysics, the trunk is
physics, and the branches growing out of this trunk are all the
other sciences, which may be reduced to three chief ones, namely,
medicine, mechanics, and ethics. Of these I regard ethics as the
highest and most perfect and, since it presupposes a knowledge of
the other sciences, as the highest degree of wisdom.

But fruits are picked not from the roots, nor from the trunks of
trees; they are picked only from the extremity of the branches. In
the same way the most useful parts of philosophy are those which
can only be learned last. Although I am ignorant in all these parts,
I had printed ten or twelve years since some essays of the things
which I thought I had learned, because of my great desire to serve
the public. The first of these essays was a discourse on how to
reason rightly and discover truth in the sciences, where I summar-
ized the chief rules of logic and of an imperfect system of ethics to
be followed provisionally until a better one could be learned.
[The Discourse on Method, 1537] The other essays were three
treatises, one on dioptrics, one on meteors [atmospheric pheno-
mena], and the last on geometry. In the *Dioptrics* I intended to
show that it was possible to go far enough in philosophy to
discover with its aid knowledge of such arts as are useful to human
existence; and the invention of telescopes, which I there discuss, is
one of the most difficult discoveries ever made. In the *Meteors* I
wanted to show the difference between philosophy as I pursue it
and as it is taught in the schools where the same subject is custom-
arily treated. Lastly, in the *Geometry* I claim to demonstrate that

I have found out several things hitherto unknown, which gives rise to the belief that several other new things might also be discovered and hence incites all men to search for the truth. Since that time, forseeing the difficulty experienced by many in reaching an understanding of the fundamentals of metaphysics, I have tried to explain its chief points in a book of *Meditations*—not a large book, but the size has been much increased and the subject much clarified by the objections sent to me by many learned persons and by my replies to them. Then when it seemed to me that these earlier works had sufficiently prepared readers' minds to receive the principles of philosophy, I published these principles also. And I have divided my *Principles of Philosophy* into four parts, the first of which contains the principles of knowledge, what may be denominated "first philosophy" or metaphysics; this is why it is well, in order to understand it properly, to begin by reading my *Meditations*, written on the same subject. The other three parts contain all the most general subjects in physics, that is to say, the explanation of the basic laws or principles of nature: the way in which the heavens, fixed stars, the planets, comets, and generally the whole world is constituted; then, in particular, the nature of the earth, and of air, water, fire, the magnet—that is, the commonest bodies around us—and of the properties noticeable in these bodies—light, heat, weight, and so on. In this way I think I have begun to explain all philosophy in due order, without omitting anything which ought at any point to precede what I have described.

In order to carry this plan to its conclusion, I ought afterwards to explain in the same fashion the nature of every one of the individual bodies found on earth—minerals, plants, animals, and above all, man. Then I ought to treat more fully medicine, ethics, and mechanics. This is what I should have done to give mankind a complete course of philosophy; and I do not yet feel myself so old, I do not yet so despair of my strength, I do not yet feel so remote from the understanding of the remainder as not to venture to try to complete this plan, if only I had means to make all the experiments which I need to support and justify my arguments. But seeing that this would require a great expenditure of money, which a private individual like myself cannot provide without public assistance, and seeing that I was not likely to receive that assistance, I thought I ought henceforward to limit myself to studying for my own private instruction and that posterity would forgive me if in future I failed to work for its instruction.

However, so that it may appear how far I have already served posterity, I shall give here the advantages which may be derived from my *Principles*. The first is the satisfaction to be taken in finding here many truths hitherto unknown; for although truth often does not touch our imagination as much as falsehood and deception do (because truth seems simpler and less striking), nevertheless it gives much more solid and durable satisfaction. The second advantage is that when the reader studies these *Principles* he becomes slowly more skilled at judging the matters he meets with and so becomes wiser. This effect is the opposite of what happens with ordinary philosophy, for it is easily noticed that so-called pedants are made less able to think that they would have been if they had never learned this philosophy. The third advantage is that the truths in my *Principles* are so clear and certain that they remove all cause for dispute and thus incline men's minds toward gentleness and harmony, quite unlike the controversies of the schools which, since they imperceptibly made those who learned them more opinionated and inclined to cavil, are very likely the chief cause of the heresies and dissensions now at work in the world. The last, and chief, advantage of these *Principles* is that those who cultivate them will discover many truths which I have not set out, and thus, passing gradually from one truth to another, will in time acquire a perfect knowledge of all philosophy and will rise to the highest point of wisdom. We see that all the arts begin crude and imperfect, yet if they have any trace of truth in them, whose effect is demonstrated by experience, they gradually become more nearly perfect with practice; in the same way, those guided in philosophy by principles which are true cannot fail by following them to encounter other truths here and there. Indeed, there is no better proof of the falsity of Aristotelian principles than that it is noticeable that no progress has been made by their means during all the centuries in which they have been followed.

I recognize that there are minds so hasty and careless that they can build nothing certain upon even the most solid foundations; and these being ordinarily the quickest to produce books are those who could very soon spoil everything that I have done and introduce uncertainty and doubt about my method of philosophizing (from which I have tried most carefully to exclude them) if their writings were to be received as mine or as containing my opinions. I had this experience recently with a person [Regius, a Dutch professor], who seemed to follow me most exactly, and of whom I had somewhere written that I relied so much upon his understanding that I did not believe that he had any opinion which I should not be glad to hold as my own. Yet last year he published a book called *Fundamentals of Nature* in which there seemed to be nothing about physics or medicine which was not drawn from my writings (both these which I have published, and another, still incomplete, about the nature of animals, which fell into his hands), but, because he had badly transcribed my words and changed the order and denied certain truths of metaphysics upon which all physics ought to be grounded, I was obliged to disown him entirely and here beg all readers to ascribe to me only those opinions which they find expressly set out in my own works, as well as to accept nothing as true—whether in my writings or elsewhere—which they do not see to be very clearly deduced from true principles.

I am also aware that several centuries may pass before all the truths which can be drawn from these principles have been deduced, especially since the majority of those still to be discovered depend upon certain specific experiments which cannot be obtained by chance but which must be sought out with care and expense by very intelligent men. Now it is very rarely that those who have the skill to make use of such experiments have the ability to make them; and most of the best minds have taken such a poor view of all philosophy, on account of the defects which they have observed in that which has commonly obtained up to the present, that they are not able to try to find a better. But if at last the difference which they perceive between these *Principles* and those of all other philosophers, and the grand sequence of truths which can be deduced from them, teaches such minds how important it is to continue the search for these truths, and to how great a degree of wisdom, what improvements of life, and what happiness these truths may lead, I venture to believe that there will be none who will not try to take up such a profitable study or who at least will fail to favor and try to assist with all their might those who work fruitfully therein. I hope that the next generation will witness its success, and so on.

9. The New Learning

The Preface to the Ingenious Reader

Dioptrical Glasses (which are now wrought up to that height and curiosity we see) are but a modern invention: Antiquity gives us not the least hint thereof, neither do their records furnish us with anything that does antedate our late discoveries of the telescope or microscope. The want of which incomparable artifice made them not only err in their fond celestial hypothesis and crystalline wheelwork of the heavens above us, but also in their nearer observations of the minute bodies and smallest sort of creatures about us, which have been by them but slightly and perfunctorily described, as being the disregarded pieces and hustlement of the Creation; when (alas!) those sons of sense were not able to see how curiously the minutest things of the world are wrought, and with what eminent signatures of Divine Providence they were enriched and embellished, without our dioptrical assistance. Neither do I think that the aged world stands now in need of spectacles, more than it did in its primitive strength and lustre: for howsoever though the faculties of the soul of our primitive father Adam might be more quick and perspicacious in apprehension, than those of our lapsed selves; yet certainly the constitution of Adam's organs was not diverse from ours, nor different from those of his fallen self, so that he could never discern those distant, or minute objects by natural vision, as we do by the artificial advantages of the telescope and microscope. So that certainly the secondary planets of Saturn and Jupiter and his ansulary [handled] appearances,[1] the sunspots, and lunations of the inferior planets, were as

SOURCE: Henry Power, *Experimental Philosophy, in three books containing New Experiments Microscopical, Mercurial, Magnetical. With some deductions, and probable hypotheses, raised from them, in Avouchment and Illustration of the now famous Atomical Hypothesis* (London, 1664).

1. Galileo had been the first to view Saturn in a telescope and see this planet as a disk with "handles" or "ears". Although Huygens had published his discovery that this appearance was caused by Saturn's rings in 1659, not all scientists immediately accepted this conclusion.

obscure to him as unknown to his posterity; only what he might ingeniously guess at by the analogy of things in nature, and some other advantageous circumstances.

And as those remote objects were beyond the reach of his natural optics, so doubtless the minute atoms and particles of nature were as unknown to him, as they are yet unseen by us: for certainly both his and our eyes were framed by providence in analogy to the rest of our senses, and as might best manage this particular engine we call the body, and best agree with the place of our habitation (the earth and elements we were to converse with) and not to be critical spectators, surveyors, and adequate judges of the immense universe: and therefore it hath often seemed to me beyond an ordinary probability, and something more than fancy (how paradoxical soever the conjecture may seem) to think, that the least bodies we are able to see with our naked eyes, are but middle proportionals (as it were) betwixt the greatest and smallest bodies in nature, which two extremes lie equally beyond the reach of human sensation. For as on the one side they are but narrow souls, and not worthy the name of philosophers, that think any body can be too great or too vast in its dimensions; so likewise are they as inapprehensive, and of the same litter with the former, that on the other side think the particles of matter may be too little, and that nature is stinted at an atom, and must have a non ultra [no further] of her subdivisions.

Such, I am sure, our modern engine (the microscope) will ocularly evince and unlearn them their opinions again: for herein you may see what a subtle divider of matter nature is; herein we can see what the illustrious wits of the atomical and corpuscularian philosophers durst but imagine, even the very atoms and their reputed indivisibles and least realities of matter, nay, the curious mechanism and organical contrivance of those minute animals, with their distinct parts, colour, figure and motion, whose whole bulk were to them almost invisible: so that were Aristotle now alive, he might write a new history of animals; for the first tome of Zoography [zoology] is still wanting, the naturalists hitherto having only described unto us the larger and more voluminous sort of animals, as bulls, bears, tigers etc. whilst they have regardlessly passed by the insectile automata (those living exiguities) with only a bare mention of their names, whereas in these pretty engines (by an incomparable stenography of providence) are lodged all the perfections of the largest animals; they have the same organs of body, multiplicity of parts, variety of motions, diversity of figures, severality of functions with those of the largest size: and that which augments the miracle, is that all these in so narrow a room neither interfere nor impede one another in their operations. Who therefore with the learned Doctor,[2] admires not Regiomontanus his fly beyond his eagle,[3] and wonders not more at the operation of two souls in those minute bodies, than but one in the trunk of Cedar? Ruder heads stand amazed at those prodigious and collossean pieces of nature, as whales, elephants and dromedaries, but in these narrow engines there is more curious mathematics, and the architecture of these little fabrics more neatly set forth the wisdom of their maker.

Now as matter may be great or little, yet never shrink by sub-division into nothing; so, it is not probable, that motion also may be indefinitely swift or slow, and yet never come to a quiescency? and so consequently there can be no rest in nature, more than a vacuity in matter. The following observations seem to make out, that the minute particles of most (if not all) bodies are constantly in some kind of motion, and that motion may be both invisibly and unintelligibly slow, as well as swift, and probably is as un-separable an attribute to bodies, as well as extension is.

And indeed, if the very nature of fluidity consist in the intestine [internal] motion of that body called fluid, as Descartes happily supposed, and Mr Boyle has more happily demonstrated,[4] why may we not be bold to think and say, that there is no such thing in the world as an absolute quiescence? for 1. the greatest part of the world (viz. the aetherial medium, wherein all the stars and planets swim) is now confessed by all to be fluid, and so, consequently, in a perpetual motion. 2. All the fixed lights of heaven are generally concluded to be pure fire, and so consequently fluid also, and then subconsequentially in motion also; not to mention the dinetical [rotary] rotations of their whole bodies, which every one is sup-posed to have, as well as our Sun: and as for the opaque and

2. Dr. [Thomas] Browne, *Religio Medici*—Power's note.
3. Regiomontanus, a fifteenth-century astronomer and humanist was, ac-cording to legend, the constructor of various mechanical toys, including a tiny self-propelled fly.
4. Boyle's essays on "fluidity and firmness," with a wealth of illustrative experiment, were published at the end of his *Certain Physiological Essays* (London, 1661).

planetary bodies of the universe, they are all porous, and the aetherial matter is continually streaming through them, their internal fire and heat constantly subliming atoms out of them, the magnetical atoms continually playing about them: not to mention also their dinetical motions about their own axes and circum-revolutions about their central suns: so that, is it not, I say, more than probable, that rest and quiescency is a more Peripatetical [Aristotelian] notion, and that the Supreme Being (who is activity itself) never made anything inactive or utterly devoid of motion?

Hence will unavoidably follow some other principles of the ever-to-be-admired Descartes:

1. That as matter is made greater or less, by the addition or subduction [subtraction] of parts, so is motion made swifter or slower by addition given to the movent, by other contiguous bodies more swiftly moving, or by subduction of it by bodies slowlier moved.

2. As the parts of matter can be transferred from one body to another, and as long as they remain united, would remain so forever: so motion may be translated from one body to another: but when it is not transferred, it would remain in that body forever.

But these sublime speculations I shall with more confidence treat of in another place; the speculation of motion and its origin being, as I conceive, one of the obscurest things in nature.

And therefore at present we shall keep within the compass of the microscope, and look at nothing further than what we can discover therein: the knowledge of man (saith the learned Verulam [Bacon]) hath hitherto been determined by the view or sight, so that whatsoever is invisible, either in respect of the fineness of the body itself, or the smallness of the parts, or of the subtlety of its motion, is little enquired; and yet these be the things that govern nature principally. How much therefore are we obliged to modern industry, that of late hath discovered this advantageous artifice of glasses, and furnished our necessities with such artificial eyes, that now neither the fineness of the body, nor the smallness of the parts, nor the subtlety of its motion can secure them from our discovery? And indeed, if the dioptrics further prevail, and that darling art could but perform what the theorists in conic sections demonstrate, we might hope, ere long, to see the magnetical effluviums of the loadstone, the solary atoms of light (or aetherial globules of the renowned Descartes), the springy particles of air, the constant and tumultuary motion of the atoms of all fluid bodies, and those infinite, insensible corpuscles (which daily produce these prodigious, though common, effects amongst us). And though these hopes be vastly hyperbolical, yet who can tell how far mechanical industry may prevail; for the process of art is indefinite, and who can set a non ultra [limit] to her endeavours? I am sure, if we look backwards at what the dioptrics hath already performed, we cannot but conclude such prognostics to be within the circle of possibilities, and perhaps not out of the reach of futurity to exhibit. However, this I am sure of, that without some such mechanical assistance, our best philosophers will but prove empty conjecturalists, and their profoundest speculations herein, but glossed outside fallacies: like our stage scenes, or perspectives[5], that show things inward, when they are but superficial paintings.

For, to conclude with that doubly honourable (both for his parts and parentage) [6] Mr. Boyle, "When a writer," saith he, "acquaints me only with his own thoughts or conjectures, without enriching his discourse with any real experiment or observation, if he be mistaken in his ratiocination, I am in some danger of erring with him, and at least am like to lose my time, without receiving any valuable compensations for so great a loss; but if a writer endeavours, by delivering new and real observations or experiments, to credit his opinions, the case is much otherways: for, let his opinions be never so false (his experiments being true) I am not obliged to believe the former, and am left at my liberty to benefit myself by the latter. And though he have erroneously superstructed upon his experiments, yet the foundation being solid, a more wary builder may be very much furthered by it, in the erection of a more judicious and consistent fabric."

HENRY POWER

From New-Hall near Halifax,
1. Aug. 1661.

5. Seventeenth-century taste was fond of paintings which used the tricks of perspective to create the illusion of a three-dimensional space out of a flat surface.

6. As the son of the Earl of Cork, he was the Honorable Mr. Robert Boyle. The reference is, as Power indicates in the margin, to the initial essay of *Certain Physiological Essays;* Power quotes accurately.

The Conclusion.
To the Generous Virtuosi,[7]
and Lovers of Experimental Philosophy

Certainly this world was made not only to be inhabited, but studied and contemplated by man; and how few are there in the world that perform this homage due to their Creator? who, though He hath disclaimed all brutal, yet still accepts of a rational sacrifice; 'tis a tribute we ought to pay Him for being men, for it is reason that transpeciates our natures and makes us little lower than the angels. Without the right management of this faculty, we do not so much in our kind as beasts do in theirs, who justly obey the prescript of their natures, and live up to the height of that instinct that Providence hath given them. . . . There is a world of people indeed, and but few men in it; mankind is but preserved in a few individuals, the greatest part of humanity is lost in earth, and their souls so fixed in the grosser moiety of themselves (their bodies) that nothing can volatilize them and their reasons at liberty. The numerous rabble that seem to have the signatures of Man in their faces, are brutes in their understanding, and have nothing of the nobler part that should denominate their essences: 'tis by the favour of a metaphor we call them men, for at best they are but Descartes' Automata or Aristotle's "least men-like animals," but the moving frames and zanies of men, and have nothing but their outsides to justify their titles to rationality.

Pugs [apes] and baboons may claim a traduction [translation] from Adam as well as these, and have as great a share of reason to justify their parentage.

But it is not this numerous piece of monstrosity (the multitude only) that are enemies to themselves and Learning; there is a company of men amongst the philosophers themselves, a sort of notional heads, whose ignorance (though varnished over with a little squabbling sophistry) is as great and invincible as the former. These are they that daily stuff our libraries with their philosophi-

---

7. A word peculiar to the seventeenth century, meaning amateur or gentlemanly lover of the new scientific learning, especially the experimental philosophy. It was not very commonly applied to creative scientists.

cal romances, and glut the press with their canting loquacities. For, instead of solid and experimental philosophy, it has been held accomplishment enough to graduate a student, if he could but stiffly wrangle out a vexatious dispute of some odd Peripatetic [scholastic] qualities or the like; which (translated into English) signified no more than a heat 'twixt two oyster-wives in Billingsgate. Nay, these crimes have not only stained the common, but there are spots also to be seen even in the purple gowns of learning. For it hath been a great fault, and, indeed, a solemn piece of folly, even amongst the professors and nobler sort of philosophers, that when they have arrived to a competent height in any art or science, if any difficulty do arise that their art cannot presently [immediately] reach unto, they instantly pronounce it a thing impossible to be done; which inconsiderable and rash censure and forstallment of their endeavours, does not only stifle their own further enquiries, but also hangs, to all succeeding ages, as a scare crow to affright them for ever approaching that difficulty. Hence it is, that most arts and sciences are branded at this day with some such ignominious impossibility.

Thus came they to upbraid chemistry with the alkahest and philosophers' stone; geography with longitudes; geometry with the quadrature [squaring] of a circle; stereometry with the duplication of the cube; trigonometry with the trisection of an angle; algebra with the equation of three discontinued numbers; mechanics with a perpetual motion; and our own profession [medicine] with the incurability of cancers and quartans.[8] Nay, the spring and neaptides in natural philosophy, the doctrine of comets in astronomy, the terra incognita [undiscovered lands] in geography, the heart's motion in anatomy, the forming of conic sections in dioptrics, the various variation in magnetical philosophy, are accounted

8. These are all examples of scientific follies or failures. The alkahest is the universal solvent and the philosophers' stone the universal catalyst; there was then no means of determining longitude accurately, though the correct principles were known; squaring the circle is finding an area bounded by straight lines equal to the area of the circle—an impossibility; the problem of the trisection of an angle was to trisect it by using ruler and compass only. Similarly, the problems mentioned in the next sentence were then incompletely resolved, though great discoveries were to be made in the next twenty-five years.

as insuperable difficulties as the former, whose causes (they say) defy all human industry ever to discover them.

But besides this intestine war and civil dissension that is 'twixt men of the same denomination and principles there is one more general impediment, which is an authentic discouragement to the promotion of the arts and sciences; and that is, the universal exclamation of the world's decay and approximation to its period, that both the great and little world [macrocosm and microcosm] have long since passed the meridian, and, that the faculties of the one [the little world of man] do fade and decay as well as the fabrics and materials of the other. Which though it be a conceit that hath possessed all ages past, as nearly as ours, yet the clamour was never so high as it is now. Something, therefore, I shall here offer, that will abate and qualify the rigour of this conception. . . .

This is the age wherein (methinks) philosophy comes in with a Spring tide; and the peripatetics may as well hope to stop the current of the tide or (with Xerxes) to fetter the ocean, as hinder the overflowing of free philosophy. Methinks, I see how all the old rubbish must be thrown away, and the rotten buildings be overthrown and carried away with so powerful an inundation. These are the days that must lay a new foundation of a more magnificent philosophy, never to be overthrown, that will empirically and sensibly canvass the phenomena of nature, deducing the causes of things from such originals in nature as we observe are producible by art, and the infallible demonstration of mechanics. And certainly this is the way and no other to build a true and permanent philosophy: for art, being the imitator of nature (or nature at second hand) is but a sensible expression of effects, dependent on the same (though more remote causes), and therefore the works of one must prove the most reasonable discoveries of the other. And to speak yet more close to the point, I think it is no rhetorication to say, that all things are artificial; for nature itself is nothing else but the art of God. Then, certainly, to find the various turnings and mysterious process of this divine art, in the management of this great machine of the world, must needs be the proper office of only the experimental and mechanical philosopher. For the old dogmatists and notional speculators, that only gazed at the visible effects and last resultances of things, understood no more of nature than a rude [ignorant] country fellow does of the internal fabric of a watch, that only sees the index [hand] and horary circle and perchance hears the clock and alarm strike in it. But he that will give a satisfactory account of those phenomena, must be an artificer indeed, and one well skilled in the wheelwork and internal contrivance of such anatomical engines.

Experimental Innovation

PERHAPS THE MOST familiar aspect of the scientific revolution was its remarkable progress in experimental discovery. So much was this the case that Englishmen—long to be leading empiricists in science—commonly spoke indifferently of "the new learning" and "the new experimental philosophy" as if the two were identical (which they were not). By the 1660's experimental investigation often assumed the name "Baconian"; certainly Bacon had stressed, perhaps overstressed, the importance of experiment in science, but experiment was to be found before Bacon's influence was felt, and in other countries than England. As Bacon had rightly foreseen, experimental science was democratic and leveling; anyone could experiment, given a subject or a new instrument. It nevertheless remained the case that only the exceptional scientist made original discoveries.

In the selections below, examples have been given of various sorts of experimental investigation. The first section deals with physiology, where no novel methods were involved, but new concepts were employed, and facts of fundamental importance established. The second section deals with microscopy and shows the results, for both biology and physics, obtainable with a new instrument — the microscope. Invented by Galileo about 1615, it was not much used before the 1650's and not employed in biological investigation before 1661, when Malpighi extended Harvey's work by demonstrating the existence (in the lungs of a frog) of the capillary vessels postulated by Harvey as connecting links between the veins and arteries. The third section deals with pneumatics, which, beginning with an inquiry into the mechanism of suction pumps, produced two new instruments—the barometer and air pump—and established the physical nature of air and the atmosphere.

Physiology

The great revolution in the understanding of the workings of the human body began with the revival of anatomical study in the sixteenth century. Medieval anatomy had been superficial and routine, though dissections and post mortem examinations were freely practiced from at least as early as 1300. The great stimulus to a new, more thorough anatomy came from the rediscovery of the anatomical writing of the Greek medical writer Galen (second century A.D.) in the early sixteenth century. Galen had thoroughly understood the method and importance of the study of human anatomy, even though he himself had been forced to rely upon animal dissection, because dissection of human cadavers was prohibited in ancient Rome, though not in medieval or Renaissance Rome. The sixteenth century saw the production of a great many books on anatomy following the Galenic method, most of which took full advantage of Renaissance progress in the exact artistic representation of the human body; the most famous and most beautiful is that by Andreas Vesalius, *On the Fabric of the Human Body (1543).*

Anatomical discovery progressed rapidly, and by about 1600 most of the facts needed for an understanding of the action of the heart and blood were known—though not fully understood. Thus the fact that some blood, at least, goes from the right side of the heart through the lungs to the left side of the heart (the so-called pulmonary or lesser circulation)·had been described by Michael Servetus in an obscure theological work (1553), and in lectures and an anatomy textbook (1559) by Realdus Columbus, a lecturer in anatomy. The valves in the veins had been detected from the 1540's and most carefully described by Harvey's teacher at the University of Padua, Fabricius of Aquapendente, in 1603, though he did not understand their function. William Harvey (1578-1657) alone put together all these facts, added others of his own discovery, utilized to the full careful anatomical investigation, especially of cold-blooded animals (in which rate of heartbeat is slow, and which have no lungs), saw the significance of quantitative measurements and hydraulic analogy, and produced a full understanding of the circulation in *Anatomical Studies on the Motion of the Heart and Blood in Animals (De Motu Cordis,* 1628). Few could read the massive accumulation of evidence and still deny the existence of the circulation, even though Harvey could not, as Malpighi was to do with the aid of the microscope in 1661, discern the existence of the capillaries which he knew to be there.

Harvey achieved his success partly by separating the role of the lungs from that of the heart. It was left to the Oxford physician Richard Lower (1631-1691) to finish Harvey's work by discovering what happened to the blood when it went through the lungs and how its aeration there converted venous into arterial blood *On the Heart, Tractatus de Corde,* 1669). He also concerned himself with novel attempts to utilize the knowledge of the circulation for new medical treatment.

Microscopy

After Galileo's invention of the microscope ten years were to pass before the publication of the first plates of biological interest (of the bee, by Francesco Stelluti). These were slowly followed by observations on insects, and in 1651 Pierre Borel published a little book of forty-five pages describing, with a few crude illustrations, a wide variety of objects viewed under the microscope. Though Malpighi's discovery of the capillaries in 1661 is perhaps the most important scientific discovery made with the aid of a microscope in this period, it did not have the effect upon the development of microscopy of some work of lesser immediate significance. The books of two Englishmen, Henry Power (*Experimental Philosophy*, 1664) and Robert Hooke, (*Micrographia*, 1665) were of great importance for the spread of miscroscopical technique and for the promoting of popular interest in England. (Both were written in English and neither was ever translated into any other language, though foreigners could study the illustrations in Hooke's book.) The difference between the two works is suggested by the reaction of Samuel Pepys, who had no pretensions to scientific understanding. He "mightily" admired Hooke's *Micrographia*; when he bought a microscope he turned to Power to learn how to use it.

Neither Power's work nor Hooke's is entirely concerned with microscopy. Power's book is divided into three parts, the second and third dealing with pneumatics and magnetism respectively. Hooke presents his work under the guise of sixty Observations (the last three telescopic), but his powers of digression being great, there is much in *Micrographia* that has nothing to do with microscopy, most notably the long digression on optics, dealing with his theory of light and color (see below, Document 23). Some of the fame of *Micrographia* derives from its handsome plates (many probably drawn by Christopher Wren): the descriptions are orientated towards these plates, whose presence permits Hooke to describe the ant, say, more accurately than Power was able to do. On the whole Hooke was more interested in the relation of his observations to general scientific problems. Only in his description of "eels in vinegar" (nematode worms) did Power perform such "pretty experiments," as Hooke put it, that even he could not better them, so that Observation 57 of *Micrographia* makes dull reading in comparison with Power's Observation 30.

In the selections below, both Power's and Hooke's observations on the ant are included to provide comparison between their literary styles and scientific technique. Hooke's observation on sparks shows how the Royal Society's curator of experiments could ingeniously turn a simple observation into a scientific discussion on the nature of matter and of heat, which should be compared with the discussions on the nature of matter in Part VII below.

Pneumatics

Seventeenth-century investigations of the physical nature of air and the role of the atmosphere seem to derive, along with so much else, from Galileo. In his *Discourses upon Two New Sciences* (1638) he discussed the specific gravity of the air (following the then well-known account of the ancients since Aristotle) and spoke of the fact familiar to miners and artisans that a suction pump will not raise water over about thirty feet. Galileo thought that a column of water thirty feet in length reached breaking point, like an overloaded cord; this ingenious if erroneous explanation stimulated his Italian disciples to further investigation. To test the idea, Gasparo Berti constructed at Rome a water barometer with a sealed tube over thirty feet long; when the tube was unstoppered at the bottom the water fell to about thirty feet, leaving an apparently empty space at the top. There was much dispute both about the cause of the water's suspension and about the nature of the "empty" space. Several more experiments were tried, especially to test the conjecture that it might be the pressure of the atmosphere on the bottom of the tube which made the water stand suspended. One of these was the famous experiment by Torricelli (1608-1647), described in a letter to a friend in 1644 and reproduced below (Document 14).

News of this work reached France through Marin Mersenne, an able scientist and a man with a wide scientific correspondence. The young mathematician Blaise Pascal was immediately interested and suggested what became the famous Puy-de-Dome experiment, which was performed by his brother-in-law in 1648. Other experiments were suggested by Roberval, professor of mathematics at the Collège royale, and by Adrien Auzout, the ablest French observational astronomer of his day. In the second selection below (Document 15), these are all described by an anatomist, Jean Pecquet (1622-1674), best known for his discovery of the thoracic duct. In the course of his book (*New Anatomical Experiments*, 1651), he digressed to discuss the weight and pressure of the atmosphere and described there all the new French experiments, which further demonstrated the existence and force of atmospheric pressure. This was the source from which most scientists of the seventeenth century drew their knowledge for the analogy of the elasticity of the air and a heap of wool. The text is taken from the English translation of 1653.

. Pneumatic investigation had produced the barometer; it was next to produce the air pump. The experiments with the barometer seemed to show that air behaved like an elastic fluid; in the 1650's Otto von Guericke (1602-1686), discovered that, like any fluid, it could be pumped out of a vessel. He first tried to evacuate a wine cask by filling it with water and pumping the water out; later, with tighter vessels, he pumped the air out directly. By 1654 he was able to show the immense force of atmospheric pressure by the famous "Magdeburg hemisphere" experiment, in which he fitted together two brass hemispheres about three feet in diameter, pumped out the air, and showed that teams of horses could not drag them apart. The account was first published about ten years later by a Jesuit, Gaspar Schott, who did not accept Guericke's conclusions; the same Schott rendered pneumatics a great service by first describing Guericke's air pump in 1657 in his *Mechanica Hydraulica Pneumatica.*

It was from Schott's work that the English scientist Robert Boyle (1627-1691) first heard of the air pump. Having long been interested in pneumatics he set his two laboratory assistants—one was young Robert Hooke—to constructing a pump which should work more easily than Guericke's (which took two strong men) and should have a glass receiver into which objects could be more easily introduced than they could into the Torricellian vacuum. Hooke's mechanical ingenuity produced Boyle's first air pump (he had others later), in which were performed the experiments that were to be described in his *New Experiments Physico-Mechanical, touching the Spring of the Air and its Effects* (1660). Here Boyle conclusively demonstrated the existence of the vacuum, the fact that it was indeed the pressure of the air that made the mercury stand suspended in the Torricellian experiment, and the fact that air was necessary for the propagation of sound and for the maintenance of flame and respiration. He also hinted at—and in the second edition demonstrated—that relation between pressure and volume which we know as Boyle's law. Much interest in pneumatics followed Boyle's work (rapidly translated into Latin for foreign readers), and the new doctrine of the effects of atmosphere pressure was universally if not instantaneously established. The selections below (Document 16) are taken from the abridged edition of Boyle's works published by an early eighteenth-century chemist and physician, Peter Shaw. Shaw deleted some of Boyle's literary devices—in particular his pretense that his pneumatic experiments were written to entertain his nephew—but preserved Boyle's work and expression virtually intact.

Physiology

10. *The Circulation of the Blood*

Chapter I.

The Causes Which Moved The Author To Write

When first I applied my mind to observation, from the many dissections of living creatures as they came to hand, that by that means I might find out the use of the motion of the heart and things conducible in Creatures; I straight ways found it a thing hard to be attained, and full of difficulty, so with Fracastoro[1] I did almost believe, that the motion of the heart was known to God alone: For neither could I rightly distinguish, which way the Diastole and Systole came to be, nor when nor where the dilatation and constriction have its existence. And that by reason of the quickness of the motion, which in some creatures appeared in the twinkling of an eye, like the passing of lightning; so that sometimes the Systole did present itself to me from this place, and the Diastole from that place, sometimes just contrary, sometimes the motion was various, sometimes confused: whence I was much troubled in mind, nor did I know what to resolve upon myself, or what belief to give to others; nor wondered I at that which André

SOURCE: William Harvey, *De Motu Cordis* (Frankfurt, 1628), anonymously translated into English in 1653, the 1673 edition, pp. 15-21, 42-63, 72-78, 81.

1. An early sixteenth-century physician, humanist, and poet, best remembered for his attempt to explain contagion and for his christening the "new disease of the armed forces" (or "French disease") syphilis .

du Laurens writes, That the motion of the heart, was as the ebbing and flowing of Euripus to Aristotle. At last using daily more search and diligence, by often looking into many and several sorts of creatures, I did believe I had hit the nail on the head, unwinded and freed myself from this labyrinth, and thought I had gained both the motion and use of the heart, together with that of the arteries, which I did so much desire: Since which time I have not been afraid, both privately to my friends, and publicly in my anatomy lectures to deliver my opinion.

Which, as it commonly falls out, pleased some, and displeased others; some there were that did check me, spoke harshly, and found fault that I had departed from the precepts and belief of all anatomists: others avouching that it was a thing new, worthy of their knowledge, and exceeding profitable, required it be more plainly delivered to them. At last, moved partly by the requests of my friends, that all men might be partakers of my endeavours, and partly by the malice of some, who being displeased with what I said, and not understanding it aright, endeavoured to traduce me publicly, I was forced to recommend these things to the Press, that every man might of me, and of the thing itself, deliver his judgement freely. But so much the more willing I was to it, because Fabricius of Aquapendente[2] having learnedly and accurately set down in a particular treatise almost all the parts of living creatures, left the heart only untouched. Lastly, if any profit or advantage might by my industry in this accrue to the republic of literature, it might perchance be granted that I had done well, and others might believe that I had not spent my time altogether to no purpose, and as the old man says in the Comedy

> No man so well ere laid his count to live,
> But that things, age, and use, some new things give,
> That what you thought you knew, you shall not know,
> And what you once thought best, you shall forgo.

This may perchance fall out now in the motion of the heart, that from hence the way being thus pervious, others trusting to more pregnant wits, may take occasion to do better, and search further.

Harvey's teacher at Padua and an important anatomist and embryologist.

Chapter II.
What Manner of Motion the Heart
Has in the Dissection of Living Creatures

First then in the hearts of all creatures, being dissected whilst they are yet alive, opening the breast, and cutting up the capsule, which immediately environeth the heart, you may observe that the heart moves sometimes, sometimes rests: and that there is a time when it moves, and when it moves not.

This is more evident in the hearts of colder creatures as the toads, serpents, frogs, house-snails, shrimps, crayfish, and all manner of little fishes. For it shows itself more manifestly in the hearts of hotter bodies, as of dogs, swine, if you observe attentively till the heart begin to die, and move faintly, and life is as it were departing from it. Then you may clearly and plainly see that the motions of it are more slow, and seldom, and the restings of it of a longer continuance: and you may observe and distinguish more easily, what manner of motion it is, and which ways it is made, in the resting of it, as likewise in death, the heart is yielding, flagging, weak, and lies as it were drooping.

At the motion, and whilst it is moving, three things are chiefly to be observed.

1.   That the heart is erected and that it raises itself upwards into a point, insomuch that it beats the breast at that time, so as the pulsation is felt outwardly.

2.   That there is a contraction of it every way, especially of the sides of it, so that it appears lesser, longer and contracted. The heart of an eel, taken out and laid upon a trencher, or upon one's hand, doth evidence this: It appears likewise in the hearts of little fishes, and of those colder animals whose hearts are sharp at top and long.

3.   That the heart being grasped in one's hand whilst it is in motion, feels harder. This hardness arises from tension, like as if one take hold of the tendons on one's arm by the elbow

whilst they are moving the fingers, one shall feel them bent and more resisting.

4. 'Tis moreover to be observed in fish and colder animals which have blood, as serpents, frogs, at that time when the heart moves it becomes whitish, when it leaveth motion it appears full of sanguine colour. From hence it seemed to me, that the motion of the heart was a kind of tension in every part of it, according to the drawing and constriction of the fibres every way, because it appeared that in all its motions, it was erected, received vigour, grew lesser, and harder, and that the motion of it was like that of the muscles, where the contraction is made according to the drawing of the nervous parts and fibres, for the muscles whilst they are in motion, and in action, are envigorated, and stretched, of soft become hard, they are uplifted, and thickened, so likewise the heart.

From which observations with good reason we may gather that the heart at that time whilst it is in motion, suffers constriction, and is thickened in its outside, and so straightened in its ventricles, thrusting forth the blood contained within it: which from the fourth observation is evident, because that in the tension it become white, having thrust out the blood contained within it and presently after in its relaxation, and rest, a purple and crimson colour returns to the heart. But of this no man needs to make any further scruple, since upon the inflicting of a wound into the cavity of the ventricle, upon every motion, and pulsation of the heart, in the very tension, you shall see the blood within contained to leap out.

So then these things happen at one and the same time, the tension of the heart, the erection of the point, the beating (which is felt outwardly) by reason of its hitting against the breast, the incrassation [thickening] of the sides of it, and the forcible protrusion of the blood by constriction of the ventricles.

Hence the contrary of the commonly received opinion appears, which is, that the heart at that time when it beats against the breast, and the pulsation is outwardly felt, it is believed that the ventricles of the heart are dilated, and replete with blood, though you shall understand that it is otherwise and that when the heart is contracted it is emptied. For that motion which is commonly thought the Diastole of the heart, is really the Systole, and so the proper motion of the heart is not a Diastole but a Systole, for the heart receives no vigour in the Diastole, but in the Systole, for then it is extended, moveth and receiveth vigour.

Neither is that to be allowed, though it is confirmed by a comparison alleged by the Divine Vesalius,[3] of a wreath of osiers, meaning of many twigs, joined together in fashion of a pyramid: that the heart doth not only move by the straight fibres, and so whilst the top is brought near to the bottom, the sides of it are dilated round about, and do acquire the form of a little gourd, and so take in blood (for according to all the drawing of the fibres which it has, the heart is stiffened and gathered together). But that the outside and substance of it are rather thickened and dilated, and that whilst the fibres are stretched from the top of the corner to the bottom, the sides of the heart do not incline to an orbicular figure, but rather contrary, as every fibre circular lies placed, does in its contraction incline to straightness and as all the fibres of the muscles whilst they are contracted and shortened of their length so towards the sides they are extended, and are thickened after the same fashion as the bodies of the muscles.

To this add, that not only in the motion of the heart by erection and incrassation of the sides of it, it so falls out, that the ventricles are straightened, but moreover all the sides inwardly are girt together as it were with a noose, for expelling the blood with greater force, by reason that those fibres or little tendons, amongst which there are none but straight ones, (for those in the outside are circular) called by Aristotle, nerves, are various in the ventricles of the heart of greater creatures, whilst they are contracted together with a most admirable frame.

Neither is it true which is commonly believed, that the heart by any motion or distention of its own doth draw blood into the ventricles, but that whilst it is moved and bended, the blood is thrust forth, and when it is relaxed and falls, the blood is received in manner as follows. . . .

3. Andreas Vesalius, author of *On the Fabric of the Human Body*, published in the same year as Copernicus's *De Revolutionibus* (1543).

Chapter VII.
That the Blood Does Pass
from the Right Ventricle of the Heart,
Through the Strainer of the Lungs,
into the Vein-like Artery,
and Left Ventricle of the Heart

It is well enough known that this may be, and that there is nothing which can hinder if we consider which way the water passing through the substance of the earth, doth procreate rivulets and fountains; or if we do consider how sweat passes through the skin or how urine flows through the strainer of the reins [kidneys] : It is to be taken notice of in those that make use of the waters of the Spa,[4] or de la Madonna, as they call them in Padua or other brackish or vitriolated waters; or those who in carousing swill themselves with drink, that in an hour or two they piss all this through their bladder. This great quantity ought to stay a while in concoction, it ought to flow through the liver, (as they confess that the juice of the nourishment we receive doth twice a day) so ought it through the veins, through the strainer of the reins, and through the ureters into the bladder.

Those therefore which I hear denying, that blood, yea the whole mass of blood, may pass through the substance of the lungs, as well as the nutritive juice through the liver, as if it were impossible, and no ways to be believed; It is to be thought that those kind of men, I speak with the Poet, where they like, they easily grant, where they like not, by no means: Here where need is, they are afraid, but where no need is they are not afraid to aver. The strainer of the liver, and of the reins too, is much thicker than that of the lungs, because they are far thinner woven and of a spongeous substance, if they be compared to the liver and reins.

In the liver there is no impulsive, no strength forcing, in the lungs, the blood is thrust against them by the impulsion of the right ventricle of the heart, by which impulsion there must necessarily follow a distention of the vessels, and porosities of the lungs. Besides, the lungs in respiration rise and fall, Galen *On the Use of the Parts.* By which motion it follows of necessity, that the poro-

---

4. Spa, in what is now Belgium, was long a most famous health resort because of its mineral springs.

sities of them and their vessels are opened and shut, as it falls out in sponges, and all things of a spongy substance when they are constricted and dilated again: On the contrary, the liver is at rest, nor is it seen at any time to be so constricted and dilated. Last of all, since through the liver, there is none but affirms, that the juice of all things we receive may pass into the vena cava, both in Men, Oxen or the greatest creatures, and that for this reason, because it must pass some way into the veins if there be any nutrition, and there is no other way, and for that cause they are forced to affirm this: Why should they not likewise believe this of the passage of the blood through the lungs in men come to age, upon the same arguments? And with Columbus,[5] a most skilful and learned anatomist, believe and assert the same from the structure and largness of the lungs; because that the vein-like artery[6] and likewise the ventricle, are always full of blood, which must needs come hither out of the veins, by no other path, but through the lungs; as both he and we from our words before, our own eyesight and other arguments do believe this to be clear.

But seeing there are some such persons which admit of nothing, unless there be an authority alleged for it; let them know, that the very same truth may be proved from Galen's own words, that is to say, not only that the blood may be transfused out of the artery-like vein, into the vein-like artery and thence into the left ventricle of the heart, and afterwards transmitted into the arteries; but also that this is done by a continued pulse of the heart, and motion of the lungs, whilst we breathe. There are in the orifice of the artery-like vein three shuts, or doors, made like a Σ [sigma] or half Moon, which altogether hinder the blood sent into the artery-like vein to return to the heart, which all know.

Galen expresses the use and necessity of those shuts in these words, *On the Use of the Parts*, VI, Ch. 10. "In all (says he) there is a mutual Anastomosis or opening of the veins, together with the arteries, in their kissing,[7] and they borrow both blood and spirit

5. Realdus Columbus—cf. p. 132.
6. The "vein-like artery" is the pulmonary vein (carrying blood from the lungs to the left ventricle); the "artery-like vein" is the pulmonary artery (carrying blood from the right ventricle to the lungs). We name these vessels by function; in Harvey's day they were named by position, vessels connected to the right side of the heart being considered veins and those on the left side arteries.
7. Osculation.

from one another by invisible and very narrow passages. But if the very mouth of the artery-like veins had always stood open, and Nature had found no device to shut it; when it was requisite, and to open it again, it could never have come to pass that by those invisible and little kisses, the Thorax being contracted the blood could be transfused into the arteries. For everything is not from anything extracted and emitted after the same manner; for as that which is light is easilier [sic] attracted than that which is heavy, by dilation of the instruments, and by the constriction is squeezed out again; so anything is easier attracted through a broad passage, than through a narrow passage, and so sent forth again. But when the Thorax is contracted the vein-like arteries which are in the lungs, being on every side pulsated, and compresses together strongly, do squeeze out very quickly the spirit that is in them, and do borrow through those fine touches a part of the blood, which truly could never come to pass, if through that great opening, such as is the artery-like vein, the blood could return back to the Heart; Now the return of it through that great mouth being stopped, some of it through those small orifices does drop into the Arteries, it being pressed every way." And a little after in the following chapter "how much the more the Thorax endeavours to squeeze out the blood so much the more those membranes, that is to say those three Sigma-like doors, do closlier shut the mouth of it, and suffer nothing to return." Which he says likewise in the same tenth chapter a little before "Unless there were doors there would follow a threefold inconvenience, for so the blood should make such a long journey but in vain, by flowing in the Diastoles of the lungs, and filling all the veins in them, in the Systoles, as it were a neap tide, like Euripus reciprocating its motion again and again, hither and thither, which would not be convenient for the blood. But this may seem no greater matter, but that in the meantime it should weaken the benefit of respiration, this is no more to be counted a small business." And a little after "And likewise the third inconvenience would follow, no slight one, when in our breathing our blood should return backwards, unless our Maker had ordained the natural position of those Membranes." Whence he concludes Chap. 2 "Indeed the use of all the shuts or portals is the same, to hinder the return of the matter; and either of them have a proper use to draw matter from the heart, that they may return no more, and to draw matters into the heart, that they may go no more from thence. For Nature would not have the heart to be wearied with needless travel, nor send thither whence it was better to extract, nor extract from thence again whither it was better to send. For which cause there being four orifices only, two in either ventricle, one takes in, the other draws forth." And a little after. "Furthermore, when one of the vessels consisting but of one tunicle is implanted into the heart, and the other consisting of a double tunicle is drawn forth from it, viz." (The right ventricle Galen means, so do I the left ventricle by the same reason) "It was needful that there should be as it were a cistern to both, to which both of them belonging, that the blood might be drawn out by one, and sent out by the other."

That argument which Galen brings for the passages of the blood through the right ventricle out of the vena cava into the lungs, we may more rightly use for the passage of the blood out of the veins through the heart into the arteries, changing only the terms.

It does therefore clearly appear from the words and places of Galen, a divine man, father of Physicians, both that the blood doth pass from the artery-like vein into the little branches of the vein-like artery both by reason of the pulse of the heart and also because of the motion of the lungs and thorax. . . .

Furthermore it was necessary that the heart should receive the blood continually into the ventricles, as in a pond or cistern, and sent it forth again: and for this reason it was necessary that it should be served with four locks or doors, whereof two should serve for the intromission and two for the emission of blood, lest either the blood like an Euripus, should inconveniently be driven up and down, or go back thither from whence it were fitter to be drawn, and flow from that part to which it was needful it should have been sent, and so should be wearied with idle travel, and the breathing of the lungs be hindered. Lastly our assertion appears clearly to be true, that the blood does continually and incessantly flow through the porosities of the lungs out of the right ventricle into the left, out of the vena cava into the great artery; for seeing the blood is continually sent out of the right ventricle into the lungs through the artery-like vein and likewise is continually attracted out of the lungs into the left, which appears by that which has been spoken and the position of the portals, it cannot be, but that it must needs pass through continually.

And likewise seeing that always and without intermission the blood enters into the right ventricle of the heart, and goes out (which is likewise manifest, of the left ventricle, both by reason and sense) it is impossible but that the blood should pass continually through, out of the vena cava into the aorta.

That therefore which is apparent to be done in most, and really in all whilst they are growing to age, by dissection through most open passages, is here likewise manifest to come to pass in those when they are arrived to full age, by the hidden porosities of the lungs, and touches of its vessels both by Galen's words, and that which has been spoken: From whence it appears, that albeit one ventricle of the heart, that is the left, were sufficient for the dispensation of the blood, through the whole body, and the education of it out of the vena cava (as it is in all creatures which want lungs) yet nature desiring that the blood should be strained through the lungs, was forced to add the right ventricle, by whose pulse the blood should be forced through the very lungs out of the vena cava into the receptacle of the left ventricle, and so it is to be said that the left ventricle was made for the lungs' sake and not for nutrition only; seeing in such an abundance of victual, adding to it the help of compulsion, it is no ways to be believed that the lungs should rather want so much aliment, and that of blood so much more pure and full of spirit, as being immediately conveyed from the ventricles of the heart, than either the most pure substance of the brain, or the most resplendent and divine constitution of the eyes, or the flesh of the heart itself, which is more fitly nourished by the coronary vein.

Chapter VIII.
Of the Abundance of Blood Passing Through the Heart Out of the Veins into the Arteries, and of the Circular Motion of the Blood

Thus much of the transfusion of the blood out of the veins into the arteries, and how it is disposed of and transmitted by the pulse of the heart, to some of which those perchance that were heretofore moved by the reasons of Galen, Columbus, and others, will yield; now as concerning the abundance and increase of this blood, which doth pass through, those things which remain to be spoken of, though they be very considerable, yet when I shall mention them, they are so new and unheard of, that not only I fear mischief which may arrive to me from the envy of some persons, but I likewise doubt that every man almost will be my enemy, so much does custom and doctrine once received and deeply rooted (as if it were another nature) prevail with every one, and the venerable reverence of antiquity enforces. Howsoever, my resolution is now set down, my hope is in the candor of those which love truth, and learned spirits. Truly when I had often and seriously considered with myself what great abundance there was, both by the dissection of living things, for experiments sake and the opening of arteries, and many ways of searching, and from the symmetry and magnitude of the ventricles of the heart, and of the vessels which go into it, and go out from it, (since Nature making nothing in vain, did not allot that greatness proportionably to no purpose, to those vessels) as likewise from the continued and careful artifice of the doors and fibres, and the rest of the fabric, and from many other things, and when I had a long time considered with myself how great abundance of blood was passed through and in how short time that transmission was done whether or no the juice of the nourishment which we receive could furnish this or no; at last I perceived that the veins should be quite emptied, and the arteries on the other side be burst with too much intrusion of blood, unless the blood did pass back again by some way out of the veins into the arteries and return into the right ventricle of the heart.

I began to bethink myself it might not have a circular motion, which afterwards I found true, and that the blood was thrust forth and driven out of the heart by the arteries into the habit of the body and all parts of it, by the beating of the left ventricle of the heart, as it is driven into the lungs through the artery-like vein by the beating of the right, and that it does return through the little veins into the vena cava, and to the right ear of the heart, as likewise out of the lungs through the aforesaid vein-like artery to the left ventricle as we said before.

Which motion we may call circular, after the same manner that Aristotle says that the rain and the air do imitate the motion of the superior bodies. For the earth being wet, evaporates by the heat of the Sun, and the vapours being raised aloft are condensed, and descend in showers and wet the ground, and by this means here are generated, likewise, tempests, and the beginnings of meteors, from the circular motion of the Sun, and his approach and removal.

So in all likelihood it comes to pass in the body, that all parts are nourished, cherished, and quickened with blood, which is warm, perfect, vapourous, full of spirit, and that I may so say, alimentative: in the parts the blood is refrigerated, coagulated, and made as it were barren, from thence it returns to the heart, as to the fountain or dwelling house of the body, to recover its perfection, and there again by natural heat, powerful, and vehement, it is melted, and is dispensed again through the body from thence, being fraught with spirits, as with balsam and that all the things do depend upon the motional pulsation of the heart.

So the heart is the beginning of life, the Sun of the Microcosm, as proportionably the Sun deserves to be called the heart of the world, by whose virtue and pulsation the blood is moved, perfected, made vegetable, and is defended from corruption, and mattering; and this familiar household god doth his duty to the whole body, by nourishing, cherishing and vegetating, being the foundation of life, and author of all. But we shall speak more conveniently of these in the speculation of the final cause of this motion.

Hence it is, seeing the veins are certain ways or vessels carrying the blood, there are two sorts of them, the Cava and Aorta. Not by reason of the side, as Aristotle says, but by their function; and not, as is commonly spoken, by their constitution, seeing in many Creatures (as I have said) a vein differs not from an artery, in the thickness of the tunicle, but by their use and employment distinguishable a vein and an artery, both of them not undeservedly called veins by the Ancients, as Galen has observed, because that this viz, the artery, is a way carrying the blood from the heart into the habit of the body, the other a way carrying it from the habit of the body back again into the heart. This is the way from the heart, the other way to the heart. This contains blood rawish, unprofitable, and now made unfit for nutrition, the other blood digested perfect, and alimentative.

Chapter IX.
That There Is a Circulation of the Blood
from the Confirmation of the First Supposition

But lest any should think that we put a cheat upon them, and bring only fair assertions, without any ground, and innovate without a cause, there comes three things to be confirmed, which being set down, I think this truth must needs follow, and be apparent to all men.

1. First, that the blood is continually, and without any intermission, transmitted out of the vena cava into the arteries, in so great abundance, that it cannot be recruited by those things we take in, and insomuch that the whole mass of blood would quickly pass through.

2. In the second place, that continually, duly, and without cease, the blood is driven into every member and part, and enters by the pulse of the arteries, and that in far greater abundance than is necessary for nourishment, or than the whole mass is able to furnish.

3. And likewise thirdly, that the veins themselves do perpetually bring back this blood into the mansion of the heart.

These things being proved, I think it will appear that it doth go round, is returned, thrust forward, and comes back from the heart into the extremities, and from thence into the heart again, and so makes as it were a circular motion.

Let us suppose how much blood the left ventricle contains in its dilation when it's full, either by our thought or experiment, either 2, or 3, or 1½ oz.; I have found in a dead man above 4 oz.

Let us suppose likewise, how much less in the contraction, or when it does contract itself, the heart may contain, and how much less capacious the ventricle is, and from then how much blood is thrust out of the great artery:[8] for in the Systole there is always some thrust forth, which was demonstrated in the third Chapter, and all men acknowledge, being induced to believe it from the fabric of the vessels, by a very probable conjecture we may aver that there is sent in of this into the artery a fourth, or fifth, or sixth, at least an eighth part. So let us imagine, that in a man there is sent forth in every pulse of the heart, an ounce and a half, or three drams of blood, or one dram which by reason of hindrance of the portals cannot return to the heart.

The heart in one half hour makes above a thousand pulses, yea in some, and at some times, two, three or four thousand; now multiply the drams either a thousand times three drams, or two drams or five hundred ounces, or such a proportionate quantity of blood, transfused through the heart into the arteries, which is a greater quantity than is found in the whole body. So likewise in a sheep or a dog if there pass (I grant you) but one scruple, in one

8. Aorta

half hour there passes a thousand scruples, or about three pounds and a half of blood: in whose body for the most part is not contained above four pounds of blood, for I have tried it in a sheep.

So our account being almost laid, according to which we may guess the quantity of blood which is transmitted, counting the pulsations, it seems that the whole mass of blood does pass out of the veins into the arteries through the heart, and likewise through the lungs.

But grant that it be not done in half an hour, but in a whole hour, or in a day, be it as you will, it is manifest that more blood is continually transmitted through the heart, than either the food which we receive can furnish, or is possible to be contained in the veins. Nor is it to be said, that the heart in its contraction sometimes does thrust out, sometimes not, or as much as nothing, or something imaginary. This I refuted before, and besides it's against sense or reason. For if in the dilation of the heart it must needs come to pass that the ventricles are filled with blood, it is likewise necessary that in its contraction it should always thrust forth, and that not a little, seeing the conduits are not small; and the protrusion not seldom: it's very convenient likewise [that] in every propulsion, the proportion of the blood thrust out should be a third part, or sixth part, or eight part in proportion to that which is before contained in the ventricle, and which did fill it in the dilation, according as the proportion of the ventricle being contracted is to the proportion of it being incontracted; and as in the dilatation it never comes to pass, that it is ever filled with nothing, or something merely imaginary so in the contraction it never expells nothing, or that which is imaginary, but always something, according to the proportion of the contraction. Wherefore it is to be concluded, that if in a man, a cow or a sheep, the heart doth send forth one dram, and that there be a thousand pulses in one half hour, that it shall come to pass in the same time that there shall be ten pounds and five ounces transmitted, if at one pulse it send forth two drams, twenty pounds and 10 oz., if half an ounce forty one pounds and 8 oz., if an ounce, 83 lb and 4 oz. will come to be transfused, I say, in half an hour, out of the veins into the arteries.

But it may perchance be that I shall set down here more accurately how much is thrust out at every pulsation, when more, and when less, and for what reason, out of many observations which I have gathered.

In the meantime this I know and declare to all men, that sometimes the blood passes in less, sometimes in more abundant quantity, and the circuit of the blood is performed sometimes sooner, sometimes slower, according to the age, temperature, external and internal cause, accidents, natural or innatural, sleep, rest, food, exercise, passions of the mind, and the like.

But however, though the blood pass through the heart and the lungs, in the least quantity that may be, it is conveyed in far greater abundance into the arteries, and the whole body, than it is possible that it could be supplied by juice of nourishment which we receive, unless there were a regress made by its circuition.

This likewise appears by our sense, when we look upon the dissection of living things, not only in the apertion [opening] of the great artery, but (as Galen affirms in man himself) if any, yea the least artery be cut, all the mass of blood will be drained out of the whole body, as well out of the veins as out of the arteries, in the space of half an hour.

Likewise, butchers can well witness this, when in killing an ox, they cut the jugular arteries, they drain the whole mass of blood in less than a quarter of an hour, and empty all the vessels, which we find likewise to come to pass in cutting off members and tumours, by too much profusion of blood, sometimes in a little space.

Nor does it weaken the force of this argument, that some will say, that in slaughter, or of cutting off members, the blood flows out as much through the veins as through the arteries, seeing the business is far otherwise. For the veins, because they flap down, and that there is no out-driving force in them, and because their composition is likewise with stoppages of portals, as hereafter shall appear, they shed but a very little, but the arteries pour out the blood more largely, impetuously, by impulsion, as if it were cast out of a spout. But let the case be tried omitting the veins and cutting the jugular arteries in a sheep, or a dog, it will be wonderful to see, with how great force, how great protrusion, how quickly, you shall see all the blood to be emptied from the whole body as well from the veins as from the arteries. But it is manifest by what we have said, that the arteries receive blood nowhere else but from the veins by transmission through the heart, wherefore tieing the aorta at the root of the heart, and opening the jugular or any other artery, if you see the arteries empty, and the veins only full, it is not to be wondered at.

Hence you shall plainly see the cause in Anatomy why so much blood is found in the veins, and but a little in the arteries, why there is a great deal found in the right ventricle, and but a little in the left, (which thing perchance gave occasion of doubt to the Ancients, and of believing, that spirits alone were contained in those concavities, whilst the animal was alive) the cause perchance is: because there is no passage afforded from the veins into the arteries but through the lungs and the heart, but when the lungs have expired and leave off to move, the blood is hindered to pass from the little branches of the artery-like vein into the vein-like artery, and so into the left ventricle of the heart (as in an embryo it was before observed, that it was stopped by reason of the want of motion of the lungs, which open and shut up the touches, and hidden and invisible porosities) but seeing the heart does not leave off motion at the same time with the lungs but does beat afterwards and outlive them, it comes to pass that the left ventricle and the arteries do send forth blood into the habit of the body, and not receiving it through the lungs, do therefore appear empty.

But this likewise affords no small credit to our purpose since there can be no other cause given for this but what in our supposition we have alleged.

Besides from hence it is manifest, that how much the more, or more vehemently the arteries do beat it happens in all fluxes of blood that so much the sooner the whole body is emptied.

Hence likewise it comes to pass, that in all faintings, all fear, and the like, when the heart beats more weakly, languishing, and with no force, that it happens that all fluxes of the blood are stopped and hindered.

Hence likewise it is that in a dead body after the heart ceases to beat, you cannot out of the jugular or crural veins, and opening of the arteries by any means extract above half the mass of blood, nor can a butcher when he hath knocked the ox on the head, and stunned him, draw all the blood form him unless he cut his throat before the heart leaves beating.

Last of all, from hence we may imagine that no man hitherto has said anything aright concerning the anastomosis, where it is; how it is, and for what cause; I am now in that search.

Chapter X.
The First Supposition Concerning
the Quantity of the Blood Which Passes Through
from the Veins into the Arteries,
and that There Is a Circulation of the Blood
is Vindicated from Objections,
and Further Confirmed by Experiments

Thus far the first position is vindicated, whether the matter be to be reckoned by account, or whether we refer it to experiment, or our own eyesight, viz. that the blood continually passes out of the veins into the arteries in greater abundance than can be furnished by our nourishment, so that the whole mass in a little time passing through that way, it must necessarily follow that there should be a circulation, and that the blood should return.

But if any here can say that it can pass through in great abundance, and yet it is not needful that there should be a circulation, since it comes to be made up by what we receive, and that the increase of milk in the paps may be an instance, for a cow in one day gives three, four, or seven gallons, or more, a woman likewise gives two or three pints every day or more, in the nursing of a child or two, which is manifest to be restored by what she receives, it is to be answered, that the heart is known to send out so much in one hour or two.

But if not yet as satisfied he shall still press further, and say, that although by the dissecting of an artery, and giving and opening a way, it comes to pass besides the course of nature, that the blood is forcibly poured out, yet it does not therefore come to pass in an entire body, no outlet being given, and the arteries being full, and constituted according to Nature, that such a great quantity should pass in so short space, insomuch that there must needs be a regress: It is to be answered: That by laying of an account it appears from former reckoning, that how much the heart being filled does contain more in its dilatation, than in its constriction, so much (for the most part) at every pulsation is sent forth, and for that cause does there so much pass the body being whole, and constituted according to Nature.

But in serpents, and in some fishes, binding the veins a little beneath the heart, you shall quickly see the distance betwixt the heart and the ligature to be emptied, so that you must needs affirm the recourse of blood, unless you will deny your own eyesight. The same shall clearly appear afterwards in the confirmation of the second supposition.

Let us conclude, confirming all these with one example, that everyone may believe his own eyes: if anyone cut up a live adder, he shall see the heart beat calmly, distinctly, for a whole hour, and so contract itself (in its constriction being oblong) and thrust itself out again like a worm. That it is whitish in the Systole, and contrary in the Diastole, together with all the rest, by which I said this truth was evidently confirmed, for here the parts are longer and more distinct. But this we may more especially find, and clearer than the noon day.

The vena cava enters the lower part of the heart, the artery comes out at the upper part; now taking hold of the vena cava with a pair of pincers, or with your finger and thumb, and the course of the blood being stopped a little way beneath the heart, you shall upon the pulse perceive to be presently almost emptied that place which is betwixt your fingers and the heart, the blood being exhausted by the pulse of the heart; and that the heart will be of a far whiter colour, and that it is lesser too in its dilatation for want of blood, and at last beats more faintly, insomuch that it seems in the end as it were to die, so soon again as you untie the vein, both colour and bigness returns to the heart. Afterwards, if you do leave the veins and do grasp or bind the artery a little way from the heart, you shall on the contrary see them swell vehemently there where they are grasped and that the heart is swelled beyond measure, and does acquire a purple colour till it be blackish again, and that it is at last oppressed with blood so that you would think it would be suffocated, but untying the string, that it does return to its natural constitution, colour, and bigness.

So now there are two sorts of death, extinction, by reason of defect, and suffocation by too great quantity: here you may have the example of both before your eyes, and confirm the truth which hath been spoken concerning the heart, by your own view.

Chapter XI.
The Second Supposition is Confirmed

The second is to be confirmed by us, which that it may appear the clearer to our view, some experiments are to be taken notice of, by which it is clear, that the blood doth enter into every member through the arteries, and does return by the veins, and that the arteries are the vessels carrying the blood from the heart, and that the veins are the vessels and ways by which the blood is returned to the heart itself; and that the blood in the members and extremities does pass from the arteries into the veins (either mediately by an anastomosis, or immediately through the porosities of the flesh, or both ways) as before it did in the heart and thorax out of the veins, into the arteries: whence it is manifest, that in its circulation it moves from thence hither, and from hence thither, to wit, from the centre to the extremities, and from the extremities again to the centre.

Chapter XII.
That There Is a Circulation of the Blood, from the Confirmation of the Second Supposition

Seeing these things are so, it is certain that another thing which I said before is likewise confirmed, that the blood does continually pass through the heart. For we see in the habit of the body, that the blood flows continually out of the arteries into the veins, not out of the veins into the arteries: We see moreover, that from one arm the whole mass of blood may be exhausted, and that too by opening but one cuticular vein with a lance[t], if the ligature be handsomely made: We see besides, that it is poured out so forcibly and so abundantly, that it is certain that not only that which was comprehended in the arm beneath the ligature, before the section, is quickly and in a little time evacuated; but likewise the blood out of the whole body, as well the veins as the arteries.

Wherefore we must confess first that by strength and force it is furnished, and by force it is driven beyond the ligature (for with force it goes out, and therefore by the strength and pulse of the heart) for the force and impulsion of the blood is only from the heart.

Next, that this flux comes from the heart, and that it flows by a passage made through the heart out of the great veins, seeing below the ligature the blood enters by the arteries, not by the veins, and the arteries at no time receive blood out of the veins, unless it be out of the left ventricle of the heart. Nor could there any otherwise so great abundance be exhausted out of one vein, making a ligature above, especially so forcibly, so abundantly, so easily, so suddenly, unless the consequents were achieved by the force and impulsion of the heart, as is said.

And if these things be so, we may very openly make a computation of the quantity, and argue concerning the motion of blood. For if anyone (the blood breaking out according to its usual effusion and force) suffer it to come so for half an hour, no body needs doubt but that the greatest part of it being exhausted, faintings and soundings would follow, and not only the arteries, but the greatest veins would be likewise emptied: Therefore it stands with reason, that in the space of that half hour there passes so much out of the great vein through the heart into the aorta. Further, if you should reckon how many ounces flow through one arm, or how many ounces are thrust within the gentle ligature in 20 or 30 pulsations, truly it would minister occasion of thinking how much may pass through the other arm, both the legs, and both sides of the neck, and through all the other arteries and veins of the body: and that the flux which is made through the lungs and the ventricles of the heart, must continually furnish of necessity new blood, and so make a circuit about the veins, since so great a quantity cannot be furnished from those things we eat, and that it is far greater than is convenient for the nutrition of the parts.

It is to be observed further, that in the administration of phlebotomy this truth chances sometimes to be confirmed, for though you tie the right arm, and lance it as it should be with a convenient orifice and administer all things as they ought to be, yet if fear, or any other cause, or sounding do intervene through passion of the mind, so that the heart do beat more faintly, the blood will by no means pass through but drop after drop, especially if the ligature be made a little straiter [tighter]. The reason is, because the pulse being but faint, and the out-driving force being but weak, the enfeebled part is not able to open the passage and thrust out the blood beyond the ligature, yea nor to draw it through the lungs, or to remove it plentifully out of the veins into the arteries. So after the same manner does it come to pass that women's flowers and all other fluxes of blood are stopped. This likewise appears by the contrary, for fear being removed, and the spirit recollected, when they do return to themselves, the pulsific strength being now increased, you shall straightway see the arteries beat more vehemently in that part where they are bound, and move in the wrist, and the blood leap out farther through the orifice.

Chapter XIII.
The Third Supposition is Confirmed, and that There is a Circulation of the Blood from the Third Supposition

Hitherto concerning the quantity of blood which passes through the lungs and heart in the centre of the body, and likewise from the arteries into the veins and habit of the body; it remains that we do explain which way the blood flows back from the extremities through the veins into the heart and how the veins are the vessels that carry it from the extremities to the centre, by which means we think those three grounds propounded will be true, clear, firm and sufficient to gain credit.

But this shall be plain enough from the portals which are found in the concavities of the veins, their use, and from ocular experiments.

The most famous Fabricius of Aquapendente, a most learned anatomist, and a venerable old man, or as the most learned Riolanus[9] would have it, Jac. Silvius[10] did first of any delineate the membranal portals [valves] in the veins being in the figure of a sigma or semilunary, the most eminent and thinnest parts of the inward tunicles of the veins: Their situation is in distant places, after a various manner, in diverse persons they are connate at the sides of the veins, looking upwards towards the roots of them, and in the middle capacity both of them, (for they are for the most part two) looking towards one another, equally and duly touching

9. A well-known and conservative French physician, one of the few publicly to reject Harvey's conclusions.
10. Professor at Paris in the mid-sixteenth century. Various venuous valves had been described by many anatomists in the sixteenth century. However, Fabricius was the first to note their existence in *all* the veins and to write a complete (though small) treatise on them. But he totally misunderstood their functions, believing they regulated the quantity of flow rather than (as Harvey correctly saw) its direction.

one another, insomuch that they are apt to stick together at the extremities, and to be joined; and lest they should hinder any thing to return from the roots of the veins into the little branches, or from the greater into the less, they are so placed that the horns of the hindermost are stretched towards the middles of the body of it which is before, and so interchangeable.

The finder out of these portals did not understand the use of them, nor others who have said lest the blood by its weight should fall downward: for there are in the jugular vein those that look downwards and do hinder the blood to be carried upwards. I (as likewise others) have found in the emulgent veins and branches of the Mesentery, those which did look towards the vena cava, and vena porta; add to this moreover that there are no such in the arteries, and it is to be observed that dogs and cattle have all their portals in the dividing of the crural veins at the beginning of the os sacrum, or in the iliac branch near the Coxendix, in which there is no such thing to be feared by reason of the upright stature of man. Nor are their portals in the jugulars, as others say, for fear of Apoplexy, because the matter is apt in sleep to flow into the head through the sopral arteries.

Nor that the blood may stand still in divarications [branchings] and that the whole blood should not break in into the small branches or those which are more capacious, for they are likewise placed where there are no divarications though I confess they are more frequent where divarications are.

Nor that the motion of the blood may be retarded from the centre of the body; for it is likely that it is thrust leisurely enough of its own accord, out of the greater into the lesser branches, and so that it is separated from the mass and fountain. But the portals were merely made, lest the blood should move from the greater veins into the lesser and tear or swell them; and that it should not go from the centre of the body to the extremities, but rather from the extremities to the centre. Therefore by this motion the small portals are easily shut; and hinder anything which is contrary to them; for they are so placed and ordained, that if anything should not be sufficiently hindered in the passage by the horns of the foremost, but should escape as it were through a chink, the convexity or vault of the next might receive it, and so hinder it from passing any further.

I have often tried that in dissection if beginning at the roots of the veins I did not put in the probe towards the small branches with all the skill I could, that it could not be further driven by reason of the hindrance of the portals: On the contrary, if I did put it in outwardly from the branches towards the root, it passed very easily.

In many places two portals are so interchangeably placed and fitted, that when they were elevated in the middle of the concavity of the vein, they close with one another to a hair's breadth, and in their extremities and convexities are [so] united interchangeably that you can neither see with your eyesight nor in any way discern any crevice or conjunction: On the contrary from outwardly putting in a probe they easily give way (and like those gates or sluices by which the course of rivers is stopped) they are easily turned back to intercept the motion of the blood from the vena cava and the heart, and being closely lifted up in many places whilst they are interchangeably shut they do quite hinder and suppress, nor by any means suffer the blood to move neither upwards to the head nor downwards to the feet, nor to the sides or arms, but do stop and resist all manner of motions of the blood, which is begun in the greater veins and ends in the lesser yet do obey any which is begun by the small veins and ends in the greater, and does provide a free and open way for it. . .

Chapter XIV.
The Conclusion of the Demonstration
of the Circulation of the Blood

Now then in the last place we may bring our opinion concerning the circulation of the blood, and propound it to all men.

Seeing it is confirmed by reasons and ocular experiments, that the blood does pass through the lungs and heart by the pulse of the ventricles, and is driven in and sent into the whole body, and does creep into the veins and porosities of the flesh, and through them returns from the little veins into the greater, from the circumference to the centre, from whence it comes at last into the vena cava, and into the ear of the heart in so great abundance, with so great flux, and reflux, from hence through the arteries thither, from thence through the veins hither back again, so that it cannot be furnished by those things which we do take in, and in a far greater abundance than is competent for nourishment: It must be of necessity concluded that the blood is driven into a round by a circular motion in creatures, and that it moves perpetually; and hence does arise the action and function of the heart, which by pulsation it performs; and lastly that the motion and pulsation of the heart is the only cause.

11. The Difference Between Venous and Arterial Blood

The Anatomy of the Heart
CHAPTER I.

THE POSITION AND STRUCTURE OF THE HEART

It is of very great importance for a true knowledge of the nature and qualities of the blood to have investigated not only its circular movement, but also to know and to compare its movements, its amounts, its elements, its various changes and their causes, as well as to estimate the quantity of the same fluid thrown out in individual beats. I have, therefore, thought it worth while to give a clear and concise account of the whole matter (which has been omitted hitherto by most authors, and desired, rather than explained, by others, including even Harvey himself), so far as I shall be able to achieve this object by conjecture and by experiment.

But as the movement of the Blood depends on the movement of the Heart, and in the absence of the latter can neither be understood nor exist, I must preface my account of it by some remarks on the Position and Structure of the heart. When these have been duly considered and collated, it will be easier to grasp how carefully both its Fabric and Position are adapted for movement, and how fittingly everything is arranged for the distribution of the blood to the organs of the body as a whole. . . .

SOURCE: From Richard Lower, *Tractatus de Corde* (London, 1669), translated by K. J. Franklin, *On the Heart;* in *Early Science in Oxford,* vol. IX (Oxford, 1932).

CHAPTER III.
THE MOVEMENT AND COLOUR OF THE BLOOD
THE RAPIDITY OF THE CIRCULATION,
AND THE DIFFERENCE BETWEEN
ARTERIAL AND VENOUS BLOOD

Having thus established the nature of the heart's structure, the source of its movement, the reasons for this movement's variation, and the kind of effects and symptoms such variations produce in the blood, it remains for me next to show how rapidly the whole of the blood circulates through the heart.

Whatever statements writers before Harvey made about the movement of the blood through the ventricles of the heart are so empty and worthless that they have already spontaneously disappeared into oblivion. And however much those of his successors who have accepted Harvey's discovery of the circulation affirm, under the necessity of that hypothesis, that the whole of the blood circulates through the heart, yet they have written about the rate of its passage and about the amount of blood forced out at any beat in such a way that I must think that they have not adequately considered the structure of the heart and its movements. For the majority of them give a few drops only, or a scruple, or one drachm, a few half an ounce, as the amount of the blood expelled at each beat. I grant that, in various animals, the ventricles of the heart contain and eject more or less according to the difference in size of the animal's body; but it will be clear from what follows how ill-advised it is to state that so small a quantity is expelled at a beat in man or any larger animal.

I am convinced that the whole mass of blood is ejected by the heart not once or twice within an hour, but many times. To make this clearer, we must consider carefully how much blood flows into the ventricles of the heart each time they dilate, and how much flows out from them while they are contracting. It is apparent from simple inspection that the ventricles dilate maximally in each diastole, but that in systole, on the other hand, the sides of the heart come together and are contracted so closely, that you can scarcely compress the small finger (introduced through the cut apex) more strongly with the hand itself. This is the reason I am absolutely certain that each ventricle receives in diastole as much blood as it can hold, and, on the other hand, expels completely in systole all that it has previously received. This is very obvious and clearly visible with the naked eye in the hearts and auricles of new-born animals, frogs, eels, and snakes. The hearts of these creatures are so completely emptied in any systole by the output of all their contained blood, that they appear quite white; but in diastole (when the blood flows back into them) their colour returns again. One cannot doubt that the same thing happens in the heart of larger animals, though it cannot be seen so clearly in them owing to the thickness of the parenchyma.

If one makes this assumption and counts the pulsations, it will not be difficult to calculate how much blood passes through the heart in an hour's time. Let us assume, then, that the left ventricle in a strong, healthy, human heart holds two ounces at one time, as the great Harvey observed (though I have observed it hold much more in some people). If the whole of that amount of blood is expelled by the heart at each systole, and the counted beats are two thousand within the hour (and this is, indeed, the lowest estimate), then the heart must eject four thousand ounces within the hour. This number of ounces makes three hundred and thirty-two pounds; given, however, that there are twenty-five pounds of blood in this man (a greater amount than is conceded to most men by nature or by Anatomists, for they say that the measure of all the blood contained in the human body rarely exceeds twenty-four pounds, or is less than fifteen pounds), it will most surely follow that the whole of this man's blood circulates through the heart six times within a single hour. But as it is seldom that there is so great an amount of blood in a healthy man, or that the heart beats so few times in the passing of an hour, it is reasonable to believe that the blood in most people passes through the ventricles of the heart somewhat more often than six times.

It is so in all animals, provided they are in enjoyment of good health; but it must be supposed that the blood sometimes passes through the heart much more rapidly, for example, in fevers and in violent exercise, and during convulsive movement of the heart; for the beats are then far faster, and the blood is consequently ejected twice as often. On the other hand, the blood passes through the heart much more slowly in jaundice, scurvy, and other such diseased conditions, which cause slowing and irregularity of the heart's movements, or in cases in which the vessels and ventricles of the heart are blocked by means of chyle. Further, the pulse is liable to variations in respect of temperament, sex, and age, so that it is impossible to estimate and to define with accuracy the amount of blood which is ejected, and the circulatory changes. But it is clear enough from the very size of the chambers of the ventricles, and from the number of heart-beats, that the blood is carried along and passes through the heart in strong and healthy animals at a much faster rate than is commonly believed, or has yet been described. And it must definitely be so, if one properly considers the constituents of the blood, and their liability to separate and to clot, unless they are continuously stimulated by vigorous movement.

Let no one imagine, from what I have said, that part of the blood—namely, that which is carried along by the vessels of, or near to, the heart, and not very distant from its source—does not circulate through the heart much more quickly and more often than the rest, which is carried through the extremities. The very nearness of the vessels and organs proves otherwise. But what I do assert is that, although all parts of the blood are not carried along at the same rate or so often through the heart-chambers, yet whatever amount or quantity of blood is present circulates through the heart as often as I have stated above.

How quickly all the blood is distributed by the heart through the whole of the body can be most easily grasped by the very rapid passage through the kidneys of fluid which has been mixed with the blood. Two or three pounds of beer, taken into the stomach as a morning drink, are almost completely passed by the bladder within half an hour, or sooner if the weather is cold. This fluid, moreover, formed for the most part only half of the blood carried to the kidneys. Is it not right therefore, to say that four or six pounds of blood are passed from the heart to the kidneys through the two (so-called emulgent) arteries? But, if this large amount of blood passes in this very short space of time through vessels which are small by comparison with the others, it will not be difficult to understand how vigorously the whole of the rest of the blood-mass is carried along in the other regions of the body. This is still more obvious in those, who drink bitter mineral waters in quantities so large, that I have known some who have drunk almost two gallons of water in discrete draughts in a single morning, and they have passed almost all of it through the bladder within four hours. This amount of water is more than double the quantity of blood in most men, and it is likewise certain that it passed quite often through the two ventricles of the heart in common with the rest of the blood, before it could be secreted by the kidneys or passed by the bladder. For nature provides no other passage or route to the kidneys and bladder for any fluid from the stomach or intestine than that through the blood and the heart. But, if so much fluid (which, as stated above, is only half the amount of blood brought to the kidneys) passes through arteries of this calibre in so short a time, what, I ask, are we to think of the passage of the rest of the blood through all the larger vessels?

It is easy to show by experiments, without relying only on estimates, the rapidity of movement of this blood ejected by the heart. For, if almost the whole of the blood may flow out from an opening in one cervical artery within five minutes, may one not suppose that the whole of the blood-mass circulates through the heart in a much shorter space of time? When almost the whole of the blood escapes thus quickly through one arterial branch, how much more quickly would it flow out from the trunk of the aorta or from all its branches opened simultaneously?

But for the careful estimation of the blood-flow and its remarkable rapidity it will be sufficient to quote one single experiment. I divided both cervical arteries in a fair-sized dog, and at the same time I compressed the trunk of the aorta below the heart with a finger, which I passed through an aperture made in the left side of the chest near the heart, so that no blood should pass down the aorta; finally, I took care to constrict the brachial arteries below the axillae. As a result practically all the blood (except that passing through the vertebrals) was expelled from the heart by way of the cervical arteries; and, strange to say, it all flowed out within three minutes; so that it cannot be denied that the whole mass of it had passed through the heart in that time. One may, indeed, see in cases of trauma, which involve section of any large artery, how brief a period suffices for people so mutilated to lose their life and practically all their blood. This blood, however, must all have first circulated through the ventricles of the heart.

But here I see it objected that the heart pulsates much more quickly, and therefore ejects blood much more rapidly amid such woundings and torments. Yet, if the blood is held in check for a short while, after these incisions are made, so that all pain and fear first disappear—and this takes place rapidly in younger dogs, which are not so upset by, or so long mindful of, such lesions—the heart-beat is definitely not thus quickened and accelerated. And, though one must admit that the beat becomes more rapid, after most of the blood has escaped and the animal begins to weaken, yet that does not occur until all the vessels have been so emptied, that the blood which remains is insufficient to fill the ventricles of the heart. Hence, pari passu with the continued loss of blood, the heart-beat becomes progressively smaller and faster, until with the complete failure of blood-inflow the movement of the heart too ceases altogether.

It may, however, be objected that the blood flows out from a cut artery more easily, and therefore more rapidly, than it circulates through the body; since in the former case it is carried along in a free, full-rushing stream, and in the latter only reaches the veins through various windings and obstacles, and the narrow places and pores in the flesh—in the same way, perhaps, that a stream flows along in an open channel more quickly than it does when passing through a grating.

To this objection it will be easy to answer that, notwithstanding the narrow places of the viscera and of the body-framework, the blood flows out from open veins almost as rapidly as it does from arteries, so long as the heart's movement is strong, although, granted, the rate of flow is not absolutely so fast in the former as in the latter case. Indeed, I have often found in dogs that if, as in the previous experiment, the neck is tightly ligatured and the trunk of the aorta compressed below the heart, so as to direct the blood-flow to the head, then, when one of the two jugular veins is divided, an almost equal quantity of blood will be drawn off from this vein in a given time as from the cervical artery, at all events until most of the blood is removed. But, after this stage is reached, it must be admitted that the heart's movement slackens on account of so great a loss of blood, and the remaining fluid flows out from the vein in smaller amount and more slowly, in correspondence with its more feeble expulsion from the heart.

If, however, this rapid circulation of the blood is accepted (and I think I have given sufficient proof of it), it will be evident that there is not so much difference between arterial blood and that contained in veins, as is commonly supposed.

I have spoken elsewhere of the different returns of the two kinds of blood, and of the sources from which they are derived. I have also in the same place discussed their colour-variation, and the cause of this very noticable difference between them.[1] But as I relied more in this matter on the authority and preconceived opinion of the learned Dr. Willis than on my own experience, and confused too far the torch of life with its torchbearer; as, too, the lapse of time has now taught me differently, I shall not be loth to exchange my former view for a better one. It is not my intention to attack the beliefs and opinions of others, or to bring scorn on myself by changing my own, but what is suggested by reason and confirmed by experience carries more weight with me and will always have my allegiance.

It is certain, then, that the difference in colour, which is found between venous and arterial blood, is quite independent of the heating of the blood in the heart (even if some such heating must be conceded there); for, granted that heating does occur chiefly in the heart, then, as the function of both ventricles is the same, and they do not differ in any other respects than, as stated above, in the strength and thickness of their fibres, why should the colour not undergo a similar change in the right ventricle? But it is quite certain that blood withdrawn from the pulmonary artery is similar in all respect to venous blood, and is only reddish on the surface. Indeed, it will be shown by a very convincing experiment that this fresh red colour is not conferred on the blood by the left ventricle either. For, if the trachea is exposed in the neck and divided, a cork inserted, and the trachea ligatured tightly over it to prevent any ingress of air into the lungs, then the blood flowing from a simultaneous cut in the cervical artery (or, at least, such blood as comes out some time after the asphyxiation of the lung) will be seen to be as completely venous and dark in colour, as if it had flown from a wound in the jugular vein. I have tried this fairly often, and the same truth is more evident still from the fact that the blood within the left ventricle of the heart and the trunk of the aorta of an animal, which has been strangled or has died a natural death, and in which air is prevented from passing into the blood, is found to be entirely akin to venous blood.

1. In *Diatribae Thomae Willisii...De Febribus Vindicatio* (Oxford, 1665), a defense of the ideas of his master, Dr. Thomas Willis, on the cause of fevers. Willis was a distinguished proponent of the chemical interpretation of physiology, but he was by no means as able an experimentalist as his pupil.

Finally, to abolish any possible room for doubt, it occurred to me to make an experiment on a strangled dog, after sensation and life had completely deserted it, and to see if the still-fluid blood in the vena cava would all return equally bright in colour through the pulmonary vein, after being driven to the right ventricle and to the lungs. So I drove on the blood, and carried out a simultaneous insufflation of the perforated lungs. The result corresponded very well with my expectation, for the blood was discharged into the dish as bright-red in colour, as if it were being withdrawn from an artery in a living animal.

I have shown that the bright-red colour of arterial blood is not acquired through any heating in the heart or anywhere else at any time. In like manner also the dark colour of venous blood is independent of any extinction of its heat within the veins. For, if this were so, why should the arterial blood not take on a like colour after it has left its vessels, since it has now beyond all doubt lost its heat?

This being so, we must next see to what the blood is indebted for this deep red coloration. This must be attributed entirely to the lungs, as I have found that the blood, which enters the lungs completely venous and dark in colour, returns from them quite arterial and bright.[2] For, if the anterior part of the chest is cut away and the lungs are continously insufflated by a pair of bellows inserted into the trachea, and they are also pricked with a needle in various places to allow free passage of air through them, then, on the pulmonary vein being cut near the left auricle, the blood will flow out into a suitably placed receptacle completely bright-red in colour. And, as long as the lungs are supplied with fresh air in this way, the blood will rush out scarlet, until the whole perfusate reaches several ounces, nay pounds, just as if it were being received from a cut artery. What I had written earlier about the blood withdrawn from the pulmonary vein being like venous blood was said as a result of experimental work, but at a time when I did not yet know from experiment that one could keep life

2. This is the first clear statement of the function of the lungs and the difference between venous and arterial blood. In the following passages Lower clearly demonstrates the role of the air in causing the change of color—though, of course, he did not understand the chemical action involved. This had to wait for the development of pneumatic chemistry in the eighteenth century and Lavoisier's clear understanding of the role of oxygen in combustion and respiration.

in an animal by continuous insufflation of pricked lungs; so that all the air had been forced out of the lung before I was able to seize and to lance the pulmonary vein. I acknowledge my indebtedness to the very famous Master Robert Hooke for this experiment—by which the lungs are kept continuously dilated for a long time without meanwhile endangering the animal's life—and the opportunity thereby given me to perform this piece of work.[3]

If any one, however, argues that this bright colour of the blood is to be attributed to its fragmentation in the lungs rather than to the mixture of air with the blood, he should consider whether the blood can really be broken into fragments better in the lungs than in the muscles of the body, or even as well. For the lungs are kept constantly dilated for the right conduct of this experiment, and I fail therefore to see how the blood can undergo fragmentation save in passing through their pores, as in the rest of the body-framework.

Further, that this red colour is entirely due to the penetration of particles of air into the blood, is quite clear from the fact that, while the blood becomes red throughout its mass in the lungs (because air diffuses in them through all the particles of blood, and hence becomes more thoroughly mixed with the blood), when venous blood is received into a vessel, the surface and uppermost part of it takes on this scarlet colour through exposure to the air. If this is removed with a knife, the part lying next below will soon change to the same colour through similar contact with the air.

Indeed, if the cake of blood is turned over after remaining stationary for a long while, its outer and uppermost layer takes on the red colour in a short space of time (provided the blood is still fresh). It is a matter of common knowledge that venous blood becomes completely red when received into a dish and shaken up for a long time to cause a thorough penetration of air into it. And let no one be surprised at a loss or admixture of air causing such marked colour-changes in the blood, since we see other fluids also acquiring various colorations, according as their pores take up or refract in greater or lesser amount the rays of light.

If you ask me for the paths in the lungs, through which the

---

3. This experiment was demonstrated by Hooke in his capacity as curator of experiments to the Royal Society, assisted by Lower, on October 10, 1667; but Hooke had devised it some months earlier.

nitrous spirit of the air[4] reaches the blood, and colours it more deeply, do you in turn show me the little pores by which that other nitrous spirit, which exists in snow, passes into the drinks of gourmets and cools their summer wines. For, if glass or metal cannot prevent the passage of this spirit, how much more easily will it penetrate the looser vessels of the lungs? Finally, if we do not deny the outward passage of fumes and of serous fluid, why may we not concede an inward passage of this nitrous foodstuff into the blood through the same or similar little pores?

On this account it is extremely probable that the blood takes in air in its course through the lungs, and owes its bright colour entirely to the admixture of air. Moreover, after the air has in large measure left the blood again within the body and the parenchyma of the viscera, and has transpired through the pores of the body, it is equally consistent with reason that the venous blood, which has lost its air, should forthwith appear darker and blacker.

From this it is easy to imagine the great advantage accruing to the blood from the admixture of air, and the great importance attaching to the air taken in being always healthy and pure; one can see, too, how greatly in error are those, who altogether deny this intercourse of air and blood. Without such intercourse, any one would be able to live in as good health in the stench of a prison as among the most pleasant vegetation. Wherever, in a word, a fire can burn sufficiently well, there we can equally well breathe.

CHAPTER IV

THE TRANSFUSION OF BLOOD
FROM ONE ANIMAL TO ANOTHER. THE TIME
AND OCCASION OF ITS DISCOVERY BY THE AUTHOR

The statements made hitherto about the blood relate to its circular movement, which takes place within the sphere of a single body; we have, so to speak, compared the credit and debit accounts, and have given a strict reckoning of the measure of fluid, and of the lapse of time necessary for the passage of the blood from the veins through the ventricles of the heart to the arteries of

4. It was a widely held theory (subscribed to by Hooke, among others, but not by Boyle) that air was essential for life and for flame because of the presence in it of "nitro-aerial particles," niter (saltpeter) being of course well known for its explosive properties as the essential ingredient of gunpowder. Some (like Lower here) believed that niter also played a part in freezing (because when dissolved in water it, like salt, lowers the temperature).

the same animal. With regard to our next subject, the transfusion of blood from one animal to another, I do not know if the hope of accomplishing this, or the thought of trying it, occurred to any one earlier than three years ago.[5] For, even after it was openly suggested as likely to have great applications in medicine, most people, nevertheless, withheld their hands completely from the experiment, or moved them to it in vain, either through fear of the operative difficulty, or in discouragement at its strangeness. As in the almost forgotten fable of Pythagoras, another, even less substantial, transmigration of the soul would seem to be more desired by the ignorant than hoped for by the learned.

I wish therefore, to reveal the conduct of the whole affair, and at the same time to show by what train of thought I first reasoned it out and undertook it, and, finally, by what means and aids it was carried into effect.

For many years at Oxford I saw others at work,[6] and myself, for the sake of experiment, injected into the veins of living animals various opiate and emetic solutions, and many medicinal fluids of that sort. The technical procedure for this is now quite well known, and this is not the place to describe the individual results and outcomes of these experiments. But when, in addition, I likewise injected many nutrient solutions and had seen the blood of different animals mix quite well and harmoniously with various injections of wine and beer, it soon occurred to me to try if the blood of different animals would not be much more suitable and would mix without danger or conflict. And, because in shed blood (no matter how well coagulation should be guarded against by repeated shaking) the natural blending and texture of the parts must of necessity change, I thought it much more convenient to transfer the unimpaired blood of an animal, which was still alive and breathing, into another. I thought this would be more easily

5. This interest in transfusion, a natural outgrowth of the study of the circulation and of the injection of drugs into the blood stream, was at its height in the years 1666 to 1668. It led ultimately to attempts in England and France to practice blood transfusion on human subjects, attempts which ceased abruptly after the death in 1668 of the unfortunate French lunatic into whose veins sheep's and calf's blood had been introduced. (There was, of course, no comprehension of the fact that men, unlike animals, cannot tolerate alien blood except under carefully controlled conditions.)

6. Principally Christopher Wren (not yet more famous as an architect than as mathematician and natural philosopher) and Robert Boyle.

effected, inasmuch as the movement of blood through its vessels is so rapid and swift, that I had observed almost the whole mass of blood flow out in a few seconds, where an outlet offered. Taking hope from this, I turned mind and hands to put the matter to a practical test.

And first I tried to transfer blood from the jugular vein of one animal to the jugular vein of a second by means of tubes between the two; but, seeing the blood clot at once in the tube and block its own passage on account of the slow movement of the venous blood, I soon began to try another way, and guided, as it were, by nature herself, I finally determined to transfer blood from an artery of one animal into a vein of a second; and by this new device to extend the circulation of the blood beyond the boundaries prescribed for it.

As everything answered expectation as I wished, I finally showed this new experiment at Oxford towards the end of February 1665, in an interesting demonstration and under the most happy circumstances. There were present the learned Doctor John Wallis, Savilian Professor of Mathematics, Thomas Millington, Doctor of Medicine, and other Doctors of the same Universidy.

Having got ready the dogs, and made other preparations as required, I selected one dog of medium size, opened its jugular vein, and drew off blood, until it was quite clear from its howls and struggles that its strength was nearly gone and that convulsions were not far off. Then, to make up for the great loss of this dog by the blood of a second, I introduced blood from the cervical artery of a fairly large mastiff, which has been fastened alongside the first dog, until this latter animal by its restiveness showed in its turn that it was overfilled and burdened by the amount of the inflowing blood. I ligatured the artery from which the blood was passing, and withdrew blood again from the receiving dog. This was repeated several times in succession, until there was no more blood or life left in two fairly large mastiffs (the blood of both having been taken by the smaller dog). In the meantime blood had been repeatedly withdrawn from this smaller animal and injected into it in such amount as would equal, I imagine, the weight of its whole body, yet once its jugular vein was sewn up and its binding shackles cast off, it promptly jumped down from the table, and, apparently oblivious of its hurts, soon began to fondle its master, and to roll on the grass, to clean itself of blood; exactly as it would have done if it had merely been thrown into a stream, and with no more sign of discomfort or of displeasure.

The report of these matters soon reached London, and I was earnestly requested by the Honourable Mr. Boyle, in a letter to acquaint the Royal Society with the procedure of the whole experiment. This communication I made not so long afterwards, and it was published in the *Philosophical Transactions* of the same Society in December of the following year, 1666.[7] Thereafter talk of it wandered across to nations abroad and to France, where, attracted by the novelty of the thing, some soon began to follow it up more thoroughly, to extend and embellish it by other further experiments, and to apply to the use of man that which I had only accomplished in animals. This is seen clearly in their writings published for the first time in the following March, 1667. So far so good, and all credit to that nation for their activity in the embellishment and extension of natural knowledge and of medicine. But, as this recent discovery of blood-transfusion is now a general subject of conversation, and a certain Denis,[8] Professor of Philosophy and Mathematics, seeks in a recently published letter to deprive me of priority in the discovery of this experiment, and to claim it for himself, let me be permitted to insert here the letter of the Honourable Mr. Boyle to me and my reply to it,[9] so that the reader may see how rightly or wrongly Denis so acted.

7. *Phil. Trans.*, no. 20 (December 17, 1666), pp. 353-58.
8. Jean Denis (d. 1704) was physician to Louis XIV and a distinguished Parisian practitioner. He was responsible for a series of French experiments on transfusion begun in 1667 and continued until the unfortunate death of the patient mentioned in footnote 5. After this event, French physicians were forbidden by law to practice transfusion without express permission from a magistrate, though Denis was never held to be guilty of his patient's death.
9. This, with further remarks by Lower, is printed in *Phil. Trans.*. It does indeed establish Lower's priority.

Microscopy

---

12. *Observations on Little Animals*

Observation XXII.
The Ant, Emmet, or Pismire

This little Animal is that great Pattern of Industry and Frugality: To this Schoolmaster did Solomon send his sluggard, who those virtues not only excells all Insects, but most men. Other excellent Observables there are in so small a fabric: As the Herculean strength of its body, that it is able to carry its triple weight and bulk; the agility of its limbs, that it runs so swiftly; the equality of its motion, that it trips so nimbly away without any saliency or leaping, without any fits or starts in its progression. Her head is large and globular, with a prominent snout: her eye is of a very fair black colour, round, globular, and prominent, of the bigness of a pea, foraminulous[1] and latticed like that of other Insects: her mouth (in which you may see something to move) is armed with a pair of pincers, which move laterally, and are indented on the inside like a saw, by which she bites, and better holds her prey; and you may often see them carry their white oblong eggs in them for better security.

Observation XXX.
Of the Little White Eels or Snigs, in Vinegar

They appear like small Silver-Eels, or little Snigs, and some of them as long as my little finger, constantly wriggling and swimming to and fro with a quick, smart, and restless motion. In which smallest of Animals these things are most remarkable.

First, they are not to be found in all sorts of Vinegar or

SOURCE: Henry Power, *Experimental Philosophy* (London, 1664), pp. 25, 32-36.

1. Full of little holes, like a sieve.

Alegar,[2] but only in such, probably, as has arrived to some peculiar temper or putrefaction, of which I can give you no characteristical signs; for, I have found them in all sorts of vinegar, both in the keenest and smartest, as well as in the weakest and most waterish vinegar; and in all these sorts, you shall sometimes find none at all; and I have both found them, and also vainly sought them, in the former liquors, at all seasons and times of the year also.

Secondly, the manner and best way of observing them is, upon a plain piece of white glass, whereon two or three drops of the said liquors are laid; and so laying that glass on the object-plate, and fitting your Microscope to it, you may distinctly see them to play and swim in those little ponds of vinegar (for so big every drop almost seems) to the very brink and banks of their fluid element.

Thirdly, nay you may see them (especially in old Alegar) with the bare eye, if you put a little of it into a clear Venice-glass,[3] especially into those pure thin white bubbles, they call essence-glasses; you may then see an infinite company of them swimming at the edges of the liquor, nay and in the body of it too, like so many shreds of the purest Dutch thread, as if the whole liquor was nothing else but a great shoal or mass of quick eels or hair worms.

I have another advantageous way of discoverance of them to the bare eye also, which is by putting a little of those liquors into a little cylinder of white glass, of a small bore and length, either sealed or closed up with cork and wax at one end: therein, if you invert this glass cylinder and often turn it topsy turvy, no liquor will fall out, only a little bubble of air will always pass and repass through the inverted liquor, and one pretty thing I have herein observed, that when this bubble has stood in the superior end of the glass (and sometimes it would do so for a pretty while together before it broke) I have seen some of those small snigs or animals on the top of it, crawling over the smooth convexity of the bubble (like so many eels over a looking glass) without breaking through the tender cuticle and film of so brittle and thin a substance.

Fourthly, that as the liquor (dropped upon your object plate)

2. Vinegar from ale; i.e., malt vinegar.
3. Venetian glass had long been reputed the best, so "Venice-glass" means glass of the highest quality. In fact, English glass was by the 1660's the best in the world for optical and scientific purposes.

spends[4] and dries up, so you shall see those little Quicks[5] to draw nearer and nearer together, and grow feebler in their motion; and when all the vinegar or alegar is dried away, they they lie all dead, twisted and complicated all together, like a knot of eels, and after a little time dry away to nothing.

Fifthly, their heads and tails are smaller than the rest of their bodies; which is best observed by the microscope, when the liquor wherein they swim is almost spent and dried up, so that their motion thereby is rendered more feeble and weak, or when they lie absolutely dead.

Sixthly, another remarkable thing is, their exceeding exiguity; for certainly of all animals they are least that can be seen by the bare eye, which is helped and advantaged also by the rarefaction of the water wherein they swim.

Seventhly, if you take a spoonful of the aforesaid vinegar and heat it over a few coals, it presently destroys all the quicks in it, so that you may see them all stretched out as their full length, like a pencil chopped small, or little bits of hairs swimming up and down the liquor, which in a short time will precipitate and all sink down to the bottom of the glass.

Nay these poor vermin are not only slain by actual heat, but by a potential one also: for putting but a few drops of the oil of vitriol[6] into an essence-glass full of that vinegar, it also shortly destroyed them in the same manner as the fire had done before.

Eighthly, now though heat hath that killing property, yet it seems that cold hath not: for I have taken a jar-glass full of the said vinegar, and by applying snow and salt to it, I have artificially frozen all the said liquor into a mass of ice (wherein all those animals it seemed lay incrystalled) though I could discover none of them in it (though I have taken the icy mass out on purpose to look at it) so that now I gave them [up] for gone for ever: yet when I came again (about two or three hours after,) to uncongeal the liquor, by keeping the glass in my warm hand, when the vinegar was again returned to its former liquidity, all my little animals made their reappearance, and danced and frisked about as lively as ever. Nay I have exposed a jar-glass full of this vinegar all night to a keen frost, and in the morning have thawed the ice again, and

4. To spend means "to become exhausted," or "waste away" (i.e., evaporate).
5. Living things.
6. Sulphuric acid.

these little vermin have appeared again and endured again that strong and long conglaciation without any manifest injury done to them; which is both a pretty and a strange experiment.

Ninthly, I have filled an essence-glass half with the said vinegar, and half with oil, (which floated on the vinegar) in a distinct region by itself, and I have observed that in frosty weather when the vinegar has been congealed, that all the little eels have run up into the superincumbent oil to preserve themselves there, and would not return till some warmth was applied to the vinegar again, and then they would always presently[7] return down into their native liquor again.

Tenthly, their motion is very remarkable, which is restless and constant, with perpetual undulations and wavings, like eels or snakes; so that it seems, that animals that come nearest the class of plants, have the most restless motions.

Eleventhly, the innumerable number and complicated motion of these minute animals in vinegar, may very neatly illustrate the doctrine of the incomparable Descartes, touching fluidity: (viz.) That the particles of all fluids are in a continual and restless motion, and therein consists the true nature of fluidity: for by this ocular example, we see there may be an intestine restless motion in a liquor, notwithstanding that the unassisted eye can discover no such matter.

13. Philosophical Microscopy

Observation XLIX.
Of an Ant or Pismire

This was a creature, more troublesome to be drawn, than any of the rest, for I could not, for a good while, think of a way to make it suffer its body to lie quiet in a natural posture; but whilst it was alive, if its feet were fettered in wax or glue, it would so twist and wind its body that I could not any ways get a good view of it; and

SOURCE: Robert Hooke, *Micrographia* (1665), pp. 203-5, 44-47.

7. In the seventeenth century this meant "immediately."

if I killed it, its body was so little, that I did often spoil the shape
of it before I could thoroughly view it: for this is the nature of
these minute bodies, that as soon, almost, as ever their life is
destroyed, their parts immediately shrivel and lose their beauty;
and so it is also with small plants, as I instanced before in the
description of Moss. And thence also is the reason of the variations
in the beards of wild oats,[1] and in those of musk-grass seed, that
their bodies, being exceedingly small, those small variations which
are made in the surfaces of all bodies, almost upon every change of
air, especially if the body be porous, do here become sensible,
where the whole body is so small, that it is almost nothing but
surface; for as in vegetable substances I see no great reason to
think that the moisture of the air (that, sticking to a wreathed
beard, does make it untwist) should evaporate or exhale away any
faster than the moisture of other bodies, but rather that the avola-
tion from, or access of moisture to, the surfaces of bodies being
much the same, those bodies become most sensible of it, which
have the least proportion of body to their surface. So is it also
with animal substances; the dead body of an ant or such little
creature does almost instantly shrivel and dry and your object
shall be quite another thing before you can half delineate it, which
proceeds not from the extraordinary exhalation, but from the
small proportion of body and juices to the usual drying of bodies
in the air, especially if warm. For which inconvenience, where I
could not otherwise remove it, I thought of this expedient.

I took the creature I had designed to delineate and put it into a
drop of very well rectified spirit of wine;[2] this I found would
presently dispatch, as it were, the animal, and being taken out of it
and laid on a paper, the spirit of wine would immediately fly away
[evaporate] and leave the animal dry, in its natural posture, or at
least in a constitution, that might easily with a pin be placed in
what posture you desired to draw it, and the limbs would so
remain without either moving or shrivelling. And thus I dealt with
this ant, which I have here delineated, which was one of many of a
very large kind that inhabited under the roots of a tree, from
whence they would sally out in great parties and make most griev-

1. The awns of the flower heads; Hooke used these as simple hygroscopes.
2. Alcohol. This may perhaps have some connection with Boyle's use of
alcohol to preserve anatomical specimens.

ous havoc of the flowers and fruits in the ambient [surrounding] garden and return back again very expertly, by the same ways and paths they went.

It was more than half the bigness of an earwig, of a dark brown or reddish colour, with long legs, on the hinder of which it would stand up and raise its head as high as it could above the ground, that it might stare the further about it, just after the same manner as I have also observed a hunting spider to do; and putting my finger towards them, they have at first all run towards it, till almost at it, and then they would stand round about it, at a certain distance, and smell, as it were, and consider whether they should any of them venture any further, till one more bold than the rest venturing to climb it, all the rest, if I would have suffered them, would have immediately followed: many such other seemingly rational actions I have observed in this little vermin with much pleasure, which would be too long to be here related; those that desire more of them may satisfy their curiosity in Ligon's *History of the Barbadoes.*[3]

Having ensnared several of these into a small box, I made choice of the tallest grown among them, and separating it from the rest, I gave it a gill of brandy or spirit of wine, which after a while e'en knocked him down dead drunk, so that he became moveless, though at first putting in he struggled for a pretty while very much, till at last, certain bubbles issuing out of its mouth, it ceased to move; this (because I had before found them quickly to recover again if they were taken out presently) I suffered to lie above an hour in the spirit; and after I had taken it out and put its body and legs into a natural posture [it] remained moveless about an hour; but then, upon a sudden, as if it had been awakened out of a drunken sleep, it suddenly revived and ran away; being caught and served as before, he for a while continued struggling and striving, till at last there issued several bubbles out of its mouth, and then, as if expiring, he remained moveless for a good while; but at length recovering, it was again redipped and suffered to lie some hours in the spirit; notwithstanding which, after it had lain dry some three or four hours, it again recovered life and motion: which kind of experiments, if prosecuted, which they highly deserve, seem to me of no inconsiderable use towards the invention

3. Richard Ligon was a mid-seventeenth-century traveler.

of the latent scheme (as the noble Verulam[4] calls it) or the hidden, unknown texture of bodies.

Of what figure this creature appeared through the microscope, the scheme[5] (though not so carefully graven as it ought) will represent to the eye, namely, that it had a large head, at the upper end of which were two protuberant eyes, pearled like those of a fly, but smaller; out of the nose or foremost part issued two horns [antennae] of a shape sufficiently differing from those of a blue fly, though indeed they seem to be both the same kind of organ, and to serve for a kind of smelling; beyond these were two indented jaws, which he opened sideways and was able to gape them asunder very wide; and the ends of them being armed with teeth, which meeting went between each other, it was able to grasp and hold a heavy body, three or four times the bulk and weight of its own body. It had only six legs, shaped like those of a fly, which, as I showed before, is an argument that it is a winged insect, and though I could not perceive any sign of them in the middle part of its body (which seemed to consist of three joints or pieces, out of which sprang two legs), yet 'tis known that there are of them that have long wings and fly up and down in the air.

The third and last part of its body was bigger and larger than the other two, unto which it was joined by a very small middle and had a kind of loose shell or another distinct part of its body, which seemed to be interposed and to keep the thorax and belly from touching.

The whole body was cased over with a very strong armour, and the belly was covered likewise with multitudes of small white shining bristles; the legs, horns, head and middle parts of its body were bestuck with hairs also, but smaller and darker.

Observation VIII.
Of the Fiery Sparks Struck from a Flint or Steel

It is a very common experiment, by striking with a flint against a steel, to make certain fiery and shining sparks to fly out from between those two compressing bodies. About eight years since,[6] upon casually reading the explication of this odd phenomenon by

4. Bacon (Lord Verulam). Hooke, though an ardent Cartesian in many things, like Power was also an ardent Baconian.
5. Omitted here; it is a faithful delineation of a large ant.
6. Probably about 1656, when he was still an undergraduate.

the most ingenious Descartes, I had a great desire to be satisfied what that substance was that gave such a shining and bright light: and to that end I spread a sheet of white paper, and on it, observing the place where several of these sparks seemed to vanish, I found certain very small, black, but glittering spots of a moveable substance, each of which examining with my microscope, I found to be a small round globule; some of which, as they looked pretty small, so did they from their surface yield a very bright and strong reflection on that side which was next the light; and each looked almost like a pretty bright iron ball, whose surface was pretty regular, such as is represented by the figure. In this I could perceive the image of the window pretty well, or of a stick which I moved up and down between the light and it. Others I found, which were, as to the bulk of the ball, pretty regularly round, but the surface of them, as it was not very smooth, but rough and more irregular, so was the reflection from it more faint and confused. . .[7] Some of these I found cleft or cracked, others quite broken in two or hollow, which seemed to be half the hollow shell of a grenade broken irregularly in pieces. Several others I found of other shapes; but [one] I observed to be a very big spark of fire, which went out upon one side of the flint that I struck fire withall, to which it stuck by the root, at the end of which small stem was fastened on a hemisphere or half a hollow ball, with the mouth of it open from the stemwards, so that it looked much like a funnel or an old fashioned bowl without a foot. This night, making many trials and observations of this experiment, I met among a multitude of globular ones which I had observed, a couple of instances which are very remarkable to the confirmation of my hypothesis.

And the first was of a pretty big ball fastened onto the end of a small sliver of iron, which compositum[8] seemed to be nothing else but a long, thin chip of iron, one of whose ends was melted into a small round globule; the other end remaining unmelted and irregular and perfectly iron.

The second instance was not less remarkable than the first; for I found when a spark went out nothing but a very small, thin, long sliver of iron or steel, unmelted at either end. So that it seems,

7. Here and elsewhere references to the figures and to other observations have been omitted.
8. Mixed body.

that some of these sparks are the slivers or chips of the iron vitrified, others are only the slivers melted into balls without vitrification, and the third kind are only small slivers of the iron made red hot with the violence of the stroke given on the steel by the flint.

He that shall diligently examine the phenomena of this experiment, will, I doubt not, find cause to believe that the reason I have heretofore given of this, is the true and genuine cause of it, namely, that the sparks appearing so bright in the falling is nothing else but a small piece of the steel or flint, but most commonly of the steel, which by the violence of the stroke is at the same time severed and heated red hot, and that sometimes to such a degree as to make it melt together into a small globule of steel; and sometimes also is that heat so very intense as further to melt it and vitrify it; but many times the heat is so gentle as to be able to make the sliver only red hot, which notwithstanding falling upon the tinder (that is only a very curious small coal made of the small threads of linen burned to coals and charred) it easily sets it on fire. Nor will any part of this hypothesis seem strange to him that considers, first, that either hammering or filing or otherwise violently rubbing of steel will presently make it so hot as to be able to burn one's fingers. Next, that the whole force of the stroke is exerted upon that small part where the flint and steel first touch: for the bodies being each of them so very hard, the pulse cannot be far communicated, that is, the parts of each can yield but very little and therefore the violence of the concussion will be exerted on that piece of steel which is cut off by the flint. Thirdly, that the filings or small parts of steel are very apt, as it were, to take fire and are presently red hot, that is, there seems to be a very combustible sulphureous body in iron or steel,[9] which the air very readily preys upon as soon as the body is a little violently heated.

And this is obvious in the filings of steel or iron cast through the flame of a candle; for even by that sudden transitus[10] of the small chips of iron, they are heated red hot, and that combustible sulphureous body is presently preyed upon and devoured by the aerial incompassing menstruum. . . .[11]

9. This is an instance of the seventeenth-century confusion over combustion and the composition of bodies. Hooke assumes that all inflammable substances must contain an ingredient akin to sulphur.
10. Going through, passage.
11. Solvent.

And in prosecution of this experiment, having taken the filings of iron and steel and with the point of a knife cast them through the flame of a candle, I observed where some conspicuous shining particles fell, and looking on them with my microscope, I found them to be nothing else but such round globules, as I formerly found the sparks struck from the steel by a stroke to be, only a little bigger; and shaking together all the filings that had fallen upon the sheet of paper underneath, and observing them with the microscope, I found a great number of small globules, such as the former, though there were also many of the parts that had remained untouched, and rough filings or chips of iron. So that, it seems, iron does contain a very combustible sulphureous body which is, in all likelihood, one of the causes of this phenomenon, and which may be perhaps very much concerned in the business of its hardening and tempering. . . .

So that, these things considered, we need not trouble ourselves to find out what kind of pores they are, both in the flint and steel, that contain the atoms of fire, nor how those atoms came to be hindered from running all out, when a door or passage in their pores is made by the concussion: nor need we trouble ourselves to examine by what Prometheus the element of fire comes to be fetched down from above the regions of the air, in what cells or boxes it is kept, and what Epimetheus lets it go: nor to consider what it is that causes so great a conflux of the atomical particles of fire which are said to fly to a flaming body like vultures or eagles to a putrifying carcass, and there to make a very great pudder [pother]. Since we have nothing more difficult in this hypothesis to conceive, first, as to the kindling of tinder, than how a large iron bullet, let fall red or glowing hot upon a heap of small coals, should set fire to those that are next to it first; nor secondly, is this last more difficult to be explicated, than that a body, as silver for instance, put into a weak menstruum, as unrectified aqua fortis,[12] when it is put in a great heat, be there dissolved by it and not before. . . . To conclude, we see by this instance how much experiments may conduce to the regulating of philosophical notions. For if the most acute Descartes had applied himself experimentally to have examined what substance it was that caused that shining of the falling sparks struck from a flint and a steel, he would certainly have a little altered his hypothesis, and

12. Dilute nitric acid, which dissolves silver when hot but not when cold.

we should have found that his ingenious principles would have admitted a very plausible explication of this phenomenon; whereas by not examining so far as he might, he has set down an explication which experiment does contradict.

But before I leave this description, I must not forget to take notice of the globular form into which each of these is most curiously formed. And this phenomenon, as I have elsewhere more largely shown, proceeds from a property which belongs to all kinds of fluid bodies more or less, and is caused by the incongruity of the ambient and included fluid, which so acts and modulates each other, that they acquire, as near as is possible, a spherical or globular form. . . .

One experiment, which does very much illustrate my present explication and is in itself exceedingly pretty, I must not pass by: and that is a way of making small globules or balls of lead or tin as small almost as these of iron or steel, and that exceedingly easily and quickly, by turning the filings or chips of those metals also into perfectly round globules. The way, in short, as I received it from the learned physician Dr. J. G.[13] is this:

Reduce the metal you would thus shape into exceedingly fine filings; the finer the filings are, the finer will the balls be. Stratify[14] these filings with the fine and well dried powder of quicklime in a crucible proportioned to the quantity you intend to make. When you have thus filled your crucible, by continual stratifications of the filings and powder, so that, as near as may be, no one of the filings touch another, place the crucible on a gradual fire, and by degrees let it be brought to a heat big enough to make all the filings that are mixed with the quicklime to melt, and no more; for if the fire be too hot, many of these filings will join and run together; whereas if the heat be proportioned, upon washing the lime-dust in fair water, all those small filings of the metal will subside to the bottom in a most curious powder, consisting all of exactly round globules, which, if it be very fine, is very excellent to make hour glasses of.

13. Presumably Jonathan Goddard (1617-1675), who was warden of Merton College, Oxford, in the 1650's when Hooke was an undergraduate at Christ Church College. He was very much interested in chemistry and actively promoted the study of experimental chemistry with Boyle and others.
14. Properly, lay in layers. It is clear however from Hooke's description that the lime is here being used merely to coat the filings, so each is kept separate.

Now though quicklime be the powder that this direction makes choice of, yet I doubt not, but that there may be much more convenient ones found out, one of which I have made trial of, and found very effectual; and were it not for discovering, by the mentioning of it, another secret, which I am not free to impart, I should have here inserted it.

Pneumatics

---

14. *The Principle of the Barometer*

Florence, June 11, 1644

My most illustrious Sir and most cherished Master: Several weeks ago I sent some demonstrations of mine on the area of the cycloid to Signor Antonio Nardi, entreating him to send them directly to you or to Signor Magiotti after he had seen them.[1] I have already intimated to you that a certain physical experiment was being performed on the vacuum; not simply to produce a vacuum, but to make an instrument which would show the changes in the air, which is at times heavier and thicker and at times lighter and more rarefied. Many have said that a vacuum cannot be produced, others that it can be produced but with repugnance on the part of Nature and with difficulty; so far, I know of no one who has said that it can be produced without effort and without resistance on the part of Nature. I reasoned in this way: if I were to find a plainly apparent cause for the resistance which is felt when one needs to produce a vacuum, it seems to me that it would be vain to try to attribute that action, which patently derives from some other cause, to the vacuum; indeed, I find that by making certain very easy calculations, the cause I have proposed (which is the weight of the air) should in itself have a greater effect than it does in the attempt to produce a vacuum. I say this because some Philosopher, seeing that he could not avoid the admission that the weight of air causes the resistance which is felt in producing a vacuum, did not say that he admitted the effect of the weight of the air, but persisted in asserting that Nature also

SOURCE: From Evangelista Torricelli, Letter to Ricci (1644) in I. H. B. and A. G. H. Spiers (eds.), *The Physical Treatises of Pascal* (New York: Columbia University Press, 1937), pp. 163-66.

1. Nardi, Magiotti, and Ricci were all friends living in Rome who met frequently for mathematical and scientific discussions. It was very common at this time to communicate scientific discoveries by letter. In this case it is probable that Torricelli had already discussed the barometric experiments of Berti and others with Ricci.

contributes at least to the abhorrence of a vacuum. We live submerged at the bottom of an ocean of the element air, which by unquestioned experiments is known to have weight,[2] and so much, indeed, that near the surface of the earth where it is most dense, it weights [volume for volume] about the four-hundredth part of the weight of water. Those who have written about twilight, moreover, have observed that the vaporous and visible air rises above us about fifty or fifty-four miles;[3] I do not, however, believe its height is as great as this, since if it were, I could show that the vacuum would have to offer much greater resistance than it does—even though there is in their favor the argument that the weight referred to by Galileo applies to the air in very low places where men and animals live, whereas that on the tops of high mountains begins to be distinctly rare and of much less weight than the four-hundredth part of the weight of water.

We have made many glass vessels like the following [fig. 3]

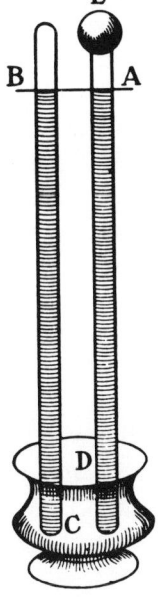

3. Torricelli's barometer.
   From Evangelista
   Torricelli, Letter to
   Ricci

2. In fact air had since antiquity been known to have weight, though it was only recently that its specific gravity had been determined with any accuracy. The question really involved was, had it enough weight to support a 30-foot column of water?
3. As, for example, the eleventh-century Moslem physicist Alhazen (Ibn al Haitham).

marked A and B with necks two cubits.[4] We filled these with quicksilver, and then, the mouths being stopped with a finger and being inverted in a basin where there was quicksilver C, they seemed to become empty and nothing happened in the vessel that was emptied; the neck AD, therefore, remained always filled to the height of a cubit and a quarter and an inch besides. To show that the vessel was perfectly empty, the underlying basin was filled with water up to D, and as the vessel was slowly raised, when its mouth reached the water, one could see the quicksilver fall from the neck, whereupon with a violent impetus the vessel was filled with water completely to the mark E. This experiment was performed when the vessel AE was empty and the quicksilver, although very heavy, was held up in the neck AD. The force which holds up that quicksilver against its nature to fall down again, has been believed hitherto to be inside of the vessel AE, and to be due either to vacuum or to that material [mercury] highly rarefied; but I maintain that it is external and that the force comes from without. On the surface of the liquid which is in the basin, there gravitates a mass of air fifty miles high; it is therefore to be wondered at if in the glass CE, where the mercury is not attracted nor indeed repelled, since there is nothing there, it enters and rises to such an extent as to come to equilibrium with the weight of this outside air which presses upon it? Water also, in a similar but much longer vessel, will rise up to almost eighteen cubits, that is, as much further than the quicksilver rises as quicksilver is heavier than water, in order to come to equilibrium with the same force, which presses alike the one and the other.

The above conclusion was confirmed by an experiment made at the same time with a vessel A and a tube B, in which the quicksilver always came to rest at the same level, AB. This is an almost certain indication that the force was not within; because if that were so, the vessel AE would have had greater force, since within it there was more rarefied material to attract the quicksilver, and a material much more powerful than that in the very small space B, on account of its greater rarefaction. I have since tried to consider from this point of view all the kinds of repulsions which are felt in the various effects attributed to vacuum, and thus far I have not

4. Torricelli regards the "vessels" as being the empty tops of the tubes not filled with mercury. The vessel A is made round (at E) to show that shape has nothing to do with the phenomenon.

encountered anything which does not go [to confirm my opinion]. I know that you will think up many objections, but I also hope that, as you think about them, you will overcome them. I must add that my principal intention—which was to determine with the instrument EC when the air was thicker and heavier and when it was more rarefied and light—has not been fulfilled; for the level AB changes from another cause (which I never would have believed), namely, on account of heat and cold; and changes very appreciably, exactly as if the vase AE were full of air.

15. Demonstrations of the Vacuum

New Anatomical Experiments of John Pecquet of Deip
By which the Hitherto Unknown Receptacle
of the Chyle, and the Transmission
from Thence to the Subclavial Veins
by the Now Discovered Lacteal Channels
of the Thorax Is Plainly Made to Appear
in Brutes . . .

CHAPTER VIII.
'TIS SHOWN BY EXPERIMENTS, THAT THERE IS
NOT ONLY A WEIGHT IN THE AIR,
BUT LIKEWISE A RAREFACTIVE ELATERY

I would treat of the Air's ponderosity, yea in its proper (as they say) place, except it were an Argument known to all. For who doth not see the Air of its own accord to descend into the Chinks and Ditches, yea even into the lowest Center of the Earth, if you delve so deep? Who doth not know that a little Bladder, the more turgid it is, is so much more heavy than it self being flaccid? Who, if he weigh a Gun burthened with condensed Air (they call it a Windgun) will not observe that its weight then is heavier than 'tis when it is discharged? To conclude, to whom is it unknown that

SOURCE: Jean Pecquent, *Experimenta nova anatomica* (Paris, 1651); the text is that of the English translation, *New Anatomical Experiments* (London, 1653), pp. 89-125.

the lightness of an Aeolipila,[1] the Air of which is rarified exactly by the force of heat, is not at all to be compared with the heaviness of it being cold?

Neither indeed believe thou that I am affrightened with the opinion of that kind of airy weight in refunding the vapours of the Atmosphere. For by the Air I understand the Atmosphere, neither for the present doth my Philosophy ascend higher.

Feign to thy self this, as a heap of spongious, or rather woolly matter encompassing the Earth-watrish Globe,[2] whose superior parts therefore are sustained by the inferior, compressing it by degrees, so that the nearer they come to the Earth, so much the more they are compressed compactly by the weight of the parts incumbent; And for that cause by its Spontaneous dilatation (which I call Elater)[3] howsoever the heaped on burthen do press them down, yet the under parts, if they have liberty, will endeavor to rarify themselves.

Hence I infer, that the lowest of those parts, as being placed under the whole burthen, so to be most condensed of them all; And for this reason the superfice[4] of the Earth-watrish Orb is pressed of the same, not by its weight alone, but also by the virtue of his Elatery, whose endeavor to rarify is most valid.

But because they are words, and I have given you leave to proceed from the Ears to the Eyes, I will prevent thy desire, and I will view the misty secrets of the Air by the search of Experience, which I at all times follow as my leader.

I will produce in the first place some Experiments concerning Vacuum, or Emptiness, which I believe as yet have not been committed to the Press; no Monuments of my own deep cunning, far be it from me to arrogate to my self the glory of those detected Miracles which were not first discovered to me: For whatsoever I did exquisitely elaborate with an exact endeavor more frequently about so weighty a matter for fear of error which troubled me, it was imitation, and not invention.

1. The aeolipile referred to here was a round vessel with a narrow aperture; when heated it supplied a blast of air and served as a bellows.
2. Or "terraequeous"—that is, the earth together with its surrounding "sphere" of water (oceans, lakes, and rivers).
3. This appears to be the first use in English of this word, commonly employed by later pneumatic writers for the expansive power or elasticity of air.
4. Or superficies—exterior surface.

I will produce Authors, not of Books, which as yet I have not so much as heard to be extant; but of those Experiments at least which follow, and whose Authority is great and Name venerable.

EXPERIMENTS PHYSICO-MATHEMATICAL OF VACUITY

The First Experiment.
A Little Bladder Being Emptied of Its Own Accord
at Falling Down of Quick-silver,
Rarifying in the High Vial of the Pipe,
Declareth the Rarefactive Elatery of the Air

Robervallius,[5] the most famous Reader of Mathematicks in Paris, in the Kings Chair, did after this manner operate, while I was present, in favour of that virtue whereby the Air of it self doth dilate it self, indeed not without success.

He had a Vial made of Glass [fig. 4], which on both ends, both

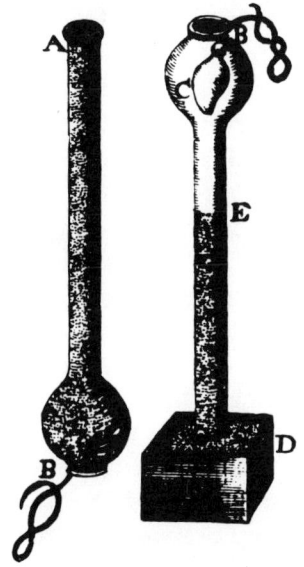

4. The first experiment. From Jean Pecquet, *New Anatomical Experiments*

on the broad Basis, and the gaping end of the neck, A, three foot long at least, was open: By the door of the Basis B, which was shut-up closely with a strait bound thread, He put and enclosed in C one of the little Bladders (or Swims) which useth to be found

5. Gilles Personne de Roberval (1602-1675), professor of mathematics at the Collège Royale at Paris.

swelling in the belly of a Carp. It was empty of air, as far as was possible; and lest that the slipperiness of its Tunicle should hinder it to express the Air, he wrested it a little without violence, being dryed, and he excluded Air out of the hole made in it with the same Ligature. And moreover, because the compacted thickness of the Tunicle of a Sow's Bladder above all other stoppeth best the passage of Quick-silver, he covered therewith the hole of the broad Basis, and filled up with Quick-silver even to the brim of the neck the whole hollowness of the Glass body, AB. He closed with his fore-finger, as with a stopple, the little door A, and holding the Vial overturned perpendicularly, he put his finger together with the mouth of A it stopped, into another Vessel wherein he had poured Quick-silver in D: withdrawing his finger easily, the highest part of the Quick-silver falling swiftly, deserted the bottle of the Vial, and after divers sparklings in the Pipe of the neck, the top of the remaining parts stood in E, neither seemed to fill the Pipe above the Superficies of the Quick-silver in the lower Vessel D, more than seven and twenty inches.

A wonderful sight appeared to the beholders; the Carp's little Bladder, which the straitness of the narrow neck kept up into the higher bottle, did of its own accord swell again. Indeed I had been amazed (except my mind had cured the errors of mine eyes) at the swelling fulness of that sudden Meteor in C, in the middle (as they say) of Vacuity. But the residue of the Air within the secret passages of the writhed indeed, but not altogether exhausted, Bladder taught that there was a Spontaneous Elatery in the Air, whilst free, and not compressed by the weight of any thing lying above it.

For 'tis but a vain reason they bring; The Air's subtility, say they, penetrateth the thickness of the glass, and being spread abroad in every place of the writhed baldder, at last entereth the same by its pores, and extends the whole to his former swelling. But let them remember that which they believe, That the Glass is on every side pervious to the entering Air, and therefore the Air is carried in all places with the same force of its rushing rays towards the Bladder, staying in the middle of the Bottle C; so that it should rather compress and oppress it, than re-acting it self against it self, have a care to raise its unprofitable dilation. Add this, that the pierced on all hands Bladder could not stay the Air, or contain it which it had received within.

Neither troubles it me, that they say that the pores pervious [and] open to the entering Air, forbid its return. For then the entered Air should be compelled, by the Valves shut on all sides, to remain within; So that the little Bladder should retain the swelling it hath once received, and should still remain full; which is demonstrated to be most false, by the lankness of the Bladder, either by inclining the engine[6] a little, or by admitting therein a little portion of the Air from abroad. And hence their opinion is ruined, which believe the Bladder swelleth again by the Spirits of the Quick-silver.

By which, seeing it clearly appears, that the Air which after writhing lay fast in the most inward places of the Bladder, did dilate the skin thereof by its insited Elatery in the falling down of the Quick-silver. Likewise it appears, that that Air which was pressed to the inward Superficies of the Vial by the Metallick weight, and did stay inclosed in the outward folds of the Vesicle, was enough by the virtue of the same Elatery to fill the whole Vial.

Neither let my Adversaries believe, that the heaviness of the Quick-silver while it filled the Vial did quite thrust forth the Air; The contiguity of dry things doth not exclude the Air, whose outrooting is only done by glue; In other things there is either Air, or both the Superficies adhere; so boys wet leather, by whose force they elevate, and weigh up, and throw forth the hanging weight of heavy stones.[7] Neither do Liquors themselves altogether go away; beat together the wet palms of your hands with as great force as you can, notwithstanding the humour remaining in the inward bosom of the skin will not evanish for any pressure; Glass by nature is smooth, nevertheless it is polished by art; neither doth the whitish silver-shining of Quick-silver argue its most perfect smoothness. Truly the weight of the Metal bruising the sides of the glassy porosities, the Air, as I have said, being freed by the falling down of the Metal, doth display it self by its proper Elatery in every place within the Pipe.

May be you will infer, The Air by the irritating of the Quick-silver falling tumultuously, doth become fit for dilation. So also would I think, except the unmoved stability of the Glassy walls should seem to gainstand her course downwards, and should rather roll down those sliding parts of the Quick-silver, which are contiguous to the Glass, towards the Center, than suffer them to fall down perpendicularly. So the brinks of the shore restrain the sides

6. That is, mechanism.
7. The damp leather shrinks enough to raise a heavy weight.

of those streams which glide apace in the midst of their Channels. So the capacity of Tunnels only granteth liberty of flowing to these whirling Liquors in the middle pool only of the falling stream, and likewise to the grain in the Millhopper, if you have taken notice of it.

I should have admired the interchangeable violence of the skipping beatings within the Pipe, except the Air which then was outwardly incumbent on the Quick-silver of the Vessel D, and whose substance is easy to be compressed, being compelled to give place to the heavier weight of the body rushing down from the top of the Vial, had not made me suspect the Air it self by the repercussion of its Elastick or dilating virtue even to reciprocate the jumps of the Quick-silver within the Pipe it self, till at last the essays of it inclosed within the Pipe be not overcome with the weight of the Quick-silver.

The Second Experiment.
The Divers Falling of Quick-silver
According to the Sundry Heights of a Hill,
Proveth that the Lower Parts of the Air
by Degrees Are More Compact than Those That Lean on Them

Neither believe thou that I obscure the matter with unskilful speeches, when I say the Earth is pressed by the Air; I have a most eminent witness, the most subtil Pascalius the son,[8] who first among our French did, not only with Quick-silver, but with other Liquors also, raise up the Experiment concerning Vacuity, scarcely being well born, and almost suffocate amongst strangers; yea with such a success of his wonderful industry he carried it up, that he did put in all the Devotaries of true wisdom through all Europe an eager desire to try the Experiments of Vacuity.

By his care there was lately an Experiment tried in Avernia, at the root of the high Mountain commonly called Le puy de Domme, near the City of Claramont, Quick-silver being poured into a Pipe four foot long, almost like that which we have described above, it did fall down to the third half line[9] of the twenty seventh inch above the superfice of the Quick-silver standing in the

---

8. Blaise Pascal (1623-1662), mathematician and writer of theological philosophy. Pecquet here describes the famous Puy-de-Dôme experiment devised by Pascal and carried out by his brother-in-law François Périer of Clermont-Ferrand in Auvergne in 1648.

9. A "line" was one-twelfth of an inch.

undermost Vessel. Ascending then the Region of the Mountain, which was about an hundred and fifty paces higher than the roots thereof, the Cylinder of the Quick-silver which remained in the Pipe possessed the length but of twenty five inches. At length coming to the top of the Hill, at least five hundred paces higher than the root, the same Cylinder of the Quick-silver found his bound of rest and height in the second line of the twenty fourth inch; and so from the lowest roots of the Mountain to the highest top thereof, it was shortened the length of three inches, and half a degree, viz. the Air which incompassed the top being lighter, by reason of less pressure, than that about the middle and root of the Mountain, ought also to hold up a shorter Cylinder of Quick-silver, according to the equality of the balance.

Neither wonder thou that this Experiment doth not agree with those that I tried in Paris by falling of Quick-silver: For both the difference of the foot of measure in Avernia from ours, which exceeds in some lines (which I exactly observed) and the divers distance of Places may be from the Center of the World (even thou being Judge) may not keep equality in these experiments.

The Third Experiment.
The Equality of Weight of the Outward Air with the Internal Cylinder of the Quick-silver Is Shown

It pleaseth me also, lest the stubborn opinion thou hast of the Ancients should murmur against my reasons, whereby I have attested the equality of weight of the external Air, and internal Quick-silver, to teach the Experiment of Vacuity in Vacuity, first happily tried by the sagacity of the acute Auzotius.[10]

Take a Vial like the former in bottle and neck AB [fig. 5] except that near the basis B it hath a little Pipe G added to it, by whose little door, when need requires, the Air hath an open entrance into the Vial. By the open door of the upturned and high Basis B, put in a four-square Vessel of equal sides C, so that its large hollowness towards the openness in the top of the Basis B, by its bottom covering horizontally the Pipe of the neck AC, placed perpendicularly under it. Truly the Vessel C, by its four Angles inwardly leaning on the Glass, shall leave a space pervious

10. Adrien Auzout (1622-1691), the leading French astronomer of the mid-seventeenth century, one of the earliest members of the Académie royale des sciences. This experiment of "le vide dans le vide" (the vacuum in a vacuum) became very famous.

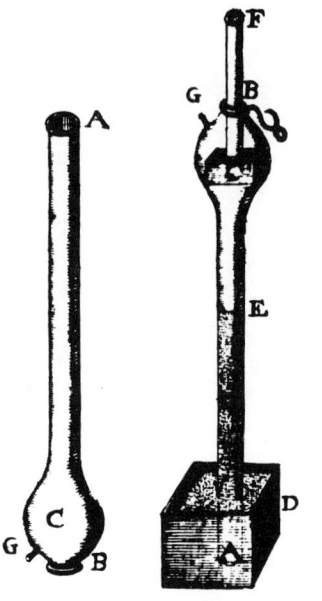

5. The third experiment. From Jean Pecquet, *New Anatomical Experiments*

between the sides of the Vessel and the Vial. Place in the Vessel C a vertical or erected Pipe of choice length likewise, made of Glass, and open at both ends CF: then with a Sow's Bladder (wherewith likewise you shall stop the Basis and little door G) you shall exactly bind it about in B to exclude the Air; then you shall command your servant, by putting his finger under, to stop the mouth of the Vial A; and let him keep it so long stopped, till with Quick-silver put in at the added entrance F, you fill the whole Engine; which entrance also being well bound up with a Sow's Bladder, if he shall withdraw his finger placed under the mouth A, and drenched, as before, in the Quick-silver standing without in the vessel D, you shall wonder to see the equal balance perpetually of the Quick-silver AE, with the outward Air; All the Pipe that stands upmost shall be emptied of its abounding Quick-silver by the sides of the inclosed Vessel C, the Pipe of the Vial AE retaining the same Quick-silver seven and twenty inches high. And then if you bore with a very small pointed Needle the Bladder of the Door G, and give entrance to a little Air, it being mixt with that which is spread abroad in the Vial, intendeth the Elatery of the same, so that now on all sides it acteth with a more strong endeavor, and oppressing the Quick-silver in the neck AE subject to it, it depresseth it not a little, and compressing that which standeth in the inward cavity of the Vessel C, it compelleth it to ascend the superior Pipe CF with a remarkable rolling; Yea according to the quantity of the entering Air you shall see it grow up twenty seven inches high towards F, the Quick-silver of the lower Pipe being altogether thrust out.

The Air presseth downward the Quick-silver in EA, by reason of the changed equal weight of the outward Air and inclosed Quick-silver, proceeding from the increase of adventitious Air entered by G: It thrusteth it upwards into CF, because the restored endeavor of the compressory virtue searcheth inwardly after an equal weight.

What then, Reader, is to be concluded from these things? The outward Air balanceth the Cylinder of the interior Quick-silver AE, therefore the Air even in its own sphere, as they call it, is weighty.

The parts of the Air in the Pipe and Carp's Bladder are distended by their own spontaneous dilatation; therefore the insited Elatery of the substance of the Air to rarify itself, doth imitate the nature of a Sponge or Wool.

And so the thicker the Air is, as hath been manifest in the Mountain Experiment, and that of Vacuity in Vacuity, insomuch it acteth on all sides with a greater and stronger Elatery invading the superfice of the Earth-watrish Globe.

The Fourth Experiment.
The Water Only by Its Weight
Compresseth the Earth-watry Globe;
but the Air Compresseth it,
Not Only by Its Weight,
but by Its Elatery

I will add an invincible Argument from an easy Experiment. Let there be a round long Pipe of Glass AB [fig. 6], at least three foot long, and four degreee, if you will, in diameter; let the end thereof B be hermetically closed exactly, that is, with the Glass it self, the other end A being open; fill the whole Glass with Quick-silver, except the seven inches CA; fill up the residue, viz. CA with water; then closing the Pipe with your finger turn it over and sustain it so long, till the water being lighter, having changed its place with the Quick-silver, at last doth ascend to the other end B: Dip both the Pipe and thy finger stopping it into the Quick-silver which standeth ready in the Vessel D; Drawing away thy finger easily, the Metal shall so flow out of the Pipe, that the Cylinder of

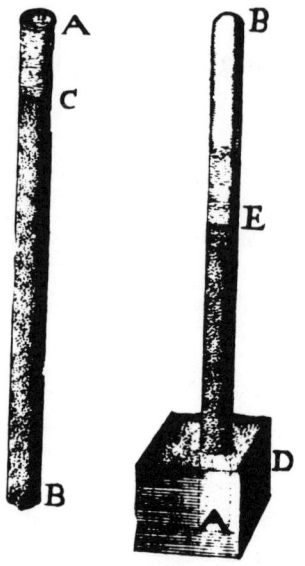

6. The fourth experiment. From Jean Pecquet, *New Anatomical Experiments*

Quick-silver AE which is to remain within the Pipe, will not remain according to its custom twenty seven inches above the standing superfice D, but it shall seem to be diminished six lines or degrees, by reason of the heaviness of the Air incumbent, and constantly remaining in seven inches measure. Neither is it any wonder; for such is the proportion between Water and Quick-silver, that half an inch of this doth weigh above seven inches of that.

Hence it clearly appears, that Water only by its weight, and by no endeavor of an Elatery, as 'tis heavy to Quick-silver within the Pipe, so it is to the superfice of the Earthly lump.

But the virtue of the Air is far more excellent; for if over again essaying the Experiment, you fill up, not with Water, but Air, the remaining space of seven inches CA, that is, if you shut up with your finger the Pipe from the Quick-silver BC, to the mouth A, being only full of Air for the space of seven inches CA, being afterwards overturned and dipped in the Engine D, you shall admire to see the Cylinder of Quick-silver AE to be lower by seven inches under the twenty seventh inch. So that it clearly appeareth, that as the subject Quick-silver, so the Earth-watry Globe is not so much pressed by the weight, as by the Elatery of the Air.

I would have you note, that when I said the Cylinder of the Quick-silver rested at the twenty seventh inch within the Pipe, I spoke then according to the most frequent Experiments; For the place of rest changeth in the Pipe according to the divers changes of the external Air in its rarefactions and condensations.

CHAPTER IX.
THE ENGINES DRAWN TO THE ASSISTANCE OF ATTRACTION
ARE DEMOLISHED

Having demonstrated by Experiments conquering all stiffness, and subduing all belief, That, more than in a wool heap, the heaping up of the intricate parts in the Air, and the Elater innate therein for rarefaction doth beat the Globe of Water and Earth; I proceed to examine their Engines who advance Attraction.

THAT THE WATER ENTERETH INTO WATERWORKS,
NOT FOR FEAR OF VACUITY,
BUT BY NECESSITY OF AN EQUAL WEIGHT IS CLEARED
BY THE PERIOD OF THE WATERS ASCENT IN THEM,
AND EXPERIMENTS

And first indeed the office of the sucker in the Pump, which they esteem the greatest, is to be considered: For they believe our opinion is overthrown, and that we have nothing to object, when they have attributed to the Heart the worthy name of the Vital Antlia.[11] For as, say they, the rising Clack[12] sucketh the Water, and falling down beateth it back; So the Heart by the gaping Diastole by drawing the Blood suppeth it up, which in the pressure of the Systole it compelleth to depart.

The mention of Antlias is so common in the Mechanicks, which some call sucking in, that none is ignorant of their office when he hears their names: But howsoever their use in drawing up the Liquour is frequent, not any as yet hath found out, in an instrument of so great use, the true inciter or genuine cause of its liquid motion.

The Inventors, and all those that afterwards have been occupied in Water-works, believed that the Water is drawn and sucked up by

11. "Antlia" originally meant bucket, then anything for drawing up liquids, as the piston of a suction pump.
12. A kind of valve.

the rising of the Clack, which whiles it riseth raiseth up the Water, and accelerateth them for fear of vacuity to rise upward against their falling nature: But I will show by the following reasons whether they think well or no.

The Earth-watry Globe is compressed on all sides, not only by the weight of the Air, as we have proved, but likewise by the most strong endeavor of the Elatery innate in it: Therefore where the heaping of the Aerial parts doth equally act in the subject Waters, it shall also be right that their emulous superficies should remain in the balance of equal height.

But if the endeavors of the Air be unequal, 'tis necessary also that the superficies of the Waters should become unequal; So that the Water should rise higher, where the pressure of the Air is lighter; and so long it should elevate its superfice, till the rising heap of Water, together with the Air which it sustains, should equally balance that other Air which is of more heavy pressure.

'Tis easy for thee to be taught this in a Bucket mechanically.

Take a Cylindrick Bucket A [fig. 7], and fill it at least half full with Water; then place it in the dish of the covering BCD, whose

Situla

7. The bucket.
From Jean Pecquet,
*New Anatomical Ex-
periments*

middle ought to be pervious D, to receive the Glass Pipe perpendicularly DE, and the covering BCD being movably contiguous to the Bucket encompassing it, should stay otherwise the eruption of the Liquor; indeed if you press down the dish with the weights FF, the Water pressed down on every side about the hole D, will fill the upright standing Pipe DE, till with its emulating heaviness it spring up to the place of equal weight G.

And this is the incitement of the ascending Water in the Pump.

Before you place in the Pipe of the Pump, standing now in the Water, the Clack, the Elastick (inforcing) weight of the Air beateth no less the inward than outward Water: Whence it comes to pass, that the Superficies of the Water both without and within the Pipe is of the same altitude: But when the Clack is fitted to the Pipe, and is lifted upwards, it together with it heaveth up the Pillar of Air which is perpendicularly stayed thereon: Hence all weight within the Pipe is taken away, nd then the Water which is urged sharplier by the perpetual pressure of the outward Air leaning thereon, rusheth into the easy Pipe; so long ascending till it come to two and thirty foot height, and there at last it is of equal weight with the circumfused Air.

Neither will the Rule of Counterposure (if we believe experience) suffer the superfice of the Waters springing within the Pipe or Pump, though the Clack should be elevate, to ascend higher.

You may likewise marvel at this in Quick-silver. For the Cylinder of Quick-silver inwardly will not ascend more than twenty seven inches above the superfice of it which standeth in the Vessel, in either Pump or Pipe: For indeed that Cylinder of Quick-silver doth counterpose a heap of water of above two and thirty feet in weight.

None of the Water ariseth above thirty two foot, none of the Quick-silver above twenty seven inches, though the Clack be drawn up with the greatest force of the Pipe: Therefore neither fear of Vacuity, who in the highest places of Nature should offer it self, moveth the Waters to follow: Neither the endeavor of the Air, which is spent with so little force of Counterposure doth oppress the Earth-watery Globe with its infused Elatery.

Hence let all Water-workers take notice, and all those Craftsmen who hope by the benefit of eared Syphons to draw Water over the tops of intermediate Hills, that the Hill be not above two and thirty foot high above the Spring, otherwise they shall lose their labour. The same judgement concerning drawing, as they call them, Pumps, is firm and stable. I omit other Pumps, and especially those in which the Clack like a Bucket is dipped in the Water to draw it up; For the elevate Water receives no period of ascent in their Pipes, except what is prescribed by the force of that power that moves the Clack: Yea they have not any business with the fear of Vacuity or of the Air.

Seeing therefore that the Water hasteth into the Pump, not by the incitement of Attraction, but by the Elastic heaviness of the spread-about Air, there is no reason why we should spend any longer time in refelling the Arguments which our Adversaries draw from this Engine to prove the attraction of the Heart.

THAT BELLOWS DO NOT DRAW THE AIR, BUT ONLY RECEIVE IT THRUST IN THEM FROM WITHOUT

Neither will the Heart more happily prove Sucking or Attraction, by playing the part of Bellows. A pair of Bellows do neither suck nor draw the Air to them, but are compelled to receive it being thrust in by external violence.

The Air lying hid within the lurking places of the folded and closed Bellows is of the same Elatery (that is, Spontaneous dilation) that the circumfused external Air is of; and so the resting of the equal weight contains both unmoved; Whiles you distend the Bellows, even then the inward Air (as was made clear in the Bladder of a Perch) of its own Spontaneous dilatation, which is therefore more weaker, it would rarify, except a more strong force of an outward Air did compell the Air next their Pipe without, and any liquid thing that is within, to equal balance.

THE FORCE OF AERIAL DILATATION, WHICH IT HATH FROM ITS PROPER ELATERY, DOTH LANGUISH, BUT BY AN EXTERNAL CAUSE BECOMETH FIRM

Neither let any man with more successful Eloquence persuade me of the Attraction of Aeolipiles and Cupping Glasses, proceeding of the effect produced by the Fire; and that thou mayst more easily understand how deservedly I oppose it, hear my opinion concerning the business of Heat.

Indeed the least Heat doth dilate the Air, as no man doubteth; But with this difference from that dilatation whereby it is sometimes of it self freely dilated, That the Spontaneous dilatation enerveth the power of the Elastic (impulsive) faculty, as the Experiments of Vacuity in Vacuity, and the Carp's Bladder do demonstrate; But the other, which is extraneous to the Air, viz. from the accession of heat, will make it firm, and will augment its growing, and is found by little and little to remit in its languishing.

THE WEATHER GLASS [BAROMETER] DOTH DEMONSTRATE, THAT NOT ONLY THE AIR BUT THAT THE WATER ALSO IS EXTENDED BY THE ACCESSION OF HEAT

So the Hand or a Coal put upon the upward Bottle of a Weather-Glass, causeth the water therein contained to descend; A great Argument that the Air which is dilated within doth in force excel the descending water which it compelleth to give it place.

Neither doth the only Air produce its own ends by the incitement of Heat; Yea the heap of water is likewise extended thereby.

This is easily tried, if you be pleased to thrust down the particle or Water C [fig. 8] sticking in the middle of the Weather-Glasses

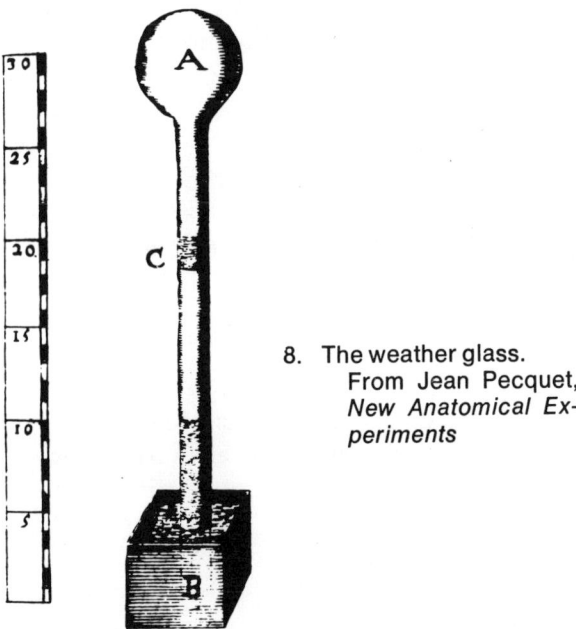

8. The weather glass. From Jean Pecquet, *New Anatomical Experiments*

stalk into the lowest seat of the Air that upholds it; For fire being put unto the upper Bottle A, doth not only compel downward the inclosed water C, but also compelleth it to grow bigger, either because it drencheth the Air which it dilateth A, in the descending water C; or because it stirreth up the Aerial parts of the Descending water to Rarefaction, viz. to the third degree, which before scarcely possessed two degrees.

I prescribe no Law to the water to be hung in C; for oft-times I have found it by chance hanging, and not a little distant from the water in B in the under Vessel, which by the diligence of my labour I could have brought to pass.

The Air growing cold in A, the water floweth upward, for the Elastick virtue of this, as it imitateth the increments of Heat, so likewise it observeth the changes of the deficiencies in the same.

16. The Air Pump

The air is so necessary to life, that most creatures which breathe cannot subsist for many minutes, without it; and most of the natural bodies we deal with, being, as well as our own, almost perpetually contiguous to it, the alterations thereof have a manifest share in many obvious effects, and particularly, in distempers: wherefore, a farther inquiry into the nature of this fluid, will, probably, show, that it concurs to exhibit abundance of phenomena, wherein it has, hitherto, seemed little concerned. So that, a true account of any new experiment, upon a thing whereof we have such a constant and necessary use, may prove advantagious to human life.

With this view, before ever I was informed that Otto Guericke, the ingenious consul of Magdeburg, had practised a way, in Germany, of emptying glass vessels of the air, I had made experiments on the same foundation, but, as that gentleman first produced considerable effects by this means, I acknowledge the assistance and encouragement which the report of his performances afforded me.

SOURCE: Robert Boyle, *New Experiments Physio-Mechanicall, touching the Spring of the Air and its Effects* (London, 1660), taken from Peter Shaw, *The Philosophical Works of the Honourable Robert Boyle, Esq., Abridged, methodized and disposed under . . . General Heads* (4 vols., London, 1725) II, 407-14, 417-18, 421-23, 438-39.

But, as few inventions happen to be complete at the first, so the engine employed by the consul, seemed very defective in its contrivance; whence but little more could be expected from it, than those very phenomena observed by the author, and related by Schottus. I, therefore, put Mr. Hooke, upon contriving an air-pump, more manageable, and convenient, that might not, like the German engine, require to be kept under water: and, after some unsuccessful attempts, he fitted me with one, consisting of two principal parts; a glass vessel, and a pump to evacuate the air. [fig 9.]

Physico-Mechanical Experiments

The first is a glass A, with a large mouth, a cover thereto, and a stop-cock fitted to the neck below. This would contain 30 quarts of water. BC, the mouth of it, is about four inches in diameter, and surrounded with a glass lip, almost an inch high, for the cover to rest on; wherein DE, is a brass ring, to cover, and be cemented on to the lip of BC. To the internal orifice of this ring, a glass stopple is fitted, to keep out the external air. In the middle of this cover is a hole HI, half an inch in diameter, incircled with a ring, or socket; to which is adapted a brass stopple K, to be turned round, without admitting the least air. In the lower end of this, is a hole 8, to admit a string, 8,9,10; which also passes through a small brass ring L, fixed to the bottom of the stopple FG, to move what is contained in the exhausted vessel, or receiver. That the stop-cock N, in the first figure, might perfectly exclude the air, we fastened a thin tin-plate, MTVW, to the shank of the cock X, all along the neck of the receiver, with a cement made of pitch, rosin, and wood ashes, poured hot into the cavity of the plate; and to prevent the cement from running in at the orifice Z, of the shank X, it was stopped with a cork fixed to a string, that it might be drawn out at the upper orifice of the receiver; and then the neck of the glass, being made warm, was pressed into the cement, which thus filled the interstices betwixt the tin-plate and the receiver, and betwixt the receiver and the shank of the cock.

The lower part of our engine consists of a sucking-pump, supported by a wooden frame, with three legs I I I so contrived, that, for the freer motion of the hand, one side of it may stand perpendicular; and across the middle of the frame we nailed a piece of board 222, to which the principal part of the pump is fixed. The pump consists of an exact strong concave cylinder of brass, four-

9. Boyle's air pump.
   From Robert Boyle, *New Experiments ... touching the Spring of the Air*

teen inches long, its cavity three inches in diameter; to which a sucker, 4455, is adapted, made up of two parts; one of which 44, is less in diameter than the cavity of the cylinder, with a thick piece of tanned leather nailed on it, whereby it excludes the air. The other part, a thick iron plate 55, is firmly joined to the middle of the former, and is a little longer than the cylinder; one edge of it being smooth, and the other indented, to receive the teeth of a small iron nut fixed by two staples to the underside of the board nailed across 22, on which the cylinder stands; and it is turned by the handle 7.

The last part of the pump is the valve R, a hole at the top of the cylinder, and taper towards the cavity; to this is fitted a brass-plug, to be taken out as occasion requires. The engine being thus contrived, some oil must be poured in at the top of the receiver upon the stop-cock, to fill up the interstices of its parts, and that the key S, may turn with the greater ease. A quantity of oil, also, must be left in the cylinder, to prevent the air from getting betwixt that and the sucker; for the like reasons, some must, likewise, be applied to the valve.

And here it is proper to observe, that when we used oil, or water, separately, for this purpose, and they have not answered the end, a mixture of the two has afterwards proved effectual. And, that the air may not enter betwixt the brass-cover and the ring, it will be convenient to lay some diachylon-plaster [adhesive tape] on their edges with a hot iron. That no air, also, may remain in the upper part of the cylinder, the handle is to be turned till the sucker rises to the top; and then, the valve being shut, it is to be drawn down to the bottom; by which means, the air being driven out of the cylinder, and a succession from without prevented, the cavity of the cylinder must be empty of air; so that, when the stop-cock is turned to afford a communication betwixt the receiver and the cylinder, part of the air before lodged in the receiver, will be drawn down into the cylinder; which, by turning back the key, is kept from entering the receiver again, and may, by unstopping the valve, and forcing up the sucker, be driven into the open air; and so, by repeated exsuctions out of the receiver, and expulsions out of the cylinder, the vessel may be exhausted as the experiment requires.

Upon drawing down the sucker of our engine, whilst the valve is shut, the cylindrical space deserted by it will be left empty of air;

and therefore, upon turning the key, the air contained in the receiver rushes into the cylinder, till in both vessels, it be brought to an equal dilatation; so that, upon shutting the receiver, turning back the key, opening the valve, and forcing up the sucker again, almost a whole cylinder of air will be driven out after this first exsuction; but after every succeeding stroke, less air will come out of the receiver into the cylinder: so that, at length, the sucker will rise almost to the top of the cylinder, before the valve need be opened. And if, when it is so exhausted, the handle of the pump be let go, and the valve be stopped, the sucker, by the force of the external air, which is an over-balance to the internal rarified air, will be forced to the upper part of the cylinder, and higher, in proportion, as the air is more exhausted. We observed, also, that, whilst any considerable quantity of air remains in the receiver, a brisk noise is immediately produced, upon turning the key.

The Spring of the Air

But to render our experiments the more intelligible, we must premise, that the air abounds in elastic particles, which being pressed together by their own weight, constantly endeavour to expand and free themselves from that force; as wool, for example, resists the hand that squeezes it, and contracts its dimensions; but recovers them as the hand opens, and endeavours at it, even whilst that is shut. It may be alleged, that though the air consists of elastic particles, yet this only accounts for the dilation of it in pneumatical engines, wherein it hath been compressed, and its spring violently bent, by an external force; upon the removal whereof, it expands, barely to recover its natural dimensions; whilst, in our experiments, the air appears not to have been compressed, before its spontaneous dilatation. But we have many experiments to prove, that our atmosphere is a heavy body, and that the upper parts of it press upon the lower. And I found a dry lamb's bladder, containing two thirds of a pint, and compressed by a pack-thread tied about it, to lose, in a very tender balance, $1\ ^1/_8$ grains of its former weight, by the recess of the air, upon pricking it. Supposing, therefore, that the air is not destitute of weight, it is easy to conceive, that the part of the atmosphere wherein we live, is greatly compressed by those directly over it, to the top of the atmosphere. And though the height of this atmosphere, according to Kepler, scarce exceeds eight miles, yet later astronomers extend it six or seven miles farther. The learned Ricciolo makes it reach fifty miles high. So that a column of air, several miles in height, pressing upon some elastic particles of the same fluid here below, may easily bend their little springs, and keep them bent; as if fleeces of wool, were piled to a vast height upon one another, the hairs of the lowest locks would, by the weight of all the incumbent parts, be strongly compressed. Hence it is, that, upon taking off the pressure of the incumbent air, from any parcel of the lower atmosphere, the particles of the latter possess more space than before. If it be farther objected against this condensation of the inferior air, that we find this fluid readily yields to the motion of flies, feathers, etc. we may reply, that as when a man squeezes wool in his hand, he feels it make a continual resistance; so each parcel of the air, about the earth, constantly endeavours to thrust away such contiguous bodies as keep it bent, and hinder the expansion of its parts; which will fly out towards the part, where they find the least resistance. And since, the corpuscles whereof the air consists, though of a springy nature, are so very small, as to compose a fluid body, it is easy to conceive, that there, as in other fluids, the component parts are in perpetual motion, whereby they become apt to yield to, or be displaced by other bodies; and that the same corpuscles are so variously moved, that, if some attempt to force a body one way, others, whose motion hath an opposite determination, as strongly press it the contrary way; whence it moves not out of its place; the pressure, on all sides, being equal. For if, by the help of our engine, the air be drawn only from one side of a body, he, who thinks to move that body, as easily as before, will, upon trial, find himself mistaken.

Thus, when our receiver is tolerably exhausted, the brass stopple in the cover, is so difficult to lift, that there seems to be some great weight fastened to the bottom of it: for the internal air being, now, very much dilated, its spring must be greatly weakened; and, consequently, it can but faintly press against the lower-end of the stopple, whilst the spring of the external air keeps it down, with its full natural force. And, as the air is gradually admitted into the receiver, the weight is manifestly felt to decrease; till, at length, the receiver being again filled with air, the stopple may be easily lifted.

It may seem surprising, that we speak of the air shut up in our receiver, as of the pressure of the atmosphere; though the glass manifestly keeps the incumbent pillar of air from pressing upon that within the vessel. But, let us consider, that if a fleece of wool, by pressure, be thus directly reduced into a narrow compass, and conveyed into a close box, though the former force ceases to bend its numerous springy parts, yet they continue as strongly bent as before; because we suppose the including box resists their expansion, as much as the force that crowded them in. Thus the air, being shut up in our glass when its parts are bent by the whole weight of the incumbent atmosphere, though that weight can no longer press upon it; yet the corpuscles of the internal air continue as forcibly bent, as before they were included. If it be said, that the continual endeavour it has to expand itself, ought then to break the glass, we must observe, that the expansive force of the internal air, is balanced by pressure of the external, which preserves the glass entire; as, by the same means, thin large bubbles, made with soapy water, will, for some time, continue whole in the open air.

And though, by help of the handle, which is a lever, the sucker may easily be drawn down to the bottom of the cylinder; yet, without such a mechanic power, the same effect could not be produced, but by a force able to surmount the pressure of the atmosphere: as in the Torricellian experiment, if the column of mercury be too high, it will subside, till its weight be a balance to the pressure of the air. Hence we need not wonder, that though the sucker move easily in the cylinder, by means of the handle, yet, if that be taken off, it will require a considerable force to raise or depress it. Nor will it seem strange, that if, when the valve, and stop-cock are exactly closed, the sucker be drawn down, and then the handle let loose, that the sucker of itself, reascends to the top of the cylinder; since the spring of the external air, finds nothing to resist its pressure upon the bottom of the sucker. And, for the same reason, when the receiver is almost emptied, though, the sucker being drawn down, the passage from the receiver to the cylinder be opened, and then stopped again, the sucker will, upon the letting go the handle, be forcibly carried up, almost to the top of the cylinder; because the air within the cylinder, being equally dilated and weakened with that of the glass, is unable to resist the pressure of the external air, till it be crowded into so little space, that both their forces are in equilibrium. So that, in this case, the sucker is drawn down with little less difficulty, than if, the cylinder being destitute of air, the stop-cock were exactly shut. It must also be observed that when the sucker hath been impelled to the top of the cylinder, and the valve is so carefully stopped, that no air remains in the cylinder, above the sucker; if, then, the sucker be drawn to the lower part of the cylinder, no greater difficulty is found to depress the sucker, when nearer the bottom of the cylinder, than when it is much farther from it. Whence it appears, that the pressure of the external air, is not increased upon the accession of the air driven out; which, to make itself room, forceth the contiguous air to a violent sub-ingression of its parts, as some suppose; for otherwise the sucker would be more resisted by the external air as it comes lower; more of the displaced air being thrust into it, to compress it.

We took a large lamb's bladder, well dried, and very limber, and leaving in it about half the air it would contain, we strongly tied the neck of it; then conveying it into the receiver, the pump was worked; and after two or three strokes, the imprisoned air began to swell in the bladder, and continued to do so, as the receiver was farther exhausted, till, at length, the bladder appeared perfectly turgid. Then, by degrees, allowing the external air to return into the receiver, the distended bladder shrunk proportionably, grew flaccid, and, at last, appeared as full of wrinkles as before.

And to try whether the actual elasticity of the fibres of the bladder, had any share in this effect, we let down to the former, two smaller bladders, of the same kind; the one not tied up at the neck, that the air it contained might pass into the receiver; the other, with its sides stretched out, and pressed together, that it might hold the less air, and then strongly tied up at the neck; and, whilst the first, upon working the pump, appeared every way distended to its full dimensions, neither of the others were remarkably swelled; and that whose neck was left loose, seemed very little less wrinkled than when first put in.

We made, likewise, a strong ligature about the middle of a long bladder, emptied of its air in part, but left open at the neck; and, upon exhausting the receiver, observed no such swelling betwixt the ligature, and the neck, as betwixt the ligature and the bottom of the bladder, where air was included.

We hung a dry bladder, well tied, and blown moderately full, in the receiver, by a string fastened to the inside of the cover; and, upon exhausting the glass, the included air first distended the bladder, and then burst it, as if it had been forcibly torn asunder.

This experiment was repeated with the like success; and the bladder bursting, long before the receiver was fully exhausted, gave a great report.

But it was often, in vain, that we tried to burst bladders, after this manner, because they were commonly grown dry, before they came to our hands; whence, if we tied them very hard, they were apt to fret, and so become unserviceable; and, if tied but moderately hard, their stiffness kept them from being closed so exactly, that the air should not get out into the receiver. We found, also, that a bladder moderately filled with air and strongly tied, being held for a while, near the fire, grew exceeding turgid; and, afterwards, being brought nearer to the fire, suddenly burst, with so loud and vehement a noise, as made us almost deaf for some time after.

Having thus found, that the air hath an elastic power, we were desirous to know how far a parcel of that fluid might be dilated by its own spring. . . .

Flame and the Air

We let down, into our receiver, a tallow-candle of a moderate size, and suspending it, so that the flame appeared in the middle of the vessel, we presently closed it up, and upon pumping found, that within little more than half a minute after, the flame went out.

At another time, the flame lasted about two minutes, though upon the first exsuction it seemed to contract itself in all its dimensions, and after two or three exsuctions, it appeared exceeding blue, and gradually receded form the tallow, till at length it seemed to possess only the very top of the wick, and there it vanished.

The same candle, being lighted again, was shut into the receiver, to try how it would burn there, without exhausting the air; and we found that it lasted much longer than formerly; and before it went out, it receded form the tallow, towards the top of the wick, though not near so much, as in the former experiment.

We took notice, that when the air was not drawn out, a considerable part of the wick remained kindled upon the extinction of the flame, which emitted a smoke, that swiftly ascended directly upwards, in a slender and uninterrupted cylinder, till it came to the top, from whence it returned, by the sides, to the lower part of the vessel; but when the flame went out, upon the exsuction of the air, we once perceived it not to be followed by any smoke at all. And at another time, the upper part of the wick, remaining kindled after the extinction of the flame, a slender steam ascended, but a very little way, and after some uncertain motions, for the greatest part, soon fell downwards.

Joining together six slender tapers of white wax, as one candle, and having lighted all the wicks, we let them down into the receiver, and made what haste we could to close it up with cement. But, though in the mean while, we left open the valve of the cylinder, the hole of the stop-cock, and that in the cover of the receiver, that some air might get in to cherish the flame, and that the smoke might have a vent; yet the air sufficed not for so great a flame, till the cover could be perfectly luted [sealed] on; so that before we were ready to employ the pump, the flame was extinguished. Wherefore, we took but one of the tapers, and having lighted it, closed it up in the receiver, to try how long a small flame, with a proportionable smoke, would continue in such a quantity of air; but we found, upon two several trials, that from our beginning to pump, the flame went out in about a minute. It appeared, indeed, that the swinging of the wire, whereby the candles hung, hastened the extinction of the flame, which seemed, by the motion of the pump, to be thrown, sometimes on one side of the wick, and sometimes on the other. But, once refraining to pump, after a very few exsuctions, the flame lasted not much longer. And lastly, closing up the same lighted taper, to discover how long it would last, without drawing out the air; we found, that it burnt vividly for a while; but afterwards, began to diminish gradually in all its dimensions, though the flame did not, as before, retire itself by little and little towards the top, but towards the bottom of the wick, so that the upper part of it manifestly appeared for some time, above the top of the flame; which, having lasted about five minutes, was succeeded by a stream of smoke, that ascended in a straight line. . . .

Magnetism in the Vacuum

We conveyed into the receiver a little pedestal of wood, in the midst of which was, perpendicularly erected, a slender iron, upon the sharp point whereof, an excited needle of steel, of about five inches long, was so placed, that, hanging in equilibrium, it could move freely every way. Then the air being pumped out, we employed a load-stone, moderately vigorous, to the outside of the glass, and found that it attracted, or repelled the ends of the needle, without any remarkable difference from what the same load-stone would have done, had none of the air been drawn away from about the needle; which, when the load-stone was removed, rested, after some tremulous vibrations, in a position north and south.

The Torricellian Experiment

A slender, and very exact cylinder of glass, near three feet in length; its bore, a quarter of an inch in diameter, being hermetically sealed, at one end, was, at the other, filled with quick-silver; care being taken, that as few bubbles as possible, should be left in the mercury. Then the tube being stopped with the finger and inverted, was opened into a long, slender, cylindrical box, half filled with quick-silver; when that in the tube subsiding, and a piece of paper being pasted level to its upper surface, the box and tube were, by strings, carefully let down into the receiver; and the cover, by means of this hole, slipped along as much of the tube, as reached above the top of the receiver: the interval left betwixt the sides of the hole, and those of the tube, being exquisitely filled up with melted diachylon; and the round chink, betwixt the cover and the receiver, likewise, very carefully closed; upon which closure, there appeared no change in the height of the mercurial cylinder: whence the air seems to bear upon the mercury, rather by virtue of its spring, than of its weight, since its weight could not be supposed to amount to above two or three ounces; which is inconsiderable, in comparison of such a cylinder of mercury as it would sustain. Now the sucker was drawn down, and immediately, upon the evacuation of a cylinder of air, out of the receiver, and the quick-silver in the tube subsided; and notice being carefully taken of the place where it stopped, we worked the pump again, and marked how low the quick-silver fell at the second exsuction: but, continuing thus, we were soon hindered from accurately marking the stages in its descent, because it presently sunk below the top of the receiver: so that we could, from hence, only mark it by the eye. And continuing pumping, for about a quarter of an hour, we could not bring the quick-silver, in the tube, totally to subside. Then we let in some air; upon which, the mercury began to reascend in the tube, and continued mounting, till having returned the key, it immediately rested at the height it had then attained. And so, by turning, and returning the key, we did, several times, impel it upwards, and check its ascent; till, at length, admitting as much of the external air as would come in, the quick-silver was impelled up, almost, to its first height; which it could not fully regain, because some little particles of air were lodged among those of the quick-silver, and rose in bubbles to the top of the tube.

It is remarkable, that having, two or three times, tried this experiment, in a small vessel; upon the very first cylinder of air that was drawn out of the receiver, the mercury fell, in the tube, 18 inches and a half; and, at another time, 19 inches and a half.

We likewise made the experiment in a tube less than two feet in length; and, when there was so much air drawn out of the receiver, that the remaining part could not counter-balance the mercurial cylinder, it fell above a span at the first stroke; and the external air being let in, impelled it up again, almost to the top of the tube: so little matters it, how heavy or light the cylinder of quick-silver be, provided its gravity overpower the pressure of as much external air, as bears upon the surface of that mercury into which it is to fall.

Lastly, we observed, that if more air were impelled up, by the pump, into the receiver, after the quick-silver had regained its usual standard in the tube, it would ascend still higher; and immediately, upon letting out that air, fall again to the height it rested at before. . . .

Sound in the Vacuum

That the air is the medium whereby sounds are conveyed to the ear, was a current opinion, till some pretended, that if a bell, with a steel clapper, be fastened to the inside of a tube, upon making the experiment de vacuo with it, the bell remaining suspended in the deserted space, at the upper end of the tube; if a vigorous load-stone be applied on the outside of the glass, it will attract the clapper; which, upon the removal of the load-stone, falling back, will strike against the bell, and thereby produce a very audible sound: whence, several have concluded, not the air, but some more subtle body, to be the medium of sounds. But suspending a watch, freed from its case, in the cavity of our receiver, by a packthread; and then, closing up the vessel with melted plaster; we listened near the sides of it, and plainly heard the balance beat, and observed, that the noise seemed to come directly in a straight line, from the watch to the ear. We found, also, a manifest difference in the noise, by holding our ears near the sides of the receiver, and near the cover of it; which seemed to proceed from the difference between the glass, the cover, and the cement, through which the sound was propagated. But, upon working the pump, the sound grew gradually fainter; so that, when the receiver was emptied as much as usual, we could not, by applying our ears to the very sides of it, hear any noise from within; though we could easily perceive, that, by the motion of the hand which marked the seconds, and by that of the balance, the watch neither stood still, nor seemed irregular. And, to satisfy ourselves farther, that it was the absence of the air about the watch that hindered us from hearing it, we let in the external air at the stop-cock; and then, though we turned the key, and stopped the valve, yet we could plainly hear the noise made by the balance; though we held our ears, sometimes, at the distance of two feet from the outside of the receiver. And this experiment, being repeated, succeeded after the like manner: which seems to prove, that the air is, at least, the principal medium of sounds. And, by the way, it is very well worth noting, that, in a vessel so exactly closed as our receiver, so weak a pulsation as that of the balance of a watch should propagate a motion to the ear, in a straight line, notwithstanding the interposition of glass, so thick as that of our receiver. We, afterwards, took a bell of about two inches in diameter at the bottom, which was supported, in the midst of the cavity of the receiver, by a bent stick, pressing with its two ends against the opposite parts of the vessel; which, being closed up, we observed the bell to sound more dead than in the open air. And yet, when we had emptied the receiver, we could not discern any considerable change in the loudness of sound; whereby it seemed, that, though the air be the principal medium of sound; yet, either a more subtle matter may be, also, a medium of it; or else that an ambient body, that contains but few particles of air, is sufficient for that purpose Whence, perhaps, in the above mentioned experiment, made with the bell and the load-stone, there might, in the deserted part of the tube, remain air enough to produce a sound.

But as, in making the experiment of firing gun-powder with a pistol in our evacuated receiver, the noise made by the flint, striking against the steel, was exceeding languid, in comparison of what it would have been in the open air: so, on several other occasions, it appeared, that the sounds produced there, if they were not lost, seemed to arrive at the ear very much weakened.

V

Scientific Societies

THE NATURAL RESULT of the success of the experimental method was the conviction that it would be even more effective if undertaken by cooperative groups, something in the manner so picturesquely described by Bacon in *The New Atlantis*. In any case, the seventeenth century was a gregarious age, and societies for the advancement of learning—especially for the study of literature and philology—had long been in existence when the first formal scientific societies took shape. The scientific societies were very much a feature of the middle of the century, languishing rather as the decades proceeded, to be revived again at the beginning of the next century.

The oldest, but the most ephemeral, of such societies was Italian, the Accademia del Cimento or Academy of Experiment. This was recognizably like the Renaissance humanist societies of an earlier age, being founded by a single wealthy man—Prince Leopold of Tuscany— interested in science as an amateur, as his ancestors had been interested in humanist learning. Leopold de Medici had been a pupil of Galileo when a boy and had a genuine interest in physical science; the scientists he assembled in his Academy were associated with the court at Florence and wholly or partly supported by Prince Leopold and his brother, the Grand Duke. The most notable members were Vicenzo Viviani, Galileo's last pupil, who, like Galileo himself, was court mathematician, and Francesco Redi, remembered for his experiments directed against spontaneous generation, who was court physician. The Academy itself was small, and little is known about its work in the years between June, 1657 (when it first met) and March, 1667 when it last met, except what is contained in the *Saggi di Naturali Esperienze fatte nell' Accademia del Cimento (Essayes of Natural Experiments made in the Academie del Cimento)*, published in 1667. The Preface to this work is reproduced below (Document 17) in the English translation by Richard Waller published in 1684. The *Saggi* describe experiments on pneumatics, freezing, mechanics, physiology, and chemistry, all rather commonplace. The Fellows of the Royal Society found it nicely written, though they thought the contents trite and old-fashioned. But it was an important piece of propaganda for the spread of experimental science.

The second experimentally based society of the seventeenth century was totally different in origin and history. The Royal Society of London, in spite of its title, was the result of long-continued and spontaneously developed groups of scientists meeting for exchange of views and news. Although there has been much controversy on the subject of its origins, there seems no reason to reject the account given (twice, 1678 and 1697) by John Wallis (1616-1703), who was a participant in its activities for all his long and vigorous life. Gresham College, founded at the very end of the sixteenth century by that great Elizabethan merchant Sir Thomas Gresham (of Gresham's Law), was a natural meeting place for scientific discussion in the gloomy days of the Civil War, when even academics were drawn to London. On the whole the Gresham professors of geometry and astronomy were far abler and more up-to-date scientists than were to be found at Oxford or Cambridge (though most had been university educated, while conversely Gresham supplied many Savilian professors to Oxford). When several of this earliest group, such as John Wilkins (a popularizer of Copernicus) and Wallis himself, went to Oxford about 1650 they naturally continued to meet for scientific discussions, which included, as they had always done, a keen interest in experiment (see Document 18).

Meanwhile the London group also continued to meet. The Oxford men gathered strength from a number of physicians, from an extraordinarily able amateur, Robert Boyle, and in time from some brilliant young men recruited as undergraduates, notably Christopher Wren, Robert Hooke, and Richard Lower.

The restoration of the monarchy took many of these men back to London, where they joined forces with returned royalists like Sir Robert Moray and Lord Brouncker. They were stimulated in part by the private scientific societies operating in Paris, of which Boyle had been fully informed by an intelligent correspondent, Henry Oldenburg (later—from 1662 to his death in 1677—to be the very active secretary of the Royal Society), they were stimulated also by the desire to revive the intellectual life of England and by the realization that the restored monarchy was more likely to promote learning and patronize science than the English government had ever before been willing to do. Accordingly, a dozen men met after a lecture by Wren at Gresham College on November 28, 1660, and agreed to form a society for the advancement of science, with a proposed membership of about fifty, to meet weekly. The emphasis was to be on the performing of experiments. During the next year the success of the Society led them to seek the advantages of a royal charter, first granted in 1662 and revised in 1663; this made them a formal society with officers, recognized laws, and rules of membership.

They were, from the beginning, orderly about keeping records; after the first charter Oldenburg became responsible for keeping minutes of all the meetings and for instituting a wide correspondence on scientific matters at home and abroad. The record of the years until 1687 was published in 1756 by the then secretary, Thomas Birch, as *The History of the Royal Society* in four volumes; the records and most of the correspondence are still preserved in the archives of the Royal Society, which has never ceased to be the leading scientific society of Great Britain. Oldenburg further advanced the work and aims of the Royal Society by founding the *Philosophical Transactions* (1665).

Initially his private venture, though published with the approval of the Society, and the aid of some of its Fellows, the *Philosophical Transactions*— the first wholly scientific journal—became so essential to scientists as a place for the publication of papers and as a source of international scientific news that it was continued after Oldenburg's death and gradually became the official journal of the Royal Society (see Document 19).

We have no such detailed knowledge of either the origin or the methods of the Académie royale des sciences (founded in 1666) as we have for the Royal Society. Much is known about the informal academies operating under the aegis of private patrons which flourished in the twenty years preceding the establishment of the Académie and which continued in many cases to exist for nearly as long after its foundation. The most famous of these is the Montmor Academy, which met at the house of Henri Louis Habert de Montmor—a government official, a minor literary figure, and member of the Académie francaise—from about 1655 to 1664. Various French scientists and intellectuals sought to interest the government in a scientific academy along the lines of the literary Académie française, as the correspondence of Colbert bears witness. The Académie royale des sciences was formally established only after Colbert had enlisted the support of various scientists by offering them pensions. Of these the most important was the distinguished Dutch physicist and astronomer Christiaan Huygens, who for twenty years was the leading figure and organizer of the French Académie. He is usually regarded as a Cartesian, as he certainly was in many respects, but his proposals for the Académie's plan of work, given below (Document 20), are curiously Baconian. The original manuscript bears Colbert's approving annotations. The Académie differed from the Royal Society in being a government department; its members were given salaries and rooms in the Louvre, provided with instruments, and alloted funds for international expeditions; they also worked cooperatively like the members of the Accademia del Cimento (cf. Document 21). But, except for Huygens, they were not such gifted experimenters as the English and, again except for Huygens, their individual scientific achievements were less notable. The Académie declined after Huygens and other Protestant scientists left France as a result of the Revocation of the Edict of Nantes (1685); it was reorganized in the 1690's and became far more active and important to the scientific life of eighteenth-century France than it had been earlier—more active too than the eighteenth-century Royal Society. Both came to dominate the scientific scene in the nineteenth century.

17. The Accademia del Cimento

The Preface to the Reader

Among the Creatures of Divine Wisdom, the birthright doubtless belongs to the Idea of Truth, which the Eternal Artificer so exactly followed in the universal fabric of nature, that no being was made with the least irregular bias of falsehood: But man afterwards (in the contemplation of so high and perfect a structure, through an extravagant desire of comprehending the admirable design, and finding out all the measures and proportions of so beautiful an order) when he aims to penetrate too deep into the truth, frames to himself an indefinite number of falsities, which proceeds from no other cause but his ambition to take those wings Nature never designed (perchance fearing to be some time or other discovered by him unwillingly in the preparation of her greater works,) yet upon these he begins to raise himself, and though charged with the weight of a material body, stretches forth these pinions to soar higher than the scale of sense leads and fixes himself upon that light, whose rays, too powerful for his eyes, dazzle and blind him. Thus we see from man's rashness, the first seeds of false notions came; from which yet it happens, not that the bright splendor of God's excellent creature is at all shaded, or by their commerce with them in the least vitiated: since all these imperfections are to be attributed to man's ignorance, whence they had their beginnings; when improperly applying the causes to the effects, he takes not from either the verity of their beings, but only delineates in his own mind a false conception of their relation to each other, and agreement; not that the sovereign beneficence of God when he creates our souls, denies them to pry, as we may say for a moment into the immense treasure of his eternal wisdom; adorning them as with the most precious jewels, with some first

SOURCE: *Saggi di Naturali Esperienze fatte nell' Academia del Cimento* (Florence, 1667); in the English translation by Richard Waller, *Essayes of Natural Experiments made in the Academie del Cimento* (London, 1684), sig. a4-b4.

sparks of truth, sufficiently evident from their retaining notions not to be acquired here, whence we must conclude, they receive them from some other place.

But it happens through our misfortune, that these rare gems, as they are but loosely set in the mind, yet too tender when she first falls into her earthly habitation, and wraps herself in this clay; as for a time they fall out of their collets [settings], are sullied, and worth nothing till by assiduous and careful study, they are again reset in their proper places. This is what the mind attempts in the search of Nature; wherefore we must confess, we have no better means than geometry, which at first essay hits the truth, and frees at once from all doubts and wearying researches. And indeed she leads into the way of philosophical speculations, but at last leaves us; not that geometry has not a large field to expatiate in, and travels not over all Nature's works; as they all submit to those mathematical laws, by which the eternal decree freely rules and commands them, but because we hitherto are unable to follow her in so long and wide a path [except] only a few steps. Now where we may not trust ourselves to go farther, we can rely on nothing with greater assurance than the faith of experience, which (like one that having several loose and scattered gems, endeavours to fix each in its proper collet) by adapting the effects to the causes, and again the causes to the effects, if not at the first essay, as geometry yet at last succeeds so happily, that by frequent trying and reject- ing, she hits the mark. We ought then to proceed with much circumspection, lest too great a reliance and trust in experience turn us out of the way, and impose upon us; since it sometimes falls out, that before the clear truth appears to us, when the first more open veils of deceit are taken off, we discover some cheating apperances that indeed have some likeness and resemblance of truth: and these are the imperfect lineaments that are seen through the last coverings that more nearly veil the lovely face of truth; through the fine web whereof she sometimes seems so plain and lively, that some might conclude, she was nakedly discovered.

Here then we ought to carry ourselves as master workmen: to discern between truth and error and the utmost perspicacity of judgment is but requisite to see well what really is from what is not: And to be the better able to perform this task, doubtless 'tis necessary to have at some time or other seen truth unveiled: an advantage they only have, who have had some taste of the studies of geometry.

Nor is it of less use to search among experiments already made, than to attempt new ones, if haply any may be found that have at all disguised the simple face of truth; wherefore 'tis aimed at in our academy, besides what has been invented by us, to try also (either for curiosity or as we light upon them by chance) those things which have been already done or written of by others: observing too well, that under this name of experiments, frequent errors have crept in and have been entertained.

This was the first motive of the perspicacious and indefatigable mind of the most serene Prince Leopold of Tuscany, who, in the recess of those daily negotiations and solicitous cares that attend his high quality, diverted into the rough path of the noblest sciences. But his Highness's discerning judgment easily foreseeing that the reputation of great authors proves too often hurtful to the studious, who through too much confidence and veneration of their names, fear to call in question what is delivered upon their authority; wherefore he judges it an undertaking worthy of his great mind to confront with the most accurate and sensible experiments the force of their assertions, and with the due rejection of errors, and embracing of realities, to make so desirable and inestimable a present to those that earnestly wish for the discovery of truth. These prudent instructions of our most serene Patron, received with due reverence and respect by our academy, has not moved us to be indiscreet censurers of the earned pains of others, nor made us bold obtruders of our own sentiments for truths and discoveries of abuses: but it is our principal intent to incite others also to repeat with the greatest severity and niceness, the same experiments, as we have now adventured to do with those of any other person: though in publishing these first essays we have, what we could, abstained therefrom, that we might, by this due respect, gain upon the adversary to believe the sincerity of our impartial and respectful thoughts.

And to the full completing of so generous and useful an undertaking, we desire only a free correspondence with those several societies that are dispersed throughout the more illustrious and noted parts of Europe: that with the same design of attaining such high ends, so profitable a commerce being in all parts round about promoted, we may all go on with equal freedom, enquiring as much as possible and participating of the truth: and for our parts we will concur to this work with the greatest simplicity and ingenuity; whereof 'tis no small argument that when we have related the experiments of others, we have still mentioned the author's name, when known to us; and that we have often freely confessed, that supposition concerning some experiments, which when put in practice we were never so successful as to bring to perfection. But above all, to prove clearly the unfeigned sincerity of our procedure, let that freedom suffice, wherewith we have still communicated the Essays and Experiments themselves to any that, travelling by our country, showed any desire or relish of such sciences, moved either by a genteel humor, esteem of learning or spur of curiosity; and that from the first time our Academy was founded in the year 1657, when the greatest part, if not all the experiments were invented whereof these *Essays* are now published. If it shall happen that among them there shall be any found, thought of before or after the time they were made here by other persons, and made public, let us not be blamed for it, since we could neither know nor see all things; so that no man ought to wonder at the lucky accord of our minds and inventions with other men's, nor indeed will we, if we find those of other men agree with ours.

Lastly, we are unwilling any should imagine, that we pretend in this publication a perfect work; or, in the least, an exact model of a large experimental history; conscious to ourselves, that more time, and greater abilities are necessary to so vast a design; as may be seen by the very title we have prefixed, only of *Essays,* which we had never put forth, had we not been much urged thereto by persons meriting from us, by their dear importunities, the sacrifice of a blush, for exposing such imperfect embryos.

And now we will close with a protestation, that we never desire to entertain controversy with any, or engage in any nice disputation, or heat of contradiction; and if sometimes, as a transition from one experiment to another, or upon what occasion soever, there shall be inserted any hints of speculation, we request they may be taken always for the thoughts and particular sense of some one of our members, but not imputed to the whole academy, whose sole design is to make experiments, and relate them. For such was our first proposal, and the intent of that great personage, who with his particular protection, and far-reaching judgment, caused us to take that method; to which sage and prudent advice we have still punctually and regularly conformed.

18. The Origins of the Royal Society

I[1] take its first ground and foundation to have been in London about the year 1645, if not sooner, when Dr. Wilkins,[2] (then chaplain to the Prince Elector Palatine, in London) and others, met weekly at a certain day and hour, under a certain penalty, and a weekly contribution for the charge of experiments, with certain rules agreed upon amongst us. When (to avoid diversion to other discourses, and for some other reasons) we barred all discourses of divinity, of state-affairs, and of news, other than what concerned our business of Philosophy. These meetings we removed soon after to the Bull Head in Cheapside, and in term-time to Gresham College, where we met weekly, at Mr. Foster's lecture (then Astonomy Professor there), and after the lecture ended, repaired, sometimes to Mr. Foster's lodgings, sometimes to some other place not far distant, where we continued such inquiries, and our numbers increased.

About the year 1648/49 some of our company were removed to Oxford; first, Dr. Wilkins, then I, and soon after, Dr. Goddard,[3]

SOURCE: John Wallis, *A Defence of the Royal Society* (London, 1678).

1. John Wallis (1616-1703), a Cambridge graduate, was a distinguished pure mathematician. During the 1640's he was employed in London as a chaplain and, being a strong Parliamentarian, to decipher intercepted royalist dispatches. He was appointed Savilian Professor of Geometry at Oxford in 1649, a position he kept for the remainder of his life, being confirmed in his post at the restoration of Charles II.
2. John Wilkins (1614-1672), an Oxford graduate, was like Wallis a private chaplain and a strong Parliamentarian, who married Cromwell's sister. He was appointed warden of Wadham College, Oxford, in 1648; at the Restoration he was deprived of his academic posts but given clerical preferment, finally becoming bishop of Chester in 1668. His early writings were all popular, mostly in support of the Copernican system but including some science fiction; he is best remembered for his work on a (symbolic) "Universal Language."
3. Jonathan Goddard (1617-1675) was a successful physician, much interested in chemical remedies, and also a Parliamentarian. He was warden of Merton College, Oxford, from 1651-1660, and Gresham Professor of Physic (medicine) from 1655 until his death.

whereupon our company divided. Those at London (and we, when we had occasion to be there,) met as before. Those of us at Oxford, with Dr. Ward,[4] Dr. Petty,[5] and many others of the most inquisitive persons in Oxford, met weekly (for some years) at Dr. Petty's lodgings, on the like account, to wit, so long as Dr. Petty continued in Oxford, and for some while after, because of the conveniences we had there (being the house of an apothecary) to view, and make use of, drugs and other like matters, as there was occasion.

Our meetings there were very numerous and very considerable. For, besides the diligence of persons studiously inquisitive, the novelty of the design made many to resort thither; who, when it ceased to be new, began to grow more remiss, or did pursue such inquiries at home. We did afterwards (Dr. Petty being gone for Ireland, and our numbers growing less) remove thence; and (some years before His Majesty's return) did meet at Dr. Wilkins's lodgings in Wadham College. In the meantime, our company at Gresham College being much again increased, by the accession of divers eminent and noble persons, upon His Majesty's return, we were (about the beginning of the year 1662) by his Majesty's grace and favour, incorporated by the name of the Royal Society.[6]

4. Seth Ward (1617-1689), a Cambridge graduate, was the most distinguished English theoretical astronomer of the first half of the seventeenth century. He was Savilian Professor of Astronomy, 1649-61. At the Restoration he, like Wilkins, sought ecclesiastical preferment, became bishop of Exeter (1662-67) and Salisbury (1667), and devoted himself to theology.

5. William Petty (1623-1687), now remembered as a political economist, was professor of anatomy at Oxford in 1651 and active in anatomical research. He soon left for Ireland to conduct a great land survey (from which he made his wealth) and after the Restoration worked mainly in economics.

6. The Royal Society was formally constituted by charter in July, 1662 and by the second charter in May, 1663. Its first suggestion as a chartered body was at a meeting at Gresham College after a lecture (exactly as Wallis describes) on November 28, 1660. During 1661 and early 1662 it met formally but without a charter or council.

19. The Aims of the Royal Society

The Philosophical Transactions, the oldest scientific journal, was at first a private venture of Henry Oldenburg, secretary of the Royal Society from 1662 (the first charter) until his death in 1677. He thought of his journal as a scientific newsletter. Naturally, with the permission of the Royal Society's council, he published more English news than enything else. Soon he was also publishing the more interesting part of his large correspondence; next, scientists began to write letters which were really papers intended for publication; finally, *The Philosophical Transactions* came to be taken for the official organ of the Royal Society, long before it was technically so. When Oldenburg died the Society found it necessary to persuade its secretaries to continue publication under supervision; in the latter part of the eighteenth century the Society finally made itself responsible for its publication.

The Introduction

Whereas there is nothing more necessary for promoting the improvement of Philosophical Matters, than the communicating to such, as apply their Studies and Endeavours that way, such things as are discovered or put in practice by others; It is therefore thought fit to employ the Press, as the most proper way to gratify those, whose engagement in such Studies, and delight in the Advancement of Learning and profitable Discoveries, doth entitle them to the knowledge of what this Kingdom, or other parts of the World, do, from time to time, afford, as well of the Progress of the Studies, Labors, and attempts of the Curious and Learned in things of this kind, as of their complete Discoveries and Performances: To the end, that such Productions being clearly and truly communicated, desires after solid and useful knowledge may be further entertained, ingenious Endeavours and Undertakings cherished, and those, addicted to and conversant in such Matters, may be invited and encouraged to search, try, and find out new things, impart their knowledge to one another, and contribute what they can to the Grand Design of improving Natural knowl-

SOURCE: *Philosophical Transactions: Giving some Accompt of the Present Undertakings, Studies, and Labours of the Ingenious in many considerable parts of the World,* vol. I, no. 1, March 6, 1664/5.

edge, and perfecting all Philosophical Arts and Sciences. All for the Glory of God, the Honour and Advantage of these Kingdoms, and the Universal Good of Mankind.

20. Huygens' Proposals for the Académie royale des sciences

Make experiments on the vacuum with the pump and otherwise and determine the weight of the air.

Examine the explosive force of gunpowder enclosed (in small amounts) in an iron or very thick copper box.

Examine in the same way the force of water rarified by fire.

Examine the force and speed of the wind and the uses made of it in navigation and machines.

Examine the force of percussion or the communication of motion in impact, of which I think I have been the first to give the true laws.

For the Scientific Meeting

The principal and most useful occupation of this group, in my opinion, is to work at a natural history, pretty much according to Bacon's plan. This history is composed of experiments and observations and is the sole method of arriving at the knowledge of the causes of all that is perceived in nature. As to learn the nature of weight, heat, cold, magnetic attraction, light, colors, the particles which compose the atmosphere, water, fire, and all other bodies, the purpose of animal respiration, how metals, stones, and plants grow—all things of which little or nothing is yet known, although nothing in the world is more desirable or useful to know.

It would be necessary, by pursuing different subjects, some of which I have just named, to establish the chapters of this history and to accumulate under these heads all the observations and experiments relating to each one individually; it is not necessary to expend so much effort on recounting obscure and difficult experi-

SOURCE: Christiaan Huygens, Memorandum for Colbert, translated by the editor from Joseph Bertrand, *L'Académie royale des sciences de 1666 à 1793* (Paris, 1869), pp. 8-10.

ments as on those which appear to be essential for the discovery of the object sought even though these are very ordinary.

The usefulness of such a faithfully prepared history extends to the whole of mankind and to all ages to come, because aside from the use for various purposes which may be derived from particular experiments, the collection of all these experiments is a solid foundation for the construction of a natural philosophy in which it is essential to proceed from the knowledge of effects to the knowledge of causes.

Chemistry and the dissection of animals are certainly necessary for this plan; but it is essential that the operations of these two subjects be such as tend to augment this history in some important point directed toward the discovery of something proposed, without losing time on various observations of certain circumstances, knowledge of which can have no consequence. This is so as not to incur the reproach which Seneca made to the ancient philosophers, "That they might perhaps have discovered useful things if they had not sought out superfluous ones."

It would be necessary to begin with the subjects judged most critical and most useful, assigning several at a time to various members of the meeting who will every week report and read what has been sent in. Thus the business will be very orderly and the results will undoubtedly be very great.

HUYGENS

21. *The Activities of the French Scientists*

Francis Vernon (1637-1677), an Oxford graduate, was an insatiable traveler. In 1668 his experience of foreign lands was responsible for his appointment to a special embassy to Sweden, and in 1669 he went to Paris as secretary to the English ambassador there. In Paris he showed himself more interested in science and learning than in politics and neglected his proper job in order to keep his learned friends in England—and especially the Royal Society—informed of what went on in Parisian intellectual circles. He was elected Fellow of the Royal Society (F.R.S.) in 1672 on his return to

SOURCE: Francis Vernon, Letter to Henry Oldenburg, May 1, 1669. From Royal Society MS. V, no. 5.

England, but he soon set off again on his travels, this time to the Middle East. He got as far as Persia, where he was murdered for the sake of his penknife.

Paris, May 11th, 1669

SR

My last unto you was the 4th current. In that I gave you an account of proceedings as much as I then could, since [when] I have had more time to informe my selfe, & consequently I can better esolve you That which you desired to know of the continuation of Sigre Cassini's Ephemerides of the satellites.[1] You may assure your selfe That they will bee continued. For on Munday night last I waited on him, & wee spoke concerning all things at leisure. hee tells mee That hee hath continued them already For two yeares & that hee thought to have printed them at Rome when this Invitation came to him From the King of France to bring him to Paris. hee hath them here with him, & I askt him if hee intended to publish them here. hee told mee Yes if hee settled here, butt as to that hee was not fully resolved be cause hee finds ye climate not soe propitious to him as Italy. as likewise be cause there are Instances made by the Pope for his returne, & the Nuntio hath orders to sollicitt it butt hee is cautious to doe it least ye King of France should disrelish it since hee is butt Newly arrived. Hee told mee that the Royall Academie are not as ours in Engld a great assembly of Gentlemen, Butt only a few Persons which are eminent, & not in number above 13. or 14. To whose conferences none are admitted of what quality soever who are not of their owne body. & these have likewise a pension From the King of 1500 Livres per annum, by vertue of that Membershipp. For the King will not only have a Titular butt an effectuall influence upon his royall Academie. They meet twice a weeke Wednesdays, Fridays & one day is deputed For Physicall, the other For Mathematicall exercises. In the Physicall Academie though they doe not strictly bind themselves to one Theame, Yet that at present they are examining the Doctrine of Coagulation. In the Mathematicall Academie, They are stating the Force of Air & water as to bodies they can beare, Their weight & such like enquiries. Monsr Auzout

1. Giovanni Domenico Cassini had recently been brought to France from Italy to take charge of the Royal Observatory then in course of construction. His *Ephemerides* were tables showing the expected positions of Jupiter's satellites, which, it was hoped, might be useful for longitude determination.

is lately returned From Naples to Rome at Naples hee made an acquaintance with Sigre Thomaso Cornelio,[2] who is indeed the most eminent Virtuoso there & hee writes that hee hath a piece entitled Systema Physicum, which hee will publish, which hee showed to him in manuscript of which hee renders a very advantageous Character. Rhedi de insectis is not yet come to Paris soe as to bee sold.[3] Monsr Colbert hath one butt shortly they expect it. Bullialdus is about Promoting Dr Wallis his Arithmetica infinitorum[4] & saith hee shall Demonstrate it, butt such things I must intreat You not to print because it is butt in conception & possibly hee would take it ill if one should divulge his thoughts. Which hee communicates only under the seale of Friendship & Familiarity.

At Abbot Bourdelotts Academie[5] the last thing that was discust was about Judiciary Astrology. where an Italian the Physitian to the Venetian Embassador Read a Thesis to Demonstrate the efficacy, & value of it. . . . All that can bee said, hee explained the dominion of the starres over humane bodies in a more sensible & Physicall manner then generally Those Fortune tellers (who have nothing butt old arabian Rules,) use to doe.

Those books you write of I shall make it my businesse in my next to give you some account of; at Present I am preparing to make a short voyage to Saumur whither I am to goe as I think next Monday & hope to bee back within a fort night. Then I shall begin againe to renew our correspondence as present. I must crave Your pardon For its interruption & desire the continuance of your kindnesse & affection which is extremely valuable to, honoured Sr,

Your most humble & affectionate servant

FRAN. VERNON

2. Auzout had left Paris when Cassini was appointed to the Observatory, and spent the rest of his life in Italy—a great loss to French astronomy. Cornelio was an occasional correspondent of Oldenburg's.

3. This was Francesco Redi's *Experiments Upon the Generation of Insects* (Florence, 1668), in which Redi showed that decaying meat did not generate maggots, but that these were produced from the eggs laid by flies which crawled upon the meat.

4. Bullialdus (or Ishmael Boulliaud) was a distinguished elderly astronomer and mathematician; he never completed his commentary upon Wallis' work.

5. This was one of many private "academies" or "salons" in France at this time; it had been in existence for about ten years at the time that Vernon wrote.

VI.

The Newtonian Triumph

THE AGE OF the scientific revolution was ushered in when men became uneasy with the accepted system of cosmology. It is no wonder that Newton, who put together the developments of his predecessors and added thereto his own discoveries to produce his monumental *Mathematical Principles of Natural Philosophy*, seemed to have solved the scientific problems of the age. This was especially so since Newton contributed equally to all the outstanding physical problems of his time—optics, mechanics, cosmology, and the theory of matter. In the sections below, some examples are given of the work of Newton and his predecessors.

Optics

By the mid-seventeenth century, mathematical optics was well developed. Following the lines laid down in Greek antiquity, most natural philosophers concerned with mathematical physics had made contributions to the mathematical description of such problems as the cause of the rainbow (brilliantly set out by Descartes, for example). The problem of refraction, left unsolved by the Greeks, received fresh stimulus from the invention of the telescope. The law of refraction was discovered by Willebrord Snel experimentally; it was derived mathematically by Fermat and deductively (as a consequence of his theory of matter) by Descartes who may, indeed, have discovered Snel's law independently. Some difficulty was experienced when, in 1670, Erasmus Bartholin announced the double refraction (associated, in fact, with polarization) of Iceland spar; this was partially explained geometrically by Huygens in his *Treatise on light (Traité de la lumière,* 1690).

In the later seventeenth century two main problems were at issue. One was the question of the way light was transmitted, which was intimately connected with the nature of light. Descartes (see below, Document 31) thought of light as a pressure transmitted instantly through the finest particles of matter, as for example would be the case with a line of marbles struck at one end; light was one form of matter. A sophisticated modification of this theory was Hooke's belief—similar to that of Grimaldi, who discovered experimentally the diffraction of light, characterized by the colored fringes produced when a narrow beam of light shines on a thin opaque object through a small hole—that light was transmitted as a vibratory motion. Descartes had been concerned with the medium: Hooke correctly concentrated on the light itself. He thought of light as being composed of a series of spherical pulses sent out by the light source, traveling in a narrow beam as a series of planes perpendicular to the beam. This was a very useful concept for explaining the other main problem of seventeenth-century optics (illustrated on p. 247, the production of color by refraction. Huygens virtually ignored this second problem; his marvelous geometrical exposition of the undulatory hypothesis of light could not adequately handle the problem of color.

Color and colored light was particularly explored by English scientists. Boyle wrote a whole work on the ways in which matter could modify the rays of light to produce color. His discussion of whiteness and blackness was brilliantly successful and is given below (Document 22); the rest of his *Experiments and Considerations Touching Colours* properly belongs to chemistry, for it was successful only where color changes (chiefly of vegetable solutions) were useful as chemical indicators—a point discussed by others besides Boyle (Power, for example) but generalized as a chemical principle by Boyle alone (see Document 36). Hooke's discussion was far more ingenious and displayed a far greater understanding of the correct approach to the problems; and in some points, as his observations on the colors of thin transparent plates, it was quite novel. But Hooke, though a keen experimentalist, could never distinguish between the ingeniously probably and the empirically necessary, and hence it was Newton, not Hooke, who saw that colored light is made up of white light and gave convincing experimental proof of this fact. Although Hooke was much influenced by Descartes in his views on light and matter, he was also a keen Baconian, and his is an excellent example of a Baconian experimental investigation.

Newton's discovery of dispersion (the breaking up of white light into its component colored parts by means of a prism) and its correct interpretation was made known by the communication of his views in a letter to the Royal Society in 1672, subsequently published in the *Philosophical Transactions* (Document 24) and followed by a series of letters in the 1670's. Newton demonstrated empirically to his own satisfaction (if not to Hooke's) that white light was the result of the mixing of the colors of the spectrum, and that colored objects resulted from the selective reflection and absorption of different colored rays (in a fashion analogous to Boyle's explanation of whiteness and blackness, but far more ingenious, sophisticated, and useful). In *Opticks* (first published in 1704, then enlarged in subsequent editions by further speculative "Queries", with their discussions of the nature of light and the structure of matter), Newton dealt fully with the phenomena "of light and colors"—with reflection, refraction, double refraction, diffraction—and tackled the difficult problem of the nature of light itself and its mode of propagation. He thought light was material and hence corporeal (which allowed him to dispense with an ether), propagated in a kind of pulse or wave (his "fits" of easy transmission and reflection); a curiously modern-sounding hypothesis too difficult of apprehension and handling ever to have been successfully incorporated into physical science. Eighteenth-century New-tonians asserted that light was corporeal and let it go at that.

Mechanics

The real triumph of the "new philosophy" was the development of the basic laws of mechanics, the mechanics of bodies in motion, in a mathematical manner which agreed with empirical observation. So successful were seventeenth-century scientists in this regard that they and their contemporaries could find no higher praise than to say of some scientific development that it was "truly mechanical." In this, as in so much else, Galileo had shown the way. As the selection above from the *Dialogue on the Two Chief Systems of the World* amply demonstrated, Galileo made most important advances in mechanics and had used these to support the Copernican system of cosmology. His *Discourses Upon Two New Sciences* of 1638—a far more technical and mathematical work than its predecessor—dealt principally with the strength of materials and ballistics, but a host of other problems were also discussed in passing. Galileo's greatest triumph was the discovery (as early as 1604) of the law of falling bodies; perhaps equally important was that he understood the nature of motion far better than any predecessor had ever done. He did not quite understand the nature of inertia, for he thought that motion could and did naturally proceed in a circle, rather than a straight line. This view was never popular. Descartes correctly stated that inertial motion is straight-line motion, and by 1660 this was so firmly and generally accepted that no one could believe that Galileo had not understood it too, as Newton's generous praise stated. Descartes, because of his belief that the universe is full of matter, saw the need for laws of impact motion, and most of his laws are concerned with cases of two bodies in collision. Huygens later stated these more accurately, as indeed did Christopher Wren and John Wallis. The striking difference between the laws of motion given by Descartes and Newton arise in part from the differences in their theories of matter. Newton, discarding the plenum, found that he could economize by ignoring the law of impact.

The Newtonian System

Galileo had used mechanics to support the reality of the Copernican system; Descartes in his *Principia Philosophiae* suggested the possibility of constructing a system of the world by describing the nature of matter and the laws of motion, from which all else could logically be derived. Newton, profoundly and rightly influenced by this superb insight into potentialities of mechanical thought, more firmly grounded in the empirical tradition, a greater mathematician and better physicist than Descartes, in the *Mathematical Principles of Natural Philosophy* (The *Principia*, 1687) successfully described his system of the world on the basis of definitions of matter, the laws of motion, and mathematically derived (though empirically related) propositions about the complex motion of bodies, together with the accumulated astronomical knowledge of his age. Books I and II of the *Principia* are highly mathematical and not immediately or obviously revelant to cosmological problems, but Book I especially is essential to the mathematical development of Book III, the system of the world—the Newtonian universe of the inverse square law of universal gravitational attraction. Here the planets follow their courses not according to some mystic formula, or because of their inner nature, but merely under the influence of gravity, which was so far from being an occult force that it could be seen everywhere, could be treated mathematically, and yielded up a supremely mechanical universe. The Rules of Reasoning in Natural Philosophy were Newton's justification for proceeding smoothly from the terrestrial world we can touch and experiment with to the celestial world we can only dimly see, and he saw no reason why this would not be universally acceptable. It was true that he could never exactly define what gravity might be, or even decide in his own mind what it was, but its effects were manifest and, as he had abundantly shown, could be reduced to the mathematical and mechanical laws firmly based upon external nature.

The selections from Newton's *Principia* are all given in the contemporary translation by Andrew Motte (a minor mathematician), first published in 1729, two years after Newton's death. Motte may have had access to Newton's manuscripts at some point, for his version often reflects Newton's private thoughts as these lay behind the printed text.

Optics

22. Whiteness and Blackness

The Nature of Whiteness

I come now to inquire into the nature of whiteness and blackness. Whiteness, considered as a quality in the object, seems, in general, chiefly to depend upon the roughness of the surface of the body called white; which gives it innumerable small superficies,[1] that acting like so many little specula[2] in various positions, they reflect the rays of light that fall on them, not towards one another, but externally, towards the spectator's eye. The sun, and other very lucid bodies, not only offend or dazzle our eyes; but if any colour is ascribed to them, it should be whiteness: for the sun

SOURCE: Robert Boyle, *Experiments and Considerations Touching Colours* (London, 1664), from the abridgment by Peter Shaw (1725), II, 27-30, 33-36.

1. Surfaces.
2. Mirrors.

at noonday, and in clear weather, when his face is serene, and his rays pass through a much less part of the atmosphere to our eyes, appears of a colour more approaching to white than when he is nearer the horizon; in which case the interposition of certain fumes and vapors make him oftentimes appear either red or yellow. And when the sun shines upon smooth water, that part of it which appears most illumined, seems far whiter than the rest. And I have sometimes found, that when the sun was veiled with a thin white cloud, though still too bright to be looked upon directly; that by casting my eyes upon still water, his body, being not far from the meridian, appeared to me exceedingly white. And though we vulgarly say, in English, a thing is red hot, to express a superlative degree of heat; yet, at the forges and furnaces of artificers, by a white heat they understand a further degree of ignition, than by a red one.

Common experience informs us, that as too much light overpowers the eye; so when the ground is covered with snow, those who have a weak sight, complain that this prospect is offensive to them. And even those who have good eyes, are from hence generally sensible of an extraordinary light in the air; and, if obliged to look very long upon the snow, find their sight injured by it. Thus Xenophon relates, that Cyrus marching his army, for many days, over mountains covered with snow, the dazzling splendour of its whiteness prejudiced the sight of a great number of his soldiers, and blinded some of them; and other accounts of the same nature may be met with in writers of good note. The like has also been affirmed to me by credible persons of my own acquaintance, and especially by one, who, during his stay in Muscovy, found his eyes much impaired, by being frequently obliged to travel in the snow: and this weakness of sight did not leave him when he left that country, but still continues, though he be a young man. I myself also, as well as others, have observed, that upon travelling by night, when the ground was all covered with snow, though it would otherwise have been dark, yet we could very well see to choose our way. But much more to my present purpose is that account given us by Olaus Magnus[3] of a way of travelling in the Northern regions during the winter, where the days of that season are very short. "In the day time," says he, "they travel twelve Italian miles,

3. His *De Gentibus Septentrionalibus* (Rome, 1555) is a description of life in Scandinavia.

but twice or thrice as far in the night, and that with ease; for the light of the moon, reflected by the snow, renders both hills and vales conspicuous, so that then they can see not only precipices afar off, but the wild beasts they would avoid."

This testimony I the less scruple to allege, because it agrees very well with what has been affirmed to me by a physician of Moscow; who informed me, that he could see things at a far greater distance, and with more clearness, when he travelled by night on the snow there, though without the assistance of the moonshine, than we, in these parts, would easily believe. 'Tis true, indeed, the intenseness of the cold might contribute something to the considerableness of the effect, by clearing the air of dark steams, which, in these more temperate climates, are usually thick in snowy weather: for this physician, and the ingenious navigator Captain James[4] agree, that in dark frosty nights, in frozen climates, they could discover more stars, and have a clearer prospect of the heavens, than we in England. I know, indeed, many learned men suppose snow thus strongly affects our eyes, not by a borrowed but a native light. I venture, however, to give it as a proof, that white bodies reflect more light than others, because having once placed a parcel of snow in a room, carefully darkened, that no celestial light might fall upon it, neither I, nor an ingenious person skilled in optics, could find it had any light besides what it received, and, 'tis usual, among such as travel in dark nights, to make their guides wear something of white to be discerned by; for there is scarce any night so dark, but that, in the free air, some light remains, though broken and debilitated, perhaps, by a thousand reflections from the opaque corpuscles, that swim in the atmosphere, and send it to one another before it arrives at the eye.

And the better to show that white bodies reflect much more light than others, I held in the darkened room, formerly mentioned, not far from the hole at which the light entered, a sheet of white paper; from whence casting the sun beams upon a white wall, it manifestly appeared both to me, and to a person I took to be witness of the experiment, to reflect a far greater light than any of the other colours; for the wall itself was not only thus notably enlightened, but also a considerable part of the room. And, fur-

---

4. Thomas James. An account of his search for a Northwest Passage, during which he discovered James's Bay and wintered (1631-32) north of Hudson's Bay, was published in 1633.

ther, to show that white bodies reflect the rays outwards, let me add, that ordinary burning glasses will not, in a great while, burn or discolour white paper, so that when I was a boy, and delighted to make trials with such glasses, I could not but wonder at this odd phenomenon, which set me very early upon guessing at the nature of whiteness; and the more, because I took notice that the image of the sun, upon white paper, was not so well defined as upon black; and because that, upon inking over the paper, the moisture would be quickly dried up, and the paper, that I could not burn before, would now presently take fire. I have also tried that by exposing my hand, with a thin black glove on it, to the warm sun, it would thereby very suddenly be more considerably heated, than if I took off the glove, and held my naked hand to its rays, or put on another glove of thin white leather. . . .

The Nature of Blackness

What we have said of whiteness may assist us to form a notion of blackness; those two qualities being sufficiently opposite to illustrate each other. And as that which makes a body white, is chiefly such a disposition of its parts, as disposes it to reflect more of the light that falls on it, than bodies of different colours; so that which renders a body black, is principally a peculiar kind of texture of its superficial particles; whereby it damps the light that falls on it, so that very little is reflected to the eye.

This texture is explicable two several ways; and first, by supposing, in the superficies of the black body, a particular kind of asperity; whence the superficial particles reflect few of the incident rays outwards, and the rest inwards, upon the body itself; as if, for instance, the surface of a black body should rise up in numberless little cylinders, pyramids, cones, etc., which, by being thick set and erect, throw the rays of light from one to another inwards, so often, that, at length, they are lost before they can come out again to the eye. The other way supposes the textures of black bodies, either to yield to the rays of light, or, upon some other account, to stifle and keep them from being reflected in any number, or with any considerable vigour outwards. According to this notion it may be said, that the corpuscles, which compose the rays of light, thrusting one another from the lucid body, and falling on black substances, meet with such a texture, that they receive into themselves, and retain almost all the motion communicated to them by those corpuscles, and consequently reflect but few of them or those but lanquidly, towards the eye; as when a ball, thrown against a floor, rebounds a great way upwards, but very little or not at all when thrown against mud or water; because the parts yield, and receive into themselves the motion which should reflect the ball back. But this last manner of accounting for blackness I barely propose, without either adopting, or absolutely rejecting it; for the hardness of touchstones, black marble, and of other bodies that are black, and solid, seems to render it somewhat improbable that they should be of so yielding a nature; unless we say that some bodies may be more disposed to yield to the impulses of the corpuscles of light, by reason of a peculiar texture, than others, which, by particular trials, appear to be softer than they. Both the solutions, however, agree in this, that black bodies reflect but little of the light which falls on them. And it is not impossible that, in some cases, both the disposition of the superficial particles as to figure and position, and the yielding of the body, or some of its parts, may jointly contribute to render a body black. The considerations which induced me to propose this notion of blackness are principally these.

Whiteness and blackness being generally reputed contrary qualities, whiteness depending, as I said, upon the disposition of the parts of the body, to reflect light plentifully, it seemed probable, that blackness might depend upon a contrary disposition of surface; but upon this I will not insist. However, if a body, of an uniform colour, be placed, part in the sunbeams and part in the shade, that part, which is not illumined, will appear nearer allied to blackness than the other, from which more light is reflected to the eye; dark colours also seem blacker, the less is the light they are viewed in: and all things seem black in the dark, when they yield no rays to make impressions upon our organs of sight; so that shadow and darkness are near akin; and shadow, we know, is but a privation of light; blackness, accordingly, seems to proceed from the want of rays reflected from the black body to the eye; though the bodies we call black, as marble, jet, etc. are not perfectly so, for if they were, we should not see them at all. But nothwithstanding the rays which fall on the sides of those erect particles we mentioned, do few of them return outwards; yet such as fall upon the points of those cylinders, cones, or pyramids, may be thence reflected to the eye, though they make but a faint impression there; because they are mixed with a great proportion of little shades. Thus, having procured a large piece of black marble to be well polished, and brought to the form of a large spherical concave speculum, the inside thereof was a kind of dark looking glass, wherein I could plainly see a little image of the sun, when it shone thereon. But this image was very far from offending or dazzling my eyes, as it would have done from another speculum, and though this were large, I could not, in a long time, set a piece of wood on fire with it; though a far less speculum of the same form, and of a more reflecting material, would presently have made it flame. And having exposed to the sun a pretty large mortar of white marble, polished on the inside, we found that it reflected a great quantity of glaring light, but so scattered that we could not make the reflected rays meet in any such conspicuous focus as that we observed in the black marble; though by holding a candle, in the night time, at a convenient distance, we were able to procure a concourse of a few reflected rays at about two inches distance from the bottom of the mortar. But we found the heat of the sunbeams so dispersedly reflected, to be very languid, even compared to the focus of the black marble; and the little picture of the sun that appeared upon the white marble as a speculum, was very faint, and exceedingly ill defined.

Taking two pieces, the one of black, and the other of white marble, whose surfaces were plain and polished; and casting on them successively the rays of the same candle, in such a manner that the adjacent superficies being shaded by an opaque and perforated body, the incident rays passed through a round hole of about half an inch in diameter, the circle of light that appeared on the white marble was, in comparison, very bright, but very ill defined; whilst that on the black marble was far less luminous, but much better defined.

When we look upon a piece of linen that has small holes in it, they appear very black; so that men are often deceived in taking holes for spots of ink: and painters, to represent holes, make use of black; the reason whereof seems to be, that the rays which fall on those holes, penetrate so deep, that none are reflected back to the eye. And in a narrow well, part of the mouth seems black, because the incident rays are reflected downwards, from one side to another, till they can no longer rebound to the eye. We may consider too, that if different parts of the same piece of black velvet be stroked opposite ways, there will appear two distinct kinds of blackness, the one far darker than the other; probably because in the less obscure part of the velvet, the little silken piles, whereof, tis composed, being inclined, there is a greater part of each of them turned to the eye; whilst in the other part the piles of silk being more erect, there are by far fewer rays sent outwards from the lateral parts of each pile; so that most of those reflected to the eye, come from the tops of the piles, which make but a small part of the whole superficies of the velvet. This explanation I propose, not that I think the blackness of the velvet proceeds from the cause assigned; since each single pile of silk is black by reason of its texture, in what position soever it be viewed; but because the greater blackness of a single tuft seems to proceed from the greater defect of rays reflected thence, and from the want of those parts of a surface that reflects rays, and the multitude of those shaded parts that reflect none. And I have often observed, that the position of particular bodies, far greater than piles of silk, may, notwithstanding each of them hath a colour of its own, make one part of their aggregate appear far darker than another. Thus a heap of carrots appear of a much darker colour when viewed with their points, than with their sides obverted to the eye.

I have observed in a darkened room, that if the sunbeams which came in at the hole were received upon white, or any other colour, and directed to a convenient part of the room, they would manifestly increase the light of that part, but if we substituted either a piece of black cloth or black velvet, it would so damp the incident rays, that the said place would be less illumined than before, when it received its light only from the weak and oblique reflections of the floor and walls of a pretty large room; over which the beams that came in at the hole, were confusedly and in a broken manner dispersed.

And to show that the rays which fall on black bodies, as they do not rebound outwards to the eye, so they are reflected towards the body itself, as the nature of those erect particles to which we have imputed blackness requires; we shall add an experiment that will at the same time confirm our doctrine of whiteness. We took, then, a broad and large tile, and having whited over one half of its superficies, and blackened the other, we exposed it to the summer sun. And having let it lie there a convenient time, we found that whilst the whited part of the tile remained cool, the blacked part of it was grown very hot. And for further satisfaction, we have sometimes left upon the surface of the tile a part that retained its native red; and exposing all to the sun, we observed the latter to have contracted a heat in comparison of the white part, but inferior to that of the black. 'Tis also remarkable, that rooms hung with black are not only darker than they would be otherwise, but warmer too. I have known a great lady of a tender constitution, complain that she commonly took cold upon going into the air, after having made any long visit to persons whose rooms were hung with black. And this is not the only lady I have heard complain of the warmth of such rooms; which, though perhaps it may partly be imputed to the effluvia of those materials wherewith the hanging were dyed, yet probably the warmth in this case depends chiefly upon the same cause with darkness; for upon exposing two pieces of silk, the one white, the other black, in the same window to the sun, I have often found the former considerable heated, when the latter has remained cool.

A virtuoso of unsuspected credit acquainted me, that in a hot climate he had, by carefully blacking the shells of eggs, and exposing them to the sun, seen them thereby well roasted, in no long time. But in England, the sun's rays seem not to be sufficiently strong to produce such an effect; for having exposed eggs in the summer season thereto, they acquired indeed a considerable degree of heat, but not enough to roast them.[5]

23. The Color of Thin Films

Observation IX.
Of the Colours Observable in Muscovy-Glass, and Other Thin Bodies

Muscovy-glass or Lapis Specularis [mica] is a Body that seems to have as many Curiosities in its Fabric as any common Mineral I have met with: for first, it is transparent to a great thickness: Next, it is compounded of an infinite number of thin flakes joined

SOURCE: Robert Hooke, *Micrographia* (London, 1665), pp. 47-51, 55-57, 64-65.
5. A series of (mainly chemical) experiments follow.

or generated one upon another so close and smooth, as with many hundreds of them to make one smooth and thin Plate of a transparent flexible substance, which with care and diligence may be slit into pieces so exceedingly thin as to be hardly perceivable by the eye, and yet even those, which I have thought the thinnest, I have with a good Microscope found to be made up of many other Plates, yet thinner; and it is probable, that, were our Microscopes much better, we might much further discover its divisibility. Nor are these flakes only regular as to the smoothness of their Surfaces; but thirdly, in many Plates they may be perceived to be terminated naturally with edges of the figure of a Rhomboid. This Figure is much more conspicuous in our English talc, much whereof is found in the Lead Mines, and is commonly called Spar, and Kauck, which is of the same kind of substance with the Selenites, but is seldom found in so large flakes as that is, nor is it altogether so tough, but is much more clear and transparent, and much more curiously shaped, and yet may be cleft and flaked like the other Selenites.[1] But fourthly, this stone has a property, which in respect of the Microscope, is more notable, and that is, that it exhibits several appearances of Colours, both to the naked Eye, but much more conspicuously to the Microscope; for the exhibiting of which, I took a piece of Muscovy-glass, and splitting or cleaving it into thin Plates, I found that up and down in several parts of them I could plainly perceive several white specks or flaws, and others diversely coloured with all the Colours of the Rainbow; and with the Microscope I could perceive, that these Colours were ranged in rings that incompassed the white speck or flaw, and were round or irregular, according to the shape of the spot which they terminated; and the position of Colours, in respect of one another, was the very same as in the Rainbow. The consecution [sequence] of those Colours from the middle of the spot outward being Blue, Purple, Scarlet, Yellow, Green; Blue, Purple, Scarlet, and so onwards, sometimes half a score repeated, that is, there appeared six, seven, eight, nine or ten several coloured rings or lines, each incircling the other, in the same manner as I have often seen a very vivid Rainbow to have four or five several Rings of Colours, that is, accounting all the Gradations between Red and Blue for one: But the order of the Colours in these Rings was quite contrary to

1. Spar or calcite. Calcium carbonate, which does indeed have rhomboidal crystals.

the primary or innermost Rainbow, and the same with those of the secondary or outermost Rainbow; these coloured Lines or Irises, as I may so call them, were some of them much brighter than others, and some of them also very much broader, they being some of them ten, twenty, nay, I believe near a hundred times broader than others; and those usually were broadish which were nearest the center of middle of the flaw. And oftentimes I found, that these Colours reached to the very middle of the flaw, and then there appeared in the middle a very large spot, for the most part, all of one colour, which was very vivid, and all the other Colours incompassing it, gradually ascending, and growing narrower towards the edges, keeping the same order, as in the secondary Rainbow, that is, if the middle were Blue, the next incompassing it would be a Purple, the third a Red, the fourth a Yellow, &c. as above; if the middle were a Red, the next without it would be a Yellow, the third a Green, the fourth a Blue, and so onward. And this order it always kept whatsoever were the middle Colour.

There was further observable in several other parts of this Body, many Lines or Threads, each of them of some one peculiar Colour, and those so exceedingly bright and vivid, that it afforded a very pleasant object through the Microscope. Some of these threads I have observed also to be pieced or made up of several short lengths of differently coloured ends (as I may so call them) as a line appearing about two inches long through the Microscope, has been compounded of about half an inch of a Peach colour, 1/8 of a lovely Grass-green, 3/4 of an inch more of a bright Scarlet, and the rest of the line of a Watchet-blue. Others of them were much otherwise coloured; the variety being almost infinite. Another thing which is very observable, is, that if you find any place where the colours are very broad and conspicuous to the naked eye, you may, by pressing that place with your finger, make the colours change places, and go from one part to another.[2]

There is one Phaenomenon more, which may, if care be used, exhibit to the beholder, as it has divers times to me, an exceeding pleasant, and not less intrusive Spectacle; And that is, if curiosity and diligence be used, you may so split this admirable Substance, that you may have pretty large Plates (in comparison of those

2. Pressure will produce the colored patterns of Newton's rings, the diffraction patterns also found in thin films.

smaller ones which you may observe in the Rings) that are perhaps an 1/8 or a 1/6 part of an inch over, each of them appearing through the Microscope most curiously, entirely, and uniformly adorned with some one vivid colour: this, if examined with the Microccope, may be plainly perceived to be in all parts of it equally thick. Two, three, or more of these lying one upon another, exhibit oftentimes curious compounded colours, which produce such a Compositum, as one would scarce imagine should be the result of such ingredients: As perhaps a faint yellow and a blue may produce a very deep purple. But when anon we come to the more strict examination of these Phaenomena, and to inquire into the causes and reasons of these productions, we shall, I hope, make it more conceivable how they are produced, and show them to be no other than the natural and necessary effects arising from the peculiar union of concurrent causes.

These Phaenomena being so various, and so truly admirable, it will certainly be very well worth our inquiry, to examine the causes and reasons of them, and to consider, whether from these causes demonstratively evidenced, may not be deduced the true causes of the production of all kinds of Colours. And I the rather now do it, instead of an Appendix or Digression to this History, than upon the occasion of examining the Colours in Peacocks, or other Feathers, because this Subject, as it does afford much better ways of examining each circumstance. And this will be made manifest to him that considers, first, that this laminated body is more simple and regular than the parts of Peacocks' feathers, this consisting only of an indefinite number of plain and smooth Plates, heaped up, or incumbent on each other. Next, that the parts of this body are much more manageable, to be divided or joined, than the parts of a Peacock's feather, or any other substance that I know. And Thirdly, because that in this we are able from a colourless body to produce several coloured bodies, affording all the variety of Colours imaginable: And several others, which the subsequent Inquiry will make manifest.

To begin, therefore, it is manifest from several circumstances, that the material cause of the apparition of these several Colours, is some Lamina or Plate of a transparent or pellucid body of a thickness very determinate and proportioned according to the greater or less refractive power of the pellucid body. And that this is so, abundance of Instances and particular Circumstances will make manifest.

As first, if you take any small piece of the Muscovy-glass, and with a Needle or some other convenient Instrument, cleave it oftentimes into thinner and thinner Laminae, you shall find, that till you come to a determinate thinness of them, they shall all appear transparent and colourless, but if you continue to split and divide them further, you shall find at last, that each Plate after it comes to such a determinate thickness, shall appear most lovely tinged or imbued with a determinate colour. If further, by any means you so flaw a pretty thick piece, that one part does begin to cleave a little from the other, and between those two there be by any means gotten some pellucid medium, those laminated pellucid bodies that fill that space, shall exhibit several Rainbows or coloured Lines, the colours of which will be disposed and ranged according to the various thicknesses of the several parts of that Plate. That this is so, is yet further confirmed by this experiment.

Take two small pieces of ground and polished Looking-glass-plate, each about the bigness of a shilling, take these two dry, and with your forefingers and thumb press them very hard and close together, and you shall find, that when they approach each other very near, there will appear several Irises or coloured Lines, in the same manner almost as in the Muscovy-glass; and you may very easily change any of the Colours of any part of the interposed body, by pressing the Plates closer and harder together, or leaving them more lax; that is, a part which appeared coloured with a red, may be presently tinged with a yellow, blue, green, purple, or the like, by altering the appropinquation [nearness] of the terminating Plates. Now that air is not necessary to be the interposed body, but that any other transparent fluid will do much the same, may be tried by wetting those approximated Surfaces with Water, or any other transparent liquor, and proceeding with it in the same manner as you did with the Air; and you will find much the like effect, only with this difference, that those compressed bodies, which differ most, in their refractive quality, from the compressing bodies, exhibit the most strong and vivid tinctures. Nor is it necessary, that this laminated and tinged body should be of a fluid substance, any other substance, provided it be thin enough and transparent, doing the same thing: this the Laminae of our

Moscovy-glass hint; but it may be confirmed by multitudes of other instances.

And first, we shall find, that even Glass itself may, by the help of a Lamp, be blown thin enough to produce these Phaenomena of Colours: which Phaenomena accidentally happening, as I have been attempting to frame small Glasses with a lamp, did not a little surprise me at first, having never heard or seen anything of it before; though afterwards comparing it with the Phaenomena I had often observed in those Bubbles which Children use to make with Soap-water, I did the less wonder; especially when upon Experiment, I found, I was able to produce the same Phaenomena in thin Bubbles made with any other transparent Substance.[3] Thus I have produced them with Bubbles of Pitch, Rosin, Colophony [rosin], Turpentine, Solutions of several Gums, as Gum Arabic in water, any glutinous liquor, as Wort, Wine, Spirit of Wine [alcohol], oil of turpentine, glare of snails, etc. . . .

It would be somewhat too long a work for this place Zetetically [by way of inquiry] to examine, and positively to prove, what particular kind of motion it is that must be the efficient of Light; for though it be a motion, yet 'tis not every motion that produces it, since we find there are many bodies very violently moved, which yet afford not such an effect; and there are other bodies, which to our senses seem not moved so much, which yet shine. Thus water and quicksilver and most other liquors heated, shine not; and several hard bodies, as Iron, Silver, Brass, Copper, Wood, etc. though very often struck with a hammer, shine not presently [immediately], though they will all of them grow exceeding hot; whereas rotten wood, rotten fish, seawater, glow-worms, etc. have nothing of tangible heat in them, and yet (where there is no stronger light to affect the Sensory) they shine some of them so vividly, that one may make a shift to read by them.

It would be too long, I say, here to insert the discursive progress by which I inquired after the properties of the motion of light, and therefore I shall only add the result.

And, first, I found it ought to be exceeding quick, such as those motions of fermentation and putrefaction, whereby, certainly, the

3. The colors of soap bubbles are described in Boyle's *Experiments and Considerations Touching Colours*, though only cursorily; Boyle was interested in the pattern of the matter which could alter light to produce color rather than, like Hooke, in any alteration in light.

parts are exceeding nimbly and violently moved; and that, because we find these motions are able more minutely to shatter and divide the body, than the most violent heats or menstruums we yet know. And that fire is nothing else but such a dissolution of the burning body, made by the most universal menstruum of all sulphureous bodies, namely, the air, we shall in another place than this Tractate endeavour to make probable. And that in all extremely hot shining bodies there is a very quick motion that causes light, as well as a more robust that causes heat, may be argued from the celerity wherewith the bodies are dissolved.

Next, it must be a vibrative motion. And for this the newly mentioned Diamond affords us a good argument, since if the motion of the parts did not return, the Diamond must, after many rubbings, decay and be wasted; but we have no reason to suspect the latter, expecially if we consider the exceeding difficulty that is found in cutting or wearing away a Diamond. And a circular motion of the parts is much more improbable, since, if that were granted, and they be supposed irregular and angular parts, I see not how the parts of the Diamond should hold so firmly together, or remain in the same sensible dimensions, which yet they do. Next, if they be globular, and moved only with a turbinated [top-like] motion, I know not any cause that can impress that motion upon the pellucid medium, which yet is done. Thirdly, any other irregular motion of the parts one amongst another, must necessarily make the body of a fluid consistence, from which it is far enough. It must therefore be a vibrating motion.

And thirdly, that it is a very short vibrating motion, I think the instances drawn from the shining of Diamonds will also make probable. For a Diamond being the hardest body we yet know in the world, and consequently the least apt to yield or bend, must consequently have also its vibrations exceeding short.

And these, I think, are the three principal properties of a motion, requisite to produce the effect called light in the Object.

The next things we are to consider, is the way or manner of the trajection of this motion through the interposed pellucid body to the eye. And here it will be easily granted:

First, that it must be a body susceptible and impartible of this motion that will deserve the name of a Transparent. And next, that the parts of such a body must be homogeneous, or of the same kind. Thirdly, that the constitution and motion of the parts must be such, that the appulse of the luminous body may be communicated or propagated through it to the greatest imaginable distance in the least imaginable time; though I see no reason to affirm, that it must be in an instant: for I know not any one experiment or observation that does prove it. And whereas it may be objected, that we see the Sun risen at the very instant when it is above the sensible horizon, and that we see a star hidden by the body of the Moon at the same instant, when the Star, the Moon, and our Eye are all in the same line; and the like observations, or rather suppositions may be urged, I have this to answer; that I can as easily deny as they affirm; for I would fain know, by what means, anyone can be assured any more of the affirmative, than I of the negative. If indeed the propagation were very slow, 'tis possible something might be discovered by eclipses of the Moon; but though we should grant the progress of the light from the Earth to the Moon, and from the Moon back to the Earth again to be full two minutes in performing, I know not any possible means to discover it; nay, there may be some instances perhaps of horizontal eclipses that may seem very much to favour this supposition of the slower progression of light than most imagine. And the like may be said of eclipses of the Sun, etc. But of this only by the by. Fourthly, that the motion is propagated every way through a homogeneous medium by direct or straight lines extended every way like rays from the center of a sphere. Fifthly, in a homogeneous medium this motion is propagated every way with equal velocity, whence necessarily every pulse of vibration of the luminous body will generate a sphere, which will continually increase, and grow bigger, just after the same manner (though indefinitely swifter) as the waves or rings on the surface of the water do swell into bigger and bigger circles about a point of it, where, by the sinking of a stone, the motion was begun, whence it necessarily follows, that all the parts of these spheres undulated through a homogeneous medium cut the rays at right angles.

But because all transparent mediums are not homogeneous to one another, therefore we will next examine how this pulse or motion will be propagated through differingly transparent mediums. And here, according to the most acute and excellent philosopher Descartes, I suppose the sine of the angle of inclination in the first medium to be to the sine of refraction in the second, as the density of the first, to the density of the second. By density, I mean not the density in respect of gravity (with which the refractions or transparency of mediums hold no proportion) but in respect only to the trajection [transmission] of the rays of light, in which respect they only differ in this: that the one propagates the pulse more easily and weakly, the other more slowly, but more strongly. But as for the pulses themselves, they will by the refraction acquire another property, which we shall now endeavour to explicate. . . .

[Hooke goes on to try to work out a geometrical theory of color, a "two-color" theory in which the other colors were produced by a mixture of red and blue, supposing light to be transmitted in pulses.]

From the considerations of the properties of these impressions, we may collect these short definitions of Colours: That Blue is an impression on the Retina of an oblique and confused pulse of light, whose weakest part precedes, and who strongest follows. And that Red is an impression on the retina of an oblique and confused pulse of light, whose strongest part precedes, and whose weakest follows.

Which properties, as they have been already manifested in the prism and falling drops of rain to be the causes of the colours there generated, may be easily found to be the efficients also of the colours appearing in thin laminated transparent bodies; for the explication of which, all this has been premised.

And that this is so, a little closer examination of the Phaenomena and the figure of the body, by this hypothesis, will make evident.

For first (as we have already observed) the laminated body must be of a determinate thickness, that is, must not be thinner than such a determinate quantity; for I have always observed, that near the edges of those which are exceeding thin, the colours disappear, and the part grows white; nor must it be thicker than another determinate quantity; for I have likewise observed, that beyond such a thickness, no colours appeared, but the plate looked white, between which two determinate thicknesses were all the coloured rings; of which in some substances I have found ten or twelve, in others not half so many, which I suppose depends much upon the transparency of the laminated body. Thus though the consecutions are the same in the scum or the skin on the top of metals, yet in those consecutions the same colour is not so often repeated as in the consecutions in thin glass, or in soap-water, or any other more transparent and glutinous liquor; for in these I have observed, red, yellow, green, blue, purple; red, yellow, green, blue, purple; red, yellow, green, blue, purple; red, yellow, etc. to succeed each other, ten or twelve times, but in the other more opaque bodies, the consecutions will not be half so many.

And therefore, secondly, the laminated body must be transparent, and this I argue from this, that I have not been able to produce any colour at all with an opaque body, though never so thin. And this I have often tried by pressing a small globule of mercury between two smooth plates of glass, whereby I have reduced that body to a much greater thinness than was requisite to exhibit the colours with a transparent body.

Thirdly, there must be a considerable reflecting body adjacent to the under or further side of the lamina or plate; for I always found, that the greater the refraction was, the more vivid were the appearing colours.

From which Observations, it is most evident, that the reflection from the under or further side of the body is the principal cause of the production of these colours. . . .

[Hooke concludes by giving a drawing of a thin plate seen under the microscope and again attempting to demonstrate geometrically how the ray is transmitted.]

24. Newton on Light and Colors.

---

*A Letter of Mr. Isaac Newton, Professor of Mathematics in the
University of Cambridge; Containing His New Theory About
Light and Colours:
Sent by the Author to the Publisher from Cambridge,
Febr. 6. 1671/72; in Order
To Be Communicated to the R. Society*

SIR,

To perform my late promise to you, I shall without further
ceremony acquaint you, that in the beginning of the year 1666 (at
which time I applied myself to the grinding of optic glasses of
other figures than spherical,) I procured a triangular glass prism, to
try therewith the celebrated phænomena of colours. And for that
purpose having darkened my chamber, and made a small hole in
my window shuts, [shutters], to let in a convenient quantity of
the sun's light, I placed my prism at his entrance, that it might be
thereby refracted to the opposite wall.[1] It was at first a very
pleasing diversion to view the vivid and intense colours produced
thereby; but after a while applying myself to consider them more
circumspectly, I was surprised to see them in an oblong form;
which according to the received laws of refraction, I expected
would have been circular.[2] They were terminated at the sides with
straight lines, but at the ends, the decay of light was so gradual, that

SOURCE: *Philosophical Transactions* (1672), in the abridgment by Charles
Hutton, George Shaw, and Richard Pearson (17 vols., London 1809),
I, 678-88.

1. Newton's extraordinary genius for experiment, combined with the luck
   inherent in all good experimental science, is abundantly plain here. Not
   everyone who experimented with prisms before him had used a darkened
   room or a narrow beam of light, and few cast the spectrum on a screen far
   enough from the prism for the phenomenon of dispersion to appear
   clearly.
2. This elongation appears to have had a great influence on Newton, who
   felt it required detailed investigation. Such investigation led him to conclude that the differently colored rays must diverge, and that therefore
   any one ray must be bent more or less than the other rays.

it was difficult to determine justly what was their figure; yet they seemed semicircular.

Comparing the length of this coloured spectrum with its breadth, I found it about five times greater; a disproportion so extravagant, that it excited me to a more than ordinary curiosity of examining from whence it might proceed. I could scarce think, that the various thickness of the glass, or the termination with shadow or darkness, could have any influence on light to produce such an effect; yet I thought it not amiss, first to examine those circumstances, and so tried what would happen by transmitting light through parts of the glass of divers thicknesses, or through holes in the window of divers sizes, or by setting the prism without, so that the light might pass through it, and be refracted before it was terminated by the hole: but I found none of those circumstances material. The fashion of the colours was in all these cases the same.

Then I suspected, whether by any unevenness in the glass, or other contingent irregularity, these colours might be thus dilated. And to try this, I took another prism like the former, and so placed it, that the light passing through them both, might be refracted contrary ways, and so by the latter returned into that course from which the former had diverted it. For, by this means, I thought the regular effects of the first prism would be destroyed by the second, but the irregular ones more augmented, by the multiplicity of refractions. The event was, that the light, which by the first prism was diffused into an oblong form, was by the second reduced into an orbicular one, with as much regularity as when it did not at all pass through them. So that, whatever was the cause of that length, it was not any contingent irregularity.

I then proceeded to examine more critically, what might be effected by the difference of the incidence of rays coming from divers parts of the sun; and to that end measured the several lines and angles, belonging to the image. Its distance from the hole or prism was 22 feet; its utmost length 13 ¼ inches; its breadth 2 $^5/_8$; the diameter of the hole ¼ of an inch; the angle, which the rays, tending towards the middle of the image, made with those lines in which they would have proceeded without refraction, was 44° 56´. And the vertical angle of the prism, 63° 12´. Also the refractions on both sides the prism, that is, of the incident and emergent rays, were as near as I could make them equal, and consequently about 54° 4′. And the rays fell perpendicularly upon the wall. Now subducting [subtracting] the diameter of the hole from the length and breadth of the image, there remains 13 inches the length, and 2 ³/₈ the breadth, comprehended by those rays, which passed through the centre of the said hole, and consequently the angle of the hole, which that breadth subtended, was about 31′, answerable to the sun's diameter; but the angle which its length subtended, was more than five such diameters, namely 2° 49′.

Having made these observations, I first computed from them the refractive power of that glass, and found it measured by the ratio of the sines, 20 to 31. And then, by that ratio, I computed the refractions of two rays flowing from opposite parts of the sun's discus, so as to differ 31′ in their obliquity of incidence, and found that the emergent rays should have comprehended an angle of about 31′, as they did, before they were incident. ³ But because this computation was founded on the hypothesis of the proportionality of the sines of incidence and refraction, which though, by my own experience, I could not imagine to be so erroneous as to make that angle but 31′, which in reality was 2° 49′; yet my curiosity caused me again to take my prism. And having placed it at my window, as before, I observed, that by turning it a little about its axis to and fro, so as to vary its obliquity to the light, more than an angle of 4 or 5 degrees, the colours were not thereby sensibly translated from their place on the wall, and consequently by that variation of incidence, the quantity of refraction was not sensibly varied. By this experiment therefore, as well as by the former computation, it was evident, that the difference of the incidence of rays, flowing from divers parts of the sun, could not make them, after a decussation [intersection], diverge at a sensibly greater angle, than that at which they before converged; which being at most but about 31 or 32 minutes, there still remained some other cause to be found out, from whence it could be 2° 49′.

Then I began to suspect whether the rays, after their trajection through the prism, did not move in curve lines, and according to their more or less curvity tend to divers parts of the wall. And it increased my suspicion, when I remembered that I had often seen a tennis ball, struck with an oblique racket, describe such a curve

3. I.e., using Snel's law of refraction.

line. For, a circular as well as a progressive motion being communicated to it by that stroke, its parts on that side, where the motions conspire, must press and beat the contiguous air more violently than on the other, and there excite a reluctancy and reaction of the air proportionably greater. And for the same reason, if the rays of light should possibly be globular bodies, and by their oblique passage out of one medium into another acquire a circulating motion, they ought to feel the greater resistance from the ambient aether, on that side where the motions conspire, and thence be continually bowed to the other. But notwithstanding this plausible ground of suspicion, when I came to examine it, I could observe no such curvity in them. And besides (which was enough for my purpose) I observed, that the difference between the length of the image and diameter of the hole, through which the light was transmitted, was proportionable to their distance.

The gradual removal of these suspicions, at length led me to the experimentum crucis,[4] which was this: I took two boards, and placed one of them close behind the prism at the window, so that the light might pass through a small hole, made in it for the purpose, and fall on the other board, which I placed at about 12 feet distance, having first made a small hole in it also, for some of that incident light to pass through. Then I placed another prism behind this second board, so that the light trajected through both the boards, might pass through that also, and be again refracted before it arrived at the wall. This done, I took the first prism in my hand, and turned it to and fro slowly about its axis, so much as to make the several parts of the image, cast on the second board, successively pass through the hole in it, that I might observe to what places on the wall the second prism would refract them. And I saw, by the variation of those places, that the light tending to that end of the image, towards which the refraction of the first prism was made, did in the second prism suffer a refraction considerably greater then the light tending to the other end. And so the true cause of the length of that image was detected to be no other, than that light consists of rays differently refrangible, which, without any respect to a difference in their incidence, were according to their degrees of refrangibility, transmitted towards divers parts of the wall.

4. Like Hooke, Newton uses this Baconian expression to mean "crucial experiment"; it is indeed an important one.

When I understood this, I left off my aforesaid glass works; for I saw, that the perfection of telescopes was hitherto limited, not so much for want of glasses truly figured according to the prescriptions of optic authors, (which all men have hitherto imagined,) as because that light itself is a heterogeneous mixture of differently refrangible rays.[5] So that, were a glass so exactly figured, as to collect any one sort of rays into one point, it could not collect those also into the same point, which having the same incidence upon the same medium are apt to suffer a different refraction. Nay, I wondered, that seeing the difference of refrangibility was so great, as I found it, telescopes should arrive to that perfection they are now at. For measuring the refractions in one of my prisms, I found, that supposing the common sine of incidence upon one of its planes was 44 parts, the sine of refraction of the utmost rays on the red end of the colours, made out of the glass into the air, would be 68 parts, and the sine of refraction of the utmost rays on the other end 69 parts: so that the difference is about a 24th or 25th part of the whole refraction; and consequently, the object glass of any telescope cannot collect all the rays which come from one point of an object, so as to make them convene at its focus in less room than in a circular space, whose diameter is the 50th part of the diameter of its aperture; which is an irregularity, some hundreds of times greater than a circularly figured lens, of so small a section as the object glasses of long telescopes are, would cause by the unfitness of its figure, were light uniform.

This made me take reflections into consideration, and finding them regular, so that the angle of reflection of all sorts of rays was equal to their angle of incidence; I understood that by their mediation optic instruments might be brought to any degree of perfection imaginable, provided a reflecting substance could be found, which would polish as finely as glass, and reflect as much light as glass transmits, and the art of communicating to it a parabolic figure be also attained. But there seemed very great difficulties, and I have almost thought them insuperable, when I further considered, that every irregularity in a reflecting superficies [surface]

5. Newton was of course wrong to think it impossible to make achromatic lenses; they were to be successfully manufactured in the late eighteenth century. But his error was a fortunate one since it led him to develop the reflecting telescope of which he sent a model to the Royal Society in 1671.

makes the rays stray 5 or 6 times more out of their due course, than the like irregularities in a refracting one: so that a much greater curiosity would be here requisite, than in figuring glasses for refraction.

Amidst these thoughts I was forced from Cambridge by the intervening plague, and it was more than two years before I proceeded further.[6] But then having thought on a tender way of polishing, proper for metal, whereby as I imagined, the figure also would be corrected to the last; I began to try what might be effected in this kind, and by degrees so far perfected an instrument (in the essential parts of it like that I sent to London,) by which I could discern Jupiter's 4 concomitants [satellites], and showed them divers times to two others of my acquaintance. I could also discern the moon-like phase of Venus, but not very distinctly, nor without some niceness in disposing the instrument.

From that time I was interrupted till this last autumn, when I made the other. And as that was sensibly better then the first (especially for day objects,) so I doubt not, but they will be still brought to a much greater perfection by their endeavours, who, as you inform me, are taking care about it at London.

I have sometimes thought to make a microscope, which in like manner should have, instead of an object glass, a reflecting piece of metal. And this I hope they will also take into consideration. For those instruments seem as capable of improvement as telescopes, and perhaps more, because but one reflective piece of metal is requisite in them, as you may perceive by the diagram [fig. 10], where AB represents the object metal, CD the eye glass, F their common focus, and O the other focus of the metal, in which the object is placed.

But to return from this digression, I told you, that light is not similar, or homogeneal, but consists of difform rays, some of which are more refrangible than others: so that of those, which are alike incident on the same medium, some shall be more refracted than others, and that not by any virtue of the glass, or other external cause, but from a predisposition, which every particular ray has to suffer a particular degree of refraction.

6. The plague year was 1665, so Newton was probably mistaken in saying at the beginning of this paper that he began these experiments in 1666. In fact there is manuscript evidence that he did some work with prisms as early as 1664, before he read Hooke's *Micrographia*.

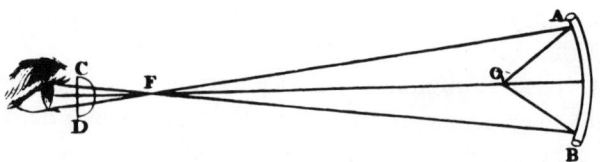

10. Newton's microscope with reflecting piece of
      metal.
      From Isaac Newton, *Philosophical Transac-
      tions*

I shall now proceed to acquaint you with another more notable
difformity in its rays, wherein the origin of colours is unfolded:
concerning which I shall lay down the doctrine first, and then, for
its examination, give you an instance or two of the experiments, as
a specimen of the rest.—The doctrine you will find comprehended
and illustrated in the following propositions:—

1. As the rays of light differ in degrees of refrangibility, so they
also differ in their disposition to exhibit this or that particular
colour. Colours are not qualifications of light, derived from refractions, or reflections of natural bodies (as it is generally believed,)
but original and connate properties, which in divers rays are
diverse.[7] Some rays are disposed to exhibit a red colour, and no
other; some a yellow, and no other; some a green, and no other,
and so of the rest. Nor are their only rays proper and particular to
the more eminent colours, but even to all their intermediate gradations.

2. To the same degree of refrangibility ever belongs the same
colour; and to the same colour ever belongs the same degree of
refrangibility. The least refrangible rays are all disposed to exhibit
a red colour, and contrarily, those rays which are disposed to
exhibit a red colour, are all the least refrangible: so the most
refrangible rays are all disposed to exhibit a deep violet-colour,
and contrarily, those which are apt to exhibit such a violet colour,
are all the most refrangible. And so to all the intermediate colours,

---

7. That is, as Boyle had said, colors are not qualities or forms, but mechanical, real properties; the colors of bodies depend upon the way they
   receive and reflect the different rays of light, now clearly shown to be a
   mixture of colored rays.

in a continued series, belong intermediate degrees of refrangibility. And this analogy betwixt colours, and refrangibility, is very precise and strict; the rays always either exactly agreeing in both, or proportionally disagreeing in both.

3. The species of colour, and degree of refrangibility proper to any particular sort of rays, is not mutable by refraction, nor by reflection from natural bodies, nor by any other cause, that I could yet observe. When any one sort of rays has been well parted from those of other kinds, it has afterwards obstinately retained its colour, notwithstanding my utmost endeavours to change it. I have refracted it with prisms, and reflected it with bodies, which in day-light were of other colours; I have intercepted it with the coloured film of air interceding two compressed plates of glass; transmitted it through coloured mediums, and through mediums irradiated with other sorts of rays, and diversely terminated it; and yet could never produce any new colour out of it. It would, by contracting or dilating, become more brisk, or faint, and by the loss of many rays, in some cases very obscure and dark; but I could never see it change in species.

4. Yet seeming transmutations of colours may be made, where there is any mixture of divers sorts of rays. For in such mixtures, the component colours appear not, but, by their mutual allaying each other, constitute a middling colour. And therefore, if by refraction, or any other of the aforesaid causes, the difform rays, latent in such a mixture, be separated, there shall emerge colours different from the colour of the composition. Which colours are not new generated, but only made apparent by being parted; for if they be again entirely mixed and blended together, they will again compose that colour, which they did before separation. And for the same reason, transmutations made by the convening of divers colours are not real; for when the difform rays are again severed, they will exhibit the very same colours, which they did before they entered the composition; as you see, blue and yellow powders, when finely mixed, appear to the naked eye green, and yet the colours of the component corpuscles are not thereby really transmuted, but only blended. For, when viewed with a good microscope, they still appear blue and yellow interspersedly.

5. There are therefore two sorts of colours. The one original and simple, the other compounded of these. The original or primary colours are, red, yellow, green, blue, and a violet-purple, together

with orange, indigo, and an indefinite variety of intermediate gradations.

6. The same colours in species with these primary ones may be also produced by composition: for a mixture of yellow and blue makes green; of red and yellow makes orange; of orange and yellowish green makes yellow. And in general, if any two colours be mixed, which in the series of those, generated by the prism, are not too far distant one from another, they by their mutual alloy compound that colour, which in the said series appears in the midway between them. But those which are situated at too great a distance, do not so. Orange and indigo produce not the intermediate green, nor scarlet and green the intermediate yellow.

7. But the most surprising and wonderful composition was that of whiteness. There is no one sort of rays which alone can exhibit this. It is ever compounded, and to its composition are requisite all the aforesaid primary colours, mixed in a due proportion[8]. I have often with admiration beheld, that all the colours of the prism being made to converge, and thereby to be again mixed as they were in the light before it was incident upon the prism, reproduced light, intirely and perfectly white, and not at all sensibly differing from a direct light of the sun, unless when the glasses, I used, were not sufficiently clear; for then they would a little incline it to their colour.

8. Hence therefore it comes to pass, that whiteness is the usual colour of light; for, light is a confused aggregate of rays indued with all sorts of colours, as they are promiscuously darted from the various parts of luminous bodies. And of such a confused aggregate, as I said, is generated whiteness, if there be a due proportion of the ingredients; but if any one predominate, the light must incline to that colour; as it happens in the blue flame of brimstone; the yellow flame of a candle; and the various colours of the fixed stars.

9. These things considered, the manner how colours are produced by the prism, is evident. For, of the rays constituting the incident light, since those which differ in colour, proportionally differ in refrangibility, they by their unequal refractions must be severed and dispersed into an oblong form in an orderly succes-

8. This simple conclusion was one which Newton's contemporaries found extraordinarily difficult to comprehend and accept.

sion, from the least refracted scarlet, to the most refracted violet. And for the same reason it is that objects, when looked upon through a prism, appear coloured. For the difform rays, by their unequal refractions, are made to diverge towards several parts of the retina, and there express the images of things coloured, as in the former case they did the sun's image upon a wall. And by this inequality of refractions they become not only coloured, but also very confused and indistinct.

10. Why the colours of the rainbow appear in falling drops of rain, is also from hence evident. For, those drops which refract the rays disposed to appear purple, in greatest quantity to the spectator's eye, refract the rays of other sorts so much less, as to make them pass beside it; and such are the drops on the inside of the primary bow, and on the outside of the secondary or exterior one. So those drops, which refract in greatest plenty the rays apt to appear red, towards the spectator's eye, refract those of other sorts so much more, as to make them pass beside it; and such are the drops on the exterior part of the primary, and interior part of the secondary bow.

11. The odd phaenomena of an infusion of lignum nephriticum[9], leaf gold, fragments of coloured glass, and some other transparently coloured bodies, appearing in one position of one colour, and of another in another, are on these grounds no longer riddles. For, those are substances apt to reflect one sort of light, and transmit another; as may be seen in a dark room, by illuminating them with similar or uncompounded light. For, then they appear of that colour only, with which they are illuminated, but yet in one position more vivid and luminous than in another, accordingly as they are disposed more or less to reflect or transmit the incident colour.

12. From hence also is manifest the reason of an unexpected experiment, which Mr. Hook, somewhere in his micrography,[10] relates to have made with two wedge-like transparent vessels, filled the one with red, the other with a blue liquor: namely, that though they were severally transparent enough, yet both together became opaque; for, if one transmitted only red, and the other only blue, no rays could pass through both.

9. Compare Boyle on this odd infusion of wood, below, p. 345.
10. *Micrographia*, Observation X, p. 74.

13. I might add more instances of this nature; but I shall conclude with this general one, that the colours of all natural bodies have no other origin than this, that they are variously qualified to reflect one sort of light in greater plenty than another. And this I have experimented in a dark room, by illuminating those bodies with uncompounded light of divers colours. For, by that means, any body may be made to appear of any colour. They have there no appropriate colour, but ever appear of the colour of the light cast upon them, but yet with this difference, that they are most brisk and vivid in the light of their own day-light colour. Minium appears there of any colour indifferently with which it is illustrated, but yet most luminous in red; and so bise appears indifferently of any colour with which it is illustrated, but yet most luminous in blue.[11] And therefore minium reflects rays of any colour, but most copiously those indued with red; and consequently when illustrated with daylight, that is, with all sorts of rays promiscusously blended, those qualified with red shall abound most in the reflected light, and by their prevalence cause it to appear of that colour. And for the same reason bise, reflecting blue most copiously, shall appear blue by the excess of those rays in its reflected light; and the like of other bodies. And that this is the entire and adequate cause of their colours, is manifest, because they have no power to change or alter the colours of any sort of rays, incident apart, but put on all colours indifferently, with which they are enlightened.

These things being so, it can be no longer disputed, whether there be colours in the dark, nor whether they be the qualities of the objects we see, no nor perhaps whether light be a body. For since colours are the qualities of light, having its rays for their entire and immediate subject, how can we think those rays qualities also, unless one quality may be the subject of and sustain another; which in effect is to call it substance. We should not know bodies for substances, were it not for their sensible qualities, and the principal of those being now found due to something else, we have as good reason to believe that to be a substance also.

Besides, whoever thought any quality to be a heterogeneous aggregate, such as light is discovered to be. But, to determine more absolutely what light is, after what manner refracted, and by what

11. Minium is red lead; bise or bice is a cobalt blue pigment.

modes or actions it produces in our minds the phantasms of colours, is not so easy. And I shall not mingle conjectures with certainties.

Reviewing what I have written, I see the discourse itself will lead to divers experiments sufficient for its examination, and therefore I shall not trouble you further, than to describe one of those which I have already insinuated.

In a darkened room make a hole in the shut of a window, whose diameter may conveniently be about a third part of an inch, to admit a convenient quantity of the sun's light; and there place a clear and colourless prism, to refract the entering light towards the further part of the room, which, as I said, will thereby be diffused into an oblong coloured image. Then place a lens of about three feet radius (suppose a broad object glass of a three-foot telescope,) at the distance of about four or five feet from thence, through which all those colours may at once be transmitted, and made by its refraction to convene at a further distance of about ten or twelve feet. If at that distance you intercept this light with a sheet of white paper, you will see the colours converted into whiteness again by being mingled. But it is requisite, that the prism and lens be placed steady, and that the paper on which the colours are cast be moved to and fro; for by such motion, you will not only find at what distance the whiteness is most perfect, but also see how the colours gradually convene, and vanish into whiteness, and afterwards having crossed one another in that place where they compound whiteness, are again dissipated and severed, and in an inverted order retain the same colours which they had before they entered the composition. You may also see, that if any of the colours at the lens be intercepted, the whiteness will be changed into the other colours. And therefore that the composition of whiteness be perfect, care must be taken that none of the colours fall beside the lens.

In the annexed design of this experiment [fig. 11.], ABC expresses the prism set endwise to sight, close by the hole F of the window EG. Its vertical angle ACB may conveniently be about 60 degrees: MN designs the lens. Its breadth 2½ or 3 inches. SF one of the straight lines, in which difform rays may be conceived to flow successively from the sun. FP and FR two of those rays unequally refracted, which the lens makes to converge towards Q, and after decussation to diverge again. And HI the paper, at divers

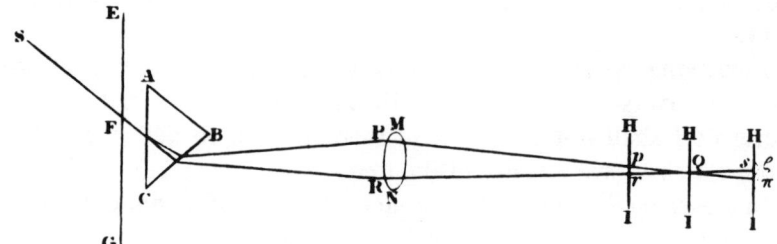

11.   Newton's experiment proving the dispersion of light.
From Isaac Newton, *Philosophical Transactions*

distances, on which the colours are projected; which in Q constitute whiteness, but are red and yellow in R, r and ρ, and blue and purple in P, p, and π.

If you proceed further to try the impossibility of changing any uncompounded colour, (which I have asserted in the 3d and 13th propositions) it is requisite that the room be made very dark, least any scattering light mixing with the colour disturb and allay it, and render it compound, contrary to the design of the experiment. It is also requisite, that there be a perfecter separation of the colours than, after the manner above described, can be made by the refraction of one single prism, and how to make such further separations, will scarcely be difficult to them that consider the discovered laws of refractions. But if trial shall be made with colours not thoroughly separated, there must be allowed changes proportionable to the mixture. Thus, if compound yellow light fall upon blue bise, the bise will not appear perfectly yellow but rather green, because there are in the yellow mixture many rays indued with green, and green being less remote from the usual blue colour of bise than yellow, is the more copiously reflected by it.

In like manner, if any one of the prismatic colours, suppose red, be intercepted, on design to try the asserted impossibility of reproducing that colour out of the others which are pretermitted; it is necessary, either that the colours be very well parted before the red be intercepted, or that together with the red the neighbouring colours, into which any red is secretly dispersed, (that is, the yellow, and perhaps green too) be intercepted, or else, that allowance be made for the emerging of so much red out of the yellow green, as may possibly have been diffused, and scatteringly blended in those colours. And if these things be observed, the new production of red, or any intercepted colour will be found impossible.

This I conceive is enough for an introduction to experiments of this kind; which if any of the Royal Society shall be so curious as to prosecute, I should be very glad to be informed with what success; that, if any thing seem to be defective, or to thwart this relation, I may have an opportunity of giving further direction about it, or of acknowledging my errors, if I have committed any.

Mechanics

---

25. The Cartesian Laws of Motion

24. Now *motion* (that is to say, locomotion, for I cannot conceive any other sort and do not think it necessary to imagine any other in nature), as commonly understood, is simply *the action by which a body passes from one place to another*. And, as noted above, just as a thing may at one and the same time change and not change its position, so at one and the same time it may be said to move and not move. For example, someone seated on the poop of a ship moved by the winds thinks of himself as moving with respect to the shore which he left and which he regards as fixed, and does not think of himself as moving with respect to the ship since he keeps the same position with respect to its parts. Thus since we commonly think that there is no motion without activity, we say that a man so seated is at rest since he feels no action within himself.

25. But if, not content with what is only based upon common usage, we wish to know what motion truly is, we must say, in order to ascribe to it a determinate nature, that it is *the transference of a part of matter or of a body from the neighborhood of those which immediately touch it and which we regard as at rest to the neighborhood of certain others*. By a body or a part of matter I mean everything that is transferred together, although this may be composed of several parts which at the same time have various other motions. And I speak of the transference and not of the force or action of transference to show that motion is always in the moved and not in the mover, for it seems to me that these two are not usually carefully enough distinguished. Moreover, I conceive that it is a *property* of the moved object and not a

SOURCE: Translated by the editor from René Descartes, *Principia Philosophiae* (French edn., Paris, 1647), part II, sections 24-40.

substantial form,[1] just as shape is a property of the object which possesses it and rest of the object at rest.

26. Now we ordinarily commit an error in thinking that more activity is required for motion than for rest, and we should consequently note here that we fell into this error early in life because we ordinarily move our bodies at will and we have an internal consciousness of this, and they are at rest only because they are attached to the earth by their weight, of which we are not aware. And as this weight and several other normally unnoticed causes tend to resist the motion of our limbs and make us tired, it has seemed to us that a larger force and greater activity is required to produce a motion than to stop it, because we have acted with an effort to move our limbs and thereby move other bodies. But we shall have no trouble in getting rid of this false notion if we note that it not only requires an effort to move bodies near to us, but also to stop their motion. Thus it requires no more activity, for example, to move a ship at rest in still water than to stop it dead when it is in motion; or if experience shows that it does in fact require a little less effort to stop it than to make it go, this is because of the weight of water it displaces as it moves and because its sluggishness (I imagine it stagnant and calm) little by little slows its motion.[2]

27. But here I am not concerned with the activity in what produces or stops motion, but rather transference or cessation of transference (rest). Clearly this transference does not exist outside the body which is moved, but rather a body differs when it is being transferred and when it is not, so that motion and rest are only two different states of the body.

28. I added that the transference of a body is made from the neighborhood of those bodies which it touches to the neighborhood of others and not from one place to another, because place may be interpreted in different ways, according to our way of thought, as said above. But when we take motion to be transference of a body from the neighborhood of contiguous bodies, then it is certain that we cannot attribute more than one motion

1. That is, motion, rest, and shape exist only in connection with the object and have no independent existence of their own as Aristotelian substantial forms, impressed upon an object, were thought to have.

2. Descartes is here working toward a true definition of inertia, which he reaches below in section 37.

to each moving object, for it can, at a given time, touch only a certain number of bodies.

29. Finally, I said that transference occurs not from the neighborhood of all kinds of bodies, but only of those taken to be at rest. For it is reciprocal, and we cannot conceive of the body AB as being transferred from the neighborhood of a body CD without also considering that CD is transferred from the neighborhood of AB, and that as much activity is required for one as for the other. Thus if we wished to attribute to motion an individual nature which could be held to be unrelated to anything else, then when we saw two bodies in contact being carried one in one direction and one in another (that is, reciprocally separated), we should have no difficulty in saying that there was as much motion in one as in the other. Now I say that this is going too far from common speech; for although we are on the earth and think that it is at rest, when we see that some of its parts which touch other, smaller bodies are removed from the neighborhood of those bodies, we yet do not for all that think that the earth is in motion.

30. This is because we think that a body does not move unless it moves as a whole and because we cannot convince ourselves that the earth as a whole moves from the sole fact that some of its parts are carried from the neighborhood of some other, smaller bodies with which they are in contact.[3] The reason for this is that we often observe around us several such transferences of bodies in opposite directions. . .with all this, we recollect that everything that is real in a moving body, by virtue of which we consider it to be moving, is equally found in those bodies which it touches, which we regard as being at rest.

31. But yet while each body in particular has only one single motion proper to it, because there are only a certain number of bodies which touch it and which are at rest with respect to it, nevertheless it can share in an infinite number of other motions insomuch as it is a part of other bodies moving in various ways. For example, if a seaman walking on a ship carried a watch, although the wheels of his watch have only a single motion proper to them, certainly they share also in the motion of the seaman

3. Descartes' definition of motion allows him to have the advantages of the Copernican system (for the earth's vortex moves about the sun) without its theological disadvantages (for his earth is at rest in the center of its vortex).

who is walking, because with him they make up a body which is transported as a unit; certainly they also share in the motion of the ship; and even in that of the sea, because they follow its course; and in that of the earth, if the earth is thought to turn on its axis, since they compose a body with it. And although it is true that all these motions reside in the wheels of the watch, nevertheless since we do not ordinarily conceive of such a large number at once and since it is not even in our power to know all those in which it shares, it suffices for us to consider in each body the motion unique to it, of which we can have certain knowledge.

32. We may even regard that unique motion which is properly attributed to each body as if it were composed of several other motions—just as we distinguish two motions in the wheels of a carriage; that is to say, a circular one made about their axle and another rectilinear one which leaves a track along the length of the road it traverses. At the same time it is clear that these two motions do not effectively differ one from another because each point of these wheels and of all other moving bodies never traces out anything except a single line. And it does not matter if this line is often bent so that it seems to have been produced by several different motions, for it is possible to imagine that this line, whatever it might be—even a straight line, the simplest of all—was described by an infinite number of such motions. . . . And although it may be useful sometimes to divide a motion into several parts so as to understand each more plainly, nevertheless, strictly speaking, we ought never to reckon more than one in each body.

33. After what has previously been demonstrated—that is, that all spaces are full of bodies and that each part of matter is truly proportional to the magnitude of the space it occupies, that it is impossible for any part to fill a larger space, or to be squeezed into a smaller, or that any other body can find room as long as it is there—we must conclude that it necessarily follows that there may always be a circle of matter or ring of body which moves all together at the same time, in such a way that, when a body leaves its place because of another which chases it, it enters into the place of another one, and this other in the place of yet another, and so down to the last one, which at the same moment occupies the space vacated by the first. . . .[4]

4. All motion for Descartes is necessarily impact motion.

34. It must be admitted that there is something in this motion which our minds conceive to be true, although they cannot understand it; that is to say, a division of certain particles of matter right down to infinity—or, better, to an indefinite division which produces so many particles that we are unable mentally to conceive any one so small that we do not think it capable of being divided into yet smaller ones. . . .

36. After having examined the nature of motion, we must consider its cause; and since it can be regarded in two ways, we shall begin with the first and more widespread way, which generally produces all the motions present in the world. The second will be considered later. . . . As for the first cause, it seems to me evident that it is God alone, who in His omnipotence created that matter with motion and rest, and who now preserves in the universe in its ordinary concourse as much motion and rest as He put into it at creation. For although motion is nothing but a mode of moved matter, it exists in a certain quantity which never increases or diminishes, although there is sometimes more and sometimes less in certain of its parts. This is why, when a particle of matter moves twice as fast as another, and when this other particle is twice as large as the first one, we ought to think that there is as much motion in the smaller as in the larger, and that when and if the motion of one particle diminishes, that of the other decreases in proportion. We also know that it is a sign of God's perfection, not only that matter is immutable by nature, but that it acts in such a way that it never changes. So that, setting aside the changes which we see in the world and those which we believe in because God has revealed them to us, and which we know occur or ought to occur naturally without any change by God, we ought not to imagine others in His works, for fear lest we attribute inconstancy to Him. From which it follows that, since He has moved the particles of matter in various different ways since He created them, and that He preserves them all in the same way and with the same laws which He made them observe at their creation, He constantly preserves a uniform quantity of motion in that matter.

37. From this, as well as the fact that God is not subject to change and acts always in the same way, we can arrive at the understanding of certain rules, which I call the laws of nature and which are the secondary causes of the different motions which we observe in all bodies. The first is that each body in particular

continues in the same state in which it is and that it never changes except by encountering others.[5] Thus we see every day that when a piece of matter is square, it remains so always unless something else occurs to change its shape, and that if it is at rest, it never begins to move by itself. But when it has once begun to move, we have also no reason to think that it ever ceases to move with the same force as long as it meets nothing which slows it or stops its motion. So that if a body has once begun to move, we ought to conclude that it continues thereafter to move and that it never stops by itself. But since we dwell on the earth, whose constitution is such that all the motions made around us cease in a little time, and often for reasons concealed from our senses, we have assumed since the beginning of our lives that the motions which thus cease because of causes unknown to us stop of themselves, and we have still at present a strong inclination to believe the same thing of all other objects in the world—that is to say, that they naturally cease of themselves and that they tend toward rest, because it seems to us that we have experience of this in various encounters. Yet this is a false assumption, which is manifestly repugnant to the laws of nature; for rest is contrary to motion and nothing naturally goes to its contrary or to self-destruction.

38. Every day we see the proof of this first rule in things pushed to a distance, for there is no other reason for their continuing to move when they have left the hand which pushed them except that, according to the laws of nature, all bodies in motion continue to move until their motion is stopped by other bodies. And it is clear that the air—and other fluids in which we see bodies move—little by little decreases the speed of their motion; we can even detect with our hands the resistance of the air, if we wave an open fan quickly enough, and there is no fluid body on earth which does not offer more manifest resistance than air to the motions of other bodies.

39. The second law I detect in nature is that every particle of matter in its own right tends to continue in motion, not in curved lines, but always in straight lines, [6] although some of these particles may often be constrained to turn aside because they meet

5. This is Newton's First Law.
6. The first published recognition of this important fact, which makes it obviously necessary (as it was not for Galileo) to *explain* circular motion, now not taken to be natural.

others in their way; and that when a body moves, it always creates a circle or ring of all the matter which moves together. This rule, like the previous one, depends on the fact that God is immutable and preserves motion in matter by a very simple operation. For He does not preserve it as it could have been some time before, but as it is exactly at the moment when He is preserving it. And although it is true that the movement is not made in an instant, nevertheless it is evident that every body which moves, moves in a straight line and not in a circular one. . .we can plainly see that every body which moves in a circle tends continually to move away from the centre of the circle it describes. And we can even detect this with our hand, when we make a stone turn on a sling, for it pulls and makes the cord stretch so that it may withdraw from our hand. This consideration is so important and will be useful so often that we must note it carefully here, and I shall explain it more fully later, when the time comes.

40. The third law which I note in nature is that if a moving body, when it meets another, has less force for continuing in a straight line than the second has to resist, it loses its determination without losing any of its motion; and if it has a greater force, it moves the second body with it and loses as much of its motion as it gives away. Thus we see that a hard body which we have pushed against another, smaller but hard and firm body, bounces back in the direction from which it came and does not lose any of its motion; but that if the body it hits is soft, it instantly stops because it transfers its motion to this other body.[7] The particular causes of the changes which occur in bodies are all comprised in this rule, at least those relating to body; for I do not, at present, reflect upon whether angels and human thoughts have the power to move bodies, a question I reserve for a treatise which I hope to write, *On Man*.

7. This and subsequent "laws" deal with the behavior of bodies upon impact, very necessary in a Cartesian world which is full of matter. In fact, the laws of impact motion are far more complex than Decartes' laws would imply, a fact he half recognized since he knew that no two bodies in his universe ever existed in isolation.

26. Newton's Laws of Motion

Axioms or Laws of Motion

LAW I.

Every body perseveres in its state of rest, or of uniform motion
in a right [straight] line, unless it is compelled to change that state
by forces impressed thereon.

Projectiles persevere in their motions, so far as they are not
retarded by the resistance of the air, or impelled downwards by
the force of gravity. A top, whose parts by their cohesion are
perpetually drawn aside from rectilinear motions, does not cease
its rotation, otherwise than as it is retarded by the air. The greater
bodies of the planets and comets, meeting with less resistance in
more free spaces, preserve their motions both progressive and circular for a much longer time.

LAW II.

The alteration of motion is ever proportional to the motive
force impressed; and is made in the direction of the right line in
which that force is impressed.

If any force generates a motion, a double force will generate
double the motion, a triple force triple the motion, whether that
force be impressed altogether and at once, or gradually and successively. And this motion (being always directed the same way
with the generating force), if the body moved before, is added to
or subducted [subtracted] from the former motion, according as
they directly conspire with or are directly contrary to each other;
or obliquely joined, when they are oblique, so as to produce a new
motion compounded from the determination of both.

SOURCE: Isaac Newton, *Principia* (London, 1687), Book I, Andrew Motte's
translation (London, 1729).

LAW III.

To every action there is always opposed an equal reaction: or the mutual actions of two bodies upon each other are always equal, and directed to contrary parts.

Whatever draws or presses another is as much drawn or pressed by that other. If you press a stone with your finger, the finger is also pressed by the stone. If a horse draws a stone tied to a rope, the horse (if I may so say) will be equally drawn back towards the stone: for the distended rope, by the same endeavour to relax or unbend itself, will draw the horse as much towards the stone, as it does the stone towards the horse, and will obstruct the progress of the one as much as it advances that of the other. If a body impinge upon another, and by its force change the motion of the other, that body also (because of the equality of the mutual pressure) will undergo an equal change, in its own motion, towards the contrary part. The changes made by these actions are equal, not in the velocities, but in the motions of bodies; that is to say, if the bodies are not hindered by any other impediments. For, because the motions are equally changed, the changes of the velocities made towards contrary parts are reciprocally proportional to the bodies. This law takes place also in attractions, as will be proved in the next scholium. . . .

The Newtonian System

27. The Possibilities of Mathematical Philosophy

The Author's Preface

Since the ancients (as we are told by Pappus)[1] made great account of the science of mechanics in the investigation of natural things; and the moderns, laying aside substantial forms and occult qualities, have endeavoured to subject the phaenomena of nature to the laws of mathematics, I have in this treatise cultivated mathematics so far as it regards philosophy. The ancients considered mechanics in a twofold respect; as rational, which proceeds accurately by demonstration; and practical. To practical mechanics all the manual arts belong, from which mechanics took its name. But, as artificers do not work with perfect accuracy, it comes to pass that mechanics is so distinguished from geometry, that what is perfectly accurate is called geometrical; what is less so, is called mechanical. But the errors are not in the art, but in the artificers. He that works with less accuracy is an imperfect mechanic; and if any could work with perfect accuracy, he would be the most perfect mechanic of all; for the description of right [straight] lines and circles, upon which geometry is founded, belongs to mechanics. Geometry does not teach us to draw these lines, but requires them to be drawn; for it requires that the learner should first be taught to describe these accurately, before he enters upon geometry; then it shows how by these operations problems may be solved. To describe right lines and circles are problems, but not geometrical problems. The solution of these problems is required from mechanics; and by geometry the use of them, when so solved, is shown; and it is the glory of geometry that from those few principles, fetched from without, it is able to produce so many things. Therefore geometry is founded in mechanical prac-

SOURCE: Isaac Newton, *Principia*, The Preface, Motte's translation.

1. A fourth-century Greek writer on mathematics.

tice, and is nothing but that part of universal mechanics which accurately proposes and demonstrates the art of measuring. But since the manual arts are chiefly conversant in the moving of bodies, it comes to pass that geometry is commonly referred to their magnitudes, and mechanics to their motion. In this sense rational mechanics will be the science of motions resulting from any forces whatsoever, and of the forces required to produce any motions, accurately proposed and demonstrated. This part of mechanics was cultivated by the ancients in the five powers[2] which relate to manual arts, who considered gravity (it not being a manual power) no otherwise than as it moved weights by those powers. Our design not respecting arts, but philosophy, and our subject not manual but natural powers, we consider chiefly those things which relate to gravity, levity, elastic force, the resistance of fluids, and the like forces, whether attractive or impulsive; and therefore we offer this work as mathematical principles of philosophy; for all the difficulty of philosophy seems to consist in this—from the phaenomena of motions to investigate the forces of nature, and then from these forces to demonstrate the other phaenomena[3] and to this end the general propositions in the first and second books are directed. In the third book, we give an example of this in the explication of the System of the World; for by the propositions mathematically demonstrated in the first book, we there derive from the celestial phaenomena the forces of gravity with which bodies tend to the sun and the several planets. Then from these forces, by other propositions which are also mathematical, we deduce the motions of the planets, the comets, the moon and the sea. I wish we could derive the rest of the phaenomena of nature by the same kind of reasoning from mechanical principles; for I am induced by many reasons to suspect that they may all depend upon certain forces[4] by which the particles of bodies, by some causes hitherto unknown, are either mutually impelled towards each other, and cohere in regular figures, or are repelled and recede from each other; which forces being unknown,

2. The five simple machines—pulley, lever, wedge, screw, inclined plane.
3. This is a true statement of the aims of the mechanical philosophy of the seventeenth century.
4. I.e., of attraction and repulsion, terms Newton hesitated to use because of their "occult" implications. (See below, Section VII, pp. 329-30.)

philosophers have hitherto attempted the search of nature in vain; but I hope the principles here laid down will afford some light either to that or some truer method of philosophy . . .

Cambridge, Trinity College   ISAAC NEWTON

May 8, 1686

28. Definitions of Matter and Motion

Definition I
The quantity of matter is the measure of the same, arising from its density and bulk conjunctly.

Thus air of a double density, in a double space, is quadruple in quantity; in a triple space, sextuple in quantity. The same thing is to be understood of snow, and find dust or powders, that are condensed by compression or liquefaction; and of all bodies that are by any causes whatever differently condensed. I have no regard in this place to a medium, if any such there is, that freely pervades the interstices between the parts of bodies.[1] It is this quantity that I mean hereafter everywhere under the name of Body or Mass. And the same is known by the weight of each body: for it is proportional to the weight, as I have found by experiments on pendulums, very accurately made, which shall be shown hereafter.

Definition II
The quantity of motion is the measure of the same, arising from the velocity and quantity of matter conjunctly.[2]

SOURCE: Isaac Newton, *Principia*, Book I, Motte's translation.

1. I.e., an ether. This definition has often been criticized as circular, since we now define density as mass times volume. Among other reasons for Newton's definition is that he regarded density and bulk (volume) as measurable quantities.
2. Newton's "quantity of motion" is the modern "momentum."

The motion of the whole is the sum of the motions of all the parts; and therefore in a body double in quantity, with equal velocity, the motion is double; with twice the velocity, it is quadruple.

Definition III

The *vis insita,* or innate force of matter, is a power of resisting, by which every body, as much as in it lies, endeavours to persevere in its present state, whether it be of rest, or of moving uniformly forward in a right line.

This force is ever proportional to the body whose force it is; and differs nothing from the inactivity of the mass, but in our manner of conceiving it. A body from the inactivity of matter, is not without difficulty put out of its state of rest or motion. Upon which account, this *vis insita*, may, by a most significant name, be called *vis inertiae* or force of inactivity. But a body exerts this force only, when another force, impressed upon it, endeavours to change its condition and the exercise of this force may be considered both as resistance and impulse: it is resistance, in so far as the body, for maintaining its present state, withstands the force impressed, it is impulse, in so far as the body, by not easily giving way to the impressed force of another, endeavours to change the state of that other. Resistance is usually ascribed to bodies at rest, and impulse to those in motion: but motion and rest as commonly conceived, are only relatively distinguished nor are those bodies always truly at rest, which commonly are taken to be so.

Definition IV

An impressed force is an action exerted upon a body, in order to change its state, either of rest, or of moving uniformly forward in a right line.

This force consists in the action only; and remains no longer in the body, when the action is over. For a body maintains every new state it acquires, by its *vis inertiae* only. Impressed forces are of different origins; as from percussion, from pressure, from centripetal force.

Definition V

A centripetal force is that by which bodies are drawn or impelled, or any way tend, towards a point as to a centre.

Of this sort is gravity, by which bodies tend to the centre of the earth; magnetism, by which iron tends to the loadstone; and that force, whatever it is, by which the planets are perpetually drawn aside from the rectilinear motions which otherwise they would pursue, and made to revolve in curvilinear orbits. A stone, whirled about in a sling, endeavours to recede from the hand that turns it; and by that endeavour, distends the sling, and that with so much the greater force, as it is revolved with the greater velocity, and as soon as ever it is let go, flies away. That force which opposes itself to this endeavour, and by which the sling perpetually draws back the stone towards the hand, and retains it in its orbit, I call the centripetal force. And the same thing is to be understood of all bodies, revolved in any orbits. They all endeavour to recede from the centres of their orbits; and were it not for the opposition of a contrary force which restrains them to, and detains them in their orbits, which I therefore call centripetal, would fly off in right lines, with an uniform motion. A projectile, if it was not for the force of gravity, would not deviate towards the earth, but would go off from it in a right line, and that with an uniform motion, if the resistance of the air was taken away. It is by its gravity that it is drawn aside perpetually from its rectilinear course, and made to deviate towards the earth, more or less, according to the force of its gravity, and the velocity of its motion. The less its gravity is, for the quantity of its matter, or the greater the velocity with which it is projected, the less will it deviate from a rectilinear course, and the farther it will go. If a leaden ball, projected from the top of a mountain by the force of gun-powder with a given velocity, and in a direction parallel to the horizon, is carried in a curved line to the distance of two miles before it falls to the ground; the same, if the resistance of the air was took away, with a double or decuple velocity, would fly twice or ten times as far. And by increasing the velocity, we may at pleasure increase the distance to which it might be projected, and diminish the curvature of the line, which it might describe, till at last it should fall at the distance of 10, 30, or 90 degrees, or even might go quite round the whole earth before it falls; or lastly, so that it might never fall to the earth, but go forward into the celestial spaces, and proceed in its motion in infinitum. And after the same manner that a projectile, by the force of gravity, may be made to revolve in an orbit, and go round the whole earth, the moon also, either by the force of gravity, if it is endued with gravity, or by any other force, that impels it towards the earth, may be perpetually drawn aside towards the earth, out of the rectilinear way, which by its innate force it would pursue; and be made to revolve in the orbit which it now describes: nor could the moon without some such force, be retained in its orbit. If this force was too small, it would not sufficiently turn the moon out of a rectilinear course: if it was too great, it would turn it too much, and draw down the moon from its orbit towards the earth. It is necessary, that the force be of a just quantity, and it belongs to the mathematicians to find the force, that may serve exactly to retain a body in a given orbit, with a given velocity; and vice versa, to determine the curvilinear way, into which a body projected from a given place, with a given velocity, may be made to deviate from its natural rectilinear way, by means of a given force.

The quantity of any centripetal force may be considered as of three kinds; absolute, accelerative, and motive.

Definition VI

The absolute quantity of a centripetal force is the measure of the same, proportional to the efficacy of the cause that propagates it from the centre, through the spaces round about.

Thus the magnetic force is greater in one loadstone and less in another, according to their sizes and strength.

Definition VII

The accelerative quantity of a centripetal force is the measure of the same, proportional to the velocity which it generates in a given time.

Thus the force of the same loadstone is greater at a less distance, and less at a greater: also the force of gravity is greater in valleys, less on tops of exceeding high mountains; and yet less (as shall be hereafter shown) at greater distances from the body of the earth; but at equal distances, it is the same everywhere; because (taking away, or allowing for, the resistance of the air) it equally accelerates all falling bodies, whether heavy or light, great or small.

Definition VIII

The motive quantity of a centripetal force, is the measure of the same, proportional to the motion which it generates in a given time.

Thus the weight is greater in a greater body, less in a less body; it is greater near to the earth, and less at remoter distances. This sort of quantity is the centripetency, or propension of the whole body towards the centre, or, as I may say, its weight; and it is ever known by the quantity of a force equal and contrary to it, that is just sufficient to hinder the descent of the body.

These quantities of forces, we may for brevity's sake call by the names of motive, accelerative, and absolute forces; and for distinction's sake consider them, with respect to the bodies that tend to the centre; to the places of those bodies; and to the centre of force towards which they tend: that is to say, I refer the motive force to the body, as an endeavour and propensity of the whole towards a centre, arising from the propensities of the several parts taken together; the accelerative force to the place of the body, as a certain power or energy diffused from the centre to all places around to move the bodies that are in them; and the absolute force to the centre, as endued with some cause, without which those motive forces would not be propagated through the spaces round about; whether that cause is some central body (such as is the loadstone, in the centre of the force of magnetism, or the earth in the centre of the gravitating force), or any thing else that does not yet appear. For I here design only to give a mathematical notion of those forces, without considering their physical causes and seats.

Wherefore the accelerative force will stand in the same relation to the motive, as celerity does to motion. For the quantity of motion arises from the celerity drawn into the quantity of matter; and the motive force arises from the accelerative force drawn into the same quantity of matter. For the sum of the actions of the accelerative force, upon the several particles of the body, is the motive force of the whole. Hence it is, that near the surface of the earth, where the accelerative gravity, or force productive of gravity, in all bodies is the same, the motive gravity or the weight is as the body: but if we should ascend to higher regions, where the accelerative gravity is less, the weight would be likewise diminished, and would always be as the product of the body, by the accelerative gravity. So in those regions, where the accelerative gravity is diminished into one half, the weight of a body two or three times less, will be four or six times less.

I likewise call attractions and impulses, in the same sense, accelerative, and motive; and use the words attraction, impulse, or propensity of any sort towards a centre, promiscuously and indifferently, one for another; considering those forces not physically, but mathematically: wherefore, the reader is not to imagine, that by those words, I anywhere take upon me to define the kind, or the manner of any action, the causes or the physical reason thereof, or that I attribute forces, in a true and physical sense, to certain centres (which are only mathematical points); when at any time I happen to speak of centres as attracting, or as endued with attractive powers.[3]

29. *Rules of Reasoning in Natural Philosophy*

Book III

In the preceding books I have laid down the principles of philosophy; principles not philosophical, but mathematical; such, to wit, as we may build our reasonings upon in philosphical enquiries. These principles are the laws and conditions of certain motions, and powers or forces, which chiefly have respect to philosophy; but, lest they should have appeared of themselve dry and barren, I have illustrated them here and there with some philosophical scholiums, giving an account of such things as are of more general nature, and which philosphy seems chiefly to be founded on; such as the density and the resistance of bodies, spaces void of all bodies, and the motion of light and sounds. It remains that, from

SOURCE: Isaac Newton, *Principia*, The Preface, Motte's translation.

3. This is part of Newton's attempt to separate the *idea* of attraction, which many scientists of his day found mentally repellent, and its *effects*, which Newton saw as inherent in the concept of gravity.

the same principles, I now demonstrate the frame of the System of the World. Upon this subject I had, indeed, composed the third book in a popular method, that it might be read by many;[1] but afterwards, considering that such as had not sufficiently entered into the principles could not easily discern the strength of the consequences, nor lay aside the prejudices to which they had been many years accustomed, therefore, to prevent the disputes which might be raised upon such accounts, I chose to reduce the substance of this book into the form of propositions (in the mathematical way), which should be read by those only who had first made themselves masters of the principles established in the preceding books: not that I would advise any one to the previous study of every proposition of those books; for they abound with such as might cost too much time, even to readers of good mathematical learning. It is enough if one carefully reads the definitions, the laws of motion, and the first three sections of the first book. He may then pass on to this book, and consult such of the remaining propositions of the first two books, as the references in this, and his occasions, shall require.

RULES OF REASONING IN PHILOSOPHY
RULE I
We are to admit no more causes of natural things than such as are both true and sufficient to explain their appearances.

To this purpose the philosophers say that Nature does nothing in vain, and more is in vain when less will serve; for Nature is pleased with simplicity, and affects not the pomp of superfluous causes.

RULE II
Therefore to the same natural effects we must, as far as possible, assign the same causes.

As to respiration in a man and in a beast; the descent of stones on Europe and in America; the light of our culinary fire and of the sun; the reflection of light in the earth, and in the planets.

1. A Latin version of the "System of the World," popular to the extent that it is nonmathematical, exists, and was annexed by Motte to his 1729 translation of the *Principia.*

RULE III

The qualities of bodies, which admit neither intension nor remission of degrees,[2] and which are found to belong to all bodies within the reach of our experiments, are to be esteemed the universal qualities of all bodies whatsoever.

For since the qualities of bodies are only known to us by experiments, we are to hold for universal all such as universally agree with experiments; and such as are not liable to diminution can never be quite taken away. We are certainly not to relinquish the evidence of experiments for the sake of dreams and vain fictions of our own devising; nor are we to recede from the analogy of Nature, which uses to be simple, and always consonant to itself. We no other way know the extension of bodies than by our senses, nor do these reach it in all bodies; but because we perceive extension in all that are sensible, therefore we ascribe it universally to all others also. That abundance of bodies are hard, we learn by experience; and because the hardness of the whole arises from the hardness of the parts, we therefore justly infer the hardness of the undivided particles not only of the bodies we feel but of all others. That all bodies are impenetrable, we gather not from reason, but from sensation. The bodies which we handle we find impenetrable, and thence conclude impenetrability to be an universal property of all bodies whatsoever. That all bodies are moveable, and endowed with certain powers (which we call the *vires inertiae)*[3] of persevering in their motion, or in their rest, we only infer from the like properties observed in the bodies which we have seen. The extension, hardness, impenetrability, mobility, and *vis inertiae* of the whole, result from the extension, hardness, impenetrability, mobility, and *vires inertiae* of the parts; and thence we conclude the least particles of all bodies to be also all extended, and hard, and impenetrable, and moveable, and endowed with their proper *vires inertiae.* And this is the foundation of all philosophy. Moreover, that the divided but contiguous particles of bodies may be

2. This is a curiously scholastic or Peripatetic expression, but presumably Newton could frame no equivalent nonscholastic phrase to differentiate between qualities like "impenetrability" (which a body either has or has not) and "heat" (which a body possesses in a greater or less degree). Rules II and III are necessary to permit the generalization which makes the law of gravitation *universal.* Rule III was enlarged in the second (1713) and third (1726) editions.

3. Inertial powers or forces—what we call inertia.

separated from one another, is matter of observation; and, in the particles that remain undivided, our minds are able to distinguish yet lesser parts, as is mathematically demonstrated. But whether the parts so distinguished, and not yet divided, may, by the powers of Nature, be actually divided and separated from one another, we cannot certainly determine. Yet, had we the proof of but one experiment that any undivided particle, in breaking a hard and solid body, suffered a division, we might by virtue of this rule conclude that the undivided as well as the divided particles may be divided and actually separated to infinity.

Lastly, if it universally appears, by experiments and astronomical observations, that all bodies about the earth gravitate towards the earth, and that in proportion to the quantity of matter which they severally contain; that the moon likewise, according to the quantity of its matter, gravitates towards the earth; that, on the other hand, our sea gravitates towards the moon; and all the planets mutually one towards another; and the comets in like manner towards the sun; we must, in consequence of this rule, universally allow that all bodies whatsoever are endowed with a principle of mutual gravitation. For the argument from the appearances concludes with more force for the universal gravitation of all bodies than for their impenetrability; of which, among those in the celestial regions, we have no experiments, nor any manner of observation. Not that I affirm gravity to be essential to bodies: by their *vis insita* I mean nothing but their *vis inertiae*. This is immutable. Their gravity is diminished as they recede from the earth.

RULE IV

In experimental philosophy we are to look upon propositions collected by general induction from phaenomena as accurately or very nearly true, notwithstanding any contrary hypotheses that may be imagined, till such time as other phaenomena occur, by which they may either be made more accurate, or liable to exceptions.

This rule we must follow, that the argument of induction may not be evaded by hypotheses.[4]

4. That is, Newton is claiming that his system which is grounded upon empirical evidence, deserves to be answered by arguments based also upon empirical evidence; he sees no reason for regarding arguments based upon *a priori* reasoning as in any sense confuting his doctrines. This rule appears only in the third edition.

VII.

The Organization of Matter

IDEAS about the structure of matter have always been investigated by both the physicist and the chemist, the former speculating about the fine structure of matter, the building up of its structure to make what we see, the latter about the behavior of matter in chemical operations. Until the seventeenth century these two aspects of the study of matter had little in common; but, in the course of the scientific revolution, both physicist and chemist learned to appreciate to some extent the advantages of a mutual apprehension of the interests and discoveries of the rational chemist and the natural philosopher. Although the two aspects are divided in the selections below, it will be obvious that some are only arbitrarily in one section or the other and could be moved without any significant disturbance.

The Mechanical Philosophy

When the seventeenth century spoke of the newer views on the structure of matter, it meant the attempt to understand the properties of bodies in terms of matter and motion. Just as writers on mechanics were showing the ability to describe the behavior of gross bodies on terms of motion, so, it was felt, it should be possible to describe the attributes of gross bodies in terms of the matter and motion of their small, component parts. Underlying all the mechanical philosophy of the seventeenth century was the conviction that matter existed primarily as small, discrete, invisible, individually indiscernible particles, presumably with characteristic sizes and shapes. These particles were influenced by, but were by no means identical with, the atoms of the ancient philosophers, of Democritos, Epicuros, and Lucretius. For, in the first place, no seventeenth-century natural philosopher cared to be saddled with the deterministic materialism of Democritos or the atheism of Epicuros and Lucretius; and secondly, seventeenth-century scientists were too philosophically and mathematically sophisticated to accept the notion of a strictly indivisible entity, however small. Hence the more usual terms are particles, small parts, or corpuscles; but the intent is the same.

Ancient atomism was widely read in the sixteenth century, and in the seventeenth century was crudely grafted onto existent theories of matter (like the Aristotelian four-element theory, or the chemist's theory of elements). It helped explain the differentiation of matter by the assumption that every kind of matter had different kinds of particles. But it was not really so much more satisfactory, in terms of economy of explanation or of susceptibility to empirical check, than the true Aristotelian theory of forms and qualities. This theory, developed in the early modern period from Aristotle's theory of causation, postulated a "form" to explain each property of a body. Thus fire was hot because it possessed the form of heat; glass had the forms of brittleness, transparency, etc.; gold was yellow, dense, malleable, and resistant to corrosion. But natural philosophers had begun to realize that this was only a semantic explanation, and further, that many properties—heat, cold, sweetness, acidity, hardness, and so on—were subjective only and did not correspond to the true state of the body. As Galileo, among others, saw, these properties, being subjective, existed only when the bodies apparently possessing them could act upon the senses; and it behooved the natural philosopher to look for the real properties of the particles which produced these. Like his successors, he thought the real properties were the size, shape, *and motion* of the particles. And these are the basic tenets of the mechanical philosophy.

There was a vast literature dealing with the mechanical philosophy in the seventeenth and eighteenth centuries; some influential examples are given below from the works of Bacon, Descartes, Boyle, Newton and finally, Dalton. Considerations of space have necessitated its presentation here in a somewhat truncated form.

The first selection (Document 30) gives some idea of Bacon's approach to the problem of "substantial forms." Bacon retained the old term with a new meaning; his forms are laws rather than occult properties, and he sought to explain such physical properties as heat in material, rational terms. His inquiry (here abridged) is at one and the same time an example of argument by induction—his new logic or organon, which he hoped would replace Aristotle's old logic—and an example of his conception of the experimental method. It was too complex for practicality, but influential in its result.

A very large part of Descartes' *Principia Philosophiae* is devoted to an investigation of the nature and properties of matter. As it is difficult to find in English, a long excerpt is given here (Document 31). This begins with Descartes' careful definition of matter as extension (a definition which precludes a vacuum) and continues with his equally careful, logical development of a particulate theory of matter. Finally, some specific examples are given, ranging from his highly successful definition of heat to the less happy definition of magnetism. Particularly influential was his use of an ether (his "subtle matter") to explain how particles were put into and kept in motion.

Boyle's corpuscular philosophy is an eclectic, empirical derivation from Baconian and Cartesian principles. Rejecting all occult forces and all *a priori* hypotheses, he tried to ground his theory upon experiment. Very many of his works combine theoretical discussion with a wealth of experimental evidence. Too long to quote here, they were immensely influential even among those who disagreed with him, for his experiments were fascinating and ingenious. The essay on *The Excellency and Grounds of the Mechanical Hypothesis* (1674) given virtually entire below as Document 32 (his eighteenth-century editor, Peter Shaw, did not much abridge this section) gives a fair sample of his turn of mind and method of argument, though not of his successful empirical explanation of such properties as heat, cold, magnetism, chemical action, and the like, which he based upon the motion and "disposition" (arrangement in space) of the particles of matter. Since he rejected all forces as occult, he was forced to assume that reacting bodies met by chance as a result of the random and ceaseless motion of their particles—a position acceptable to a Baconian empiricist perhaps, but hardly to a Cartesian rationalist.

Newton, by recognizing the existency of forces of attraction and repulsion, tried to avoid the severely empircist and skeptical attitude of Boyle without returning to the Cartesian ether and impact motion. Newton's forces could be (and in the case of gravity were) expressed in mathematical and rigorous form. Yet he came dangerously close to a return to occult forces, so carefully eschewed by both Descartes and Boyle. Newton personally had no more taste for occult forces than any of his contemporaries and often tried to explain attraction and repulsion by the use of an ether, but as he could never devise an ether which had so little density as not to interfere with the motion of terrestrial and celestial bodies and at the same time was capable of exerting pressure, he always returned in the end to recognizing the existence of forces for which he could not provide any acceptable cause.

This led him to avoid, for most of his life, extended discussions of the mechanism of attraction and repulsion or of the phenomena of matter. But he often wrote such discussions without publishing them. A famous example employing the ether was written as a letter to Boyle in 1678; publication sixty-six years later found it still topical. In the final queries appended to *Opticks* and published in 1717, four years after he decided not to write an length about the problem in the second edition of the *Principia* (1713), Newton spoke at length in the form of rhetorical questions about what attraction and repulsion might be. Query 31 is well known and easily accessible. Here (Document 33) is an earlier edition, written in 1686 or 1687 and intended as a conclusion to the *Principia.* As in *Opticks,* the style is discursive and hypothetical although there are numerous, mainly chemical, experiments; Newton rightly realized in time that this form of presentation did not suit the rigorously mathematical style of the *Principia.* He replaced it with the hints of the *Preface* (Document 27 above). But it remains a fascinating example of what he would have liked to have achieved.

The dilemma of a choice between occult forces with action at a distance or an unscientific, nonempirical ether and impulse motion was thus not resolved by Newton, and remained for discussion by his successors. True Newtonians happily chose either mode of explanation without feeling any the less Newtonian, since there was textual evidence for both views. Most Newtonian physicists and chemists were like Newton, content to exploit the existence of the forces of attraction and repulsion without inquiring too deeply what these forces might be. Others, equally ardent Newtonians, opted for one explanation or the other and constructed atomic "systems" on the basis of an ether or of attractive and repulsive forces. These had their importance, especially in the nineteenth century, but they are too complex to present here. Instead a selection (Document 34) from a work on chemistry is given which, while breaking sharply from Newtonian conceptions in many ways, in others is a fitting culmination to Newton's work.

John Dalton (1766-1844) was largely self-taught, and his main interest was in the physics of the atmosphere. His first important discovery was a physical one—that in a mixture of gases, like the atmosphere, each gas behaves as if it were independent of the others, its particles not attracting or repelling any particles of other gases. This is a rather anti-Newtonian position, yet Dalton though of himself as a good Newtonian. He was next led to study the physical composition of gases and of other substances; the result was his atomic theory, in which the atom of each elementary substance is distinguished by its weight. This weight he could only determine by means of a study of chemical composition, so he turned more and more to chemistry, without losing his interest in physics. (Most of the first volume of his *New System of Chemical Philosophy* was devoted to a physical theory of heat.) Dalton's atomic theory proved to be uniquely useful in chemistry, where it helped organize a developing science, put the coping stone upon the new French system of chemistry initiated by Lavoisier, and for the first time provided an emprirical definition of such previously vague words as "atom" and "element." After Dalton the mechanical philosophy was virtually a dead issue, and speculation about the physical structure of matter reached a low ebb in the early ninteenth century.

Chemical Theories

Whether he knows it or not, the chemist is, of necessity, interested and involved in the structure of matter. It did, indeed, take many centuries before the chemist connected his own interest in the changes in form of substances—changes of state, changes of color, changes of activity—with the underlying structure of matter. It was probably not until the sixteenth century that the growth of interest in theories of matter and the growth of interest in chemical operations began to coalesce. At first they did so uneasily. Natural philosphers were wary of chemistry, so often mystic and even overtly alchemical, and the chemists were little aware of the development of the new philosophy with its twin insistence on rationalism and empiricism. Out of one aspect of the mystic chemistry of Paracelsus (Theophrastus Bombastus Philippus Aureolus von Hohenheim Paracelsus, to give the full name he invented for himself, Bombast von Hohenheim being, improbably enough, his real name) came, ironically, a rational and useful chemistry. Paracelsus was mystically sure of the analogy between the microcosm (man) and the macrocosm (nature), and regarded chemistry and physiology as one. If the processes of the human body were chemical, then derangements in these processes could be corrected by chemical means. (It must be remembered that Paracelsus' chemistry was itself vitalistic and animalistic). Many self-styled Paracelsans of the late sixteenth and early seventeenth centuries, less mystic than their master, became out-and-out "medical chemists," mainly interested in the preparation of new, chemical drugs and the teaching of such practical chemistry to apothecaries. An excellent example is Christophe Glaser, from whose *Traité de la chymie* (1663) a selection is given below (Document 35). Glaser was both a practical chemist and a professor of chemistry at the Royal Botanic Garden in Paris. His main interest was practical, yet his theory of matter was anything but either practical or empirical, for though he claimed, like others, to find his principles experimentally, they were vague substances, not chemically precise and certainly not chemically useful. They offered a basis of argument and an apparent explanation, satisfactory enough on a verbal level of argument, but of no great use to chemists. The same was true of the famous "quantitative proof" given by van Helmont that everything is made of water—it was not a proof, for his willow tree derived subsistence from the air as well as from water, and he never utilized the concept of water as an element in his discussion of chemical reactions. Helmont's importance lay rather in his concept of chemistry as a unifying science—a very mystic science, but a science nonetheless, a point much appreciated by the excessively rational Robert Boyle.

For Boyle saw that chemistry could be a rational science just as easily as it could be a mystic science, and, early a competent and soon an original physicist, he equated science with the new natural philosophy. As the first chemical selection (Document 36) from Boyle shows, he recognized that chemistry was the ideal experimental philosophy from which confirmation and demonstration of the mechanical philosophy could best be provided; beyond this he saw that the mechanical philosophy could provide a replacement for the old speculative basis of chemical matter, real particles replacing hypothetical elements. This is the thesis of *The Sceptical Chymist* (1661), a book more famous than it deserves to be, for it deals in great detail with a now outmoded controversy. Briefly, Boyle there argued that the substances usually reckoned as elements—any or all of them—had characteristics not compatible with their supposed elementary nature, and he used a host of experiments to demonstrate this. In conclusion he said that if one took the usual contemporary definition of an element (a substance not further analyzable, which was to be found in all other nonelemental substances), then the least particles or corpuscles were the only true elements there might be. This being settled, in his own mind at least, Boyle proceeeded to concern himself with chemical analysis and the study of chemical composition and chemical reactions in terms not of elements but of simple, specific, readily indentifiable substances. He established the identity of some substances known under a variety of names (such as potassium carbonate); he established the multiplicity of some substances known under a common name (many salts and oils); and he began the method of systematic identification of compounds through tests. The most successful and famous was his classification of chemical substances into acids, alkalies, and neutral bodies (Document 37).

As true analytical chemistry developed, chemists became interested in other problems. Foremost among these were the linked problems of the nature of atmospheric air and the role of air in combustion. Documents 38 and 39 illustrate the prevailing views, on the eve of "the chemical revolution" during which the problem was solved, by two of the leading protagonists in that solution, Joseph Priestley and A. L. Lavoisier.

Air, though often considered an element, had for centuries been regarded as essentially nonchemical; that is to say, it was thought to mix with bodies, but not to combine chemically with them. Naturally, air was known to be essential to both respiration and combustion—an axiom amply proved in Boyle's air pump—but the reason for this was not understood. Sometimes it was thought to be a mechanical effect; at other times (in a popular English view held by, among others, Hooke and the physician John Mayow) because of "nitro-aerial" particles mixed with air, or because of some as yet unidentified substance mixed with the air. There was no concept of elastic fluids (what we now call gases) other than atmospheric air, supposed homogeneous. Thus Boyle, who actually produced both hydrogen and carbon dioxide—the first by the action of strong acid on steel filings, the second from fermenting fruits and vegetables—never thought he had produced anything other than atmospheric air. Nor did Stephen Hales, who in his *Vegetable Staticks* (1727) digressed from the study of sap pressure to study the air which Boyle had suggested was to be found in vegetable substances. Hales found that all vegetable substances "fixed" air, and showed how this "fixed air" could be collected and its quantity determined; but he thought his fixed air was identical with atmospheric air. The first to discover that it was not—that, in fact, there were at least two different elastic fluids, atmospheric air and fixed air—was Joseph Black. Investigating the medical properties of magnesia alba (a mineral which is a mixture of magnesium hydroxide and magnesium carbonate), he was led to examine attentively the "air" given off on the addition of acids, and by careful and patient study was led to conclude that this air was very different from atmospheric air— it would not support combustion or respiration, for example—and was identical with Hales' fixed air. For the first time, atmospheric air could no longer be regarded as unique.

Black's *Experiments upon Magnesia Alba, Quick-lime, and Some other Alcaline Substances* (1756) was a landmark in the history of chemistry, for a gas was shown to be capable of entering into another substance and taking full part in a chemical reaction. A number of chemists—Cavendish and Priestley, among others in England, and Scheele in Sweden—began to apply Black's technique to other "airs." Cavendish was the first to discover a second new gas—hydrogen—and Priestley and Scheele found numerous others. In Document 38, from *Experiments and Observations on a Different Kinds of Air* (1775), Priestley describes his discovery of oxygen (which he called "dephlogisticated air") and his view of the nature of the atmosphere. This is the culmination of the great school of pneumatic chemistry, which was completed when it was shown that water is composed of two gases, that fixed air is composed of common carbon and oxygen, and that it is the oxygen which supports respiration and combustion.

The full understanding of the role of air in combustion and respiration was discovered by Lavoisier. Not primarily a pneumatic chemist, though he was profoundly interested in effervescence, he solved the problem of combustion before he knew that the substance he needed for his antiphlogistic chemistry was oxygen. The seventeenth century had explored combustion, discovered that many substances (especially metals) gain in weight on combustion, and attributed this either to something in the air—Hooke's and Mayow's nitro-aerial particles—or, with Boyle, to fire particles. (Boyle saw no empirical evidence for nitro-aerial particles, and he thought he had performed experiments which proved that it was fire, not air, which was responsible for the gain in weight he observed on calcination of tin, copper, and other substances.) In the eighteenth century Boyle's fire was caught up in the great confusion over the material nature of heat, and hence matter of heat, caloric, phlogiston were all taken to the material principle of fire. The German chemist Stahl was the writer who codified this view, but phlogiston never was identical for all chemists. Inflammable substances were taken to be lacking in phlogiston and hence to have a great affinity for fire, while metals were thought to be rich in phlogiston, which they gave off on heating. As phlogiston was a material substance this should have meant that there was a contradiction somewhere, but the fact that all metals gained weight on calcination was not really established until just as Lavoisier was preparing to show the falsehood of the phlogiston theory.

Lavoisier built on the work of his predecessors and contemporaries, especially Stephen Hales, whom he studied thoroughly. After a series of experiments on combustion, he repeated Boyle's experiments on metals and demonstrated that Boyle was wrong in thinking that he had shown that the gain of weight could not be due to air; in fact, Lavoisier showed that it was beyond doubt derived from a part of the atmosphere. From the work of the pneumatic chemists Lavoisier was prepared to think that this "part" of the atmosphere was not an alien admixture (like nitro-aerial particles) but itself gaseous in nature—an "air", and after Priestley had described his new, eminently respirable and combustible gas to Lavoisier, the latter came slowly to see that it was this air which was involved. By 1783 Lavoisier had worked out his "anti-phlogistic theory," which was to be given great authority by its incorporation into a new system of chemical nomenclature devised by Guyton de Morveau, improved by Lavoisier and others, and first used in Lavoisier's *Traité élémentaire de chimie* (1789). Even before this, Lavoisier had carefully worked out in detail and by purely experimental means the precise relation between respiration and combustion, a relation long known to exist but never previously precisely understood. His paper is notable for its careful, precise reasoning combined with clear and logical experimental proof.

The Mechanical Philosophy

---

30. Bacon on Forms.

The Nature of Forms, and the Form of Heat

III

... Whosoever is acquainted with Forms embraces the unity of nature in substances the most unlike; and is able therefore to detect and bring to light things never yet done, and such as neither the vicissitudes of nature, nor industry in experimenting, nor accident itself would ever have brought into act, and which would never have occurred to the thought of man. From the discovery of Forms therefore results truth in speculation and freedom in operation.

IV

... The Form of a nature is such, that given the Form, the nature infallibly follows. Therefore it is always present when nature is present, and universally implies it, and is constantly inherent in it. Again, the Form is such, that if it be taken away the nature infallibly vanishes. Therefore it is always absent when the nature is absent, and implies it absence, and inheres in nothing else. Lastly, the true Form is such that it deduces the given nature from some source of being which is inherent in more natures, and which is better known in the natural order of things than the Form itself.

XX

... From a survey of the instances, all and each, the nature of which Heat is a particular case appears to be Motion. This is displayed most conspicuously in flame, which is always in motion, and in boiling or simmering liquids, which also are in perpetual motion. It is also shown in the excitement or increase of heat

SOURCE: Francis Bacon, *Novum Organum* (London, 1620); translated by R. Ellis and James Speeding, Book II, aphorisms III, IV, XX.

caused by motion, as in bellows and blasts . . . and again in other kinds of motion. . . . Again, it is shown in the extinction of fire and heat by any strong compression, which checks and stops the motion. . . . It is shown also by this, that all bodies are destroyed, or at any rate notably altered, by all strong and vehement fire and heat; whence it is quite clear that heat causes a tumult and confusion and violent motion in the internal parts of a body which perceptibly tends to its dissolution.

When I say of Motion that it is as the genus of which heat is a species, I would be understood to mean, not that heat generates motion or that motion generates heat (though both are true in certain cases), but that Heat itself, its essence and quiddity, is Motion and nothing else; limited however by the specific difference which I will presently subjoin, as soon as I have added a few cautions for the sake of avoiding ambiguity.

Sensible heat is a relative notion, and has relation to man, not to the universe, and is correctly defined as merely the effect of heat on the animal spirits. Moreover, in itself it is variable, since the same body, according as the senses are predisposed, induces a perception of cold as well as of heat. . . .

Nor again must the communication of Heat or its transitive nature, by means of which a body become hot when a hot body is applied to it, be confounded with the Form of Heat. For heat is one thing, heating another. Heat is produced by the motion of attrition without any preceding heat, an instance which excludes heating from the Form of Heat. And even when heat is produced by the approach of a hot body this does not proceed from the Form of Heat, but depends entirely on a higher and more general nature, that is, on the nature of assimilation or self-multiplication, a subject which requires a separate inquiry.

Again, our notion of fire is popular, and of no use; being made up of the combination in any body of heat and brightness, as in common flame and bodies heated to redness.

Having thus removed all ambiguity, I come at length to the true specific differences which limit Motion and constitute the Form of Heat.

The first difference then is this. Heat is an expansive motion, whereby a body strives to dilate and stretch itself to a larger sphere or dimension than it had previously occupied. This difference is most observable in flame, where the smoke or thick vapour manifestly dilates and expands itself into flame.

It is shown also in all boiling liquid, which manifestly swells, rises and bubbles; and carries on the process of self-expansion, till it turns into a body far more extended and dilated than the liquid itself, namely into vapour, smoke or air. . . .

It is shown also in the opposite nature of cold. For cold contracts all bodies and makes them shrink; insomuch that in intense frosts nails fall out from walls, brazen vessels crack, and heated glass on being suddenly placed in the cold cracks and breaks. . . .

The second difference is a modification of the former; namely, that heat is a motion expansive or towards the circumference, but with this condition, that the body has at the same time a motion upwards. . . .

The third specific difference is this; that heat is a motion of expansion not uniformly of the whole body together, but in the smaller parts of it. . . .

The fourth specific difference is a modification of the last; it is, that the preceding motion of stimulation or penetration must be somewhat rapid and not sluggish, and must proceed by particles, minute indeed, yet not the finest of all, but a degree larger. . . .

The Form or true definition of heat (heat, that is, in relation to the universe, not simply in relation to man), is in a few words as follows: Heat is a motion, expansive, restrained, and acting in its strife upon the smaller particles of bodies. . . . If in any natural body you can excite a dilating or expanding motion, and can so repress this motion and turn it back upon itself that the dilation shall not proceed equably, but have its way in one part, and be counteracted in another, you will undoubtedly generate heat.

31. Descartes' Theory of Matter

Part II.
The Principles of Material Things

Although we are thoroughly persuaded that there are truly bodies in the world, yet since we formerly doubted this and since we have listed it as one of the judgments made at the beginning of our lives, it is necessary here to search for the reasons which lead us to establish a certain knowledge of bodies.

First, we experience ourselves that all that we feel comes from something outside our minds: we cannot by ourselves experience one sensation rather than another, but this derives from something else according to the way in which it stimulates our senses. It is true that we could inquire whether God (or some other than He) was not this something else; but we feel, or rather our senses cause us often to perceive clearly and distinctly, a substance which has extension in length, breadth, and width, whose parts have shapes and various motions from whence proceed the sensations we experience of colors, odors, of pain, and so on. Now if God implanted in our souls directly from Himself the idea of that extended substance, or even if He permitted it to be caused in us by something which did not possess extension, shape, or motion, there would be nothing to prevent our believing that God took pleasure in deceiving us, for we conceive of this substance as a thing different from God and from our own minds, and it appears that the idea of matter which we possess is framed upon outside bodies, to which this idea conforms very well. Now since God does not deceive us (for that is repugnant to His nature, as has already been said), we must conclude that there exists a substance possessing extension in length, depth, and breadth, which exists in the world at this moment and has all the properties which we perceive

SOURCE: Translated by the editor from René Descartes, *Principia Philosophiae* (Paris, 1647), sections 1, 4-22, 54-56, 61-63; part III, sections 48-52; part IV, sections 1-2, 14-16, 18-20, 22-23, 28-30, 133, 207.

to belong to it. And this extended susbtance is what is properly called body or the substance of material things. . . .

. . . Now we shall learn that the nature of matter, or more generally of body, does not consist in its being hard, weighty, or colored, or in its affecting our senses in any other way, but only in its being a substance which is extended in length, breadth, and depth. As for hardness, we learn nothing by touch about bodies except that the parts of a hard body resist the movement of our hands when they encounter it. Now if whenever our hands were brought towards a place, the bodies which make up that place retreated as quickly as our hands approached them, we should certainly· never feel "hardness"; yet we have no reason to believe that our bodies which might retreat in this way would thereby lose what makes them bodies. From which it follows that their nature does not consist in the hardness which we sometimes feel on their account, nor in their weight, heat, or other qualities of this kind. For if we examine any body we may notice none of these qualities in it, and yet we clearly and distinctly perceive that it possesses everything necessary to be a body, provided that it has extension in length, breadth, and depth. From this it also follows that a body does not require those qualities in any way and that its nature consists only in its possessing extension.

5. To make this truth appear entirely plain it only remains to clear up two difficulties. The first is this: some people seeing near at hand bodies which are sometimes more and sometimes less rarefied have thought that the same body possesses more extension when it is rarefied than when it is condensed; there are even some so subtle as to try to distinguish the substance of a body from its own size, and its size from its extension. The other difficulty arises from a common manner of thought: we do not conceive that there is a body where there is said to be only extension in length, breadth, and depth, but only think of a space, even an empty space which is easily taken to be nothing.

6. As for rarefaction and condensation, anyone who reflects carefully and admits nothing on this subject but what he clearly and distinctly perceives will agree that they occur only through the change of shape of the body as it is rarefied or condensed; that is to say, when we see that a body is rarefied, we must think that it has many interstices between it particles which are filled with some other body, and in a condensed body these same particles

are closer to one another than they were before, either because the interstices between them are smaller or else have been done away with entirely, in which case it is impossible to suppose that the body could be further condensed. Yet it does not fail to have as much extension as when these same particles, being farther apart and as it were scattered in various branches, occupy a greater space. For we must not ascribe to it the extension in its pores or interstices which its own particles do not occupy, because this belongs to the other bodies which fill these pores, just as when we see a sponge full of water or of some other fluid we do not think that on this account the sponge has a greater extension, but only that there are pores or interstices between its particles which are larger than when it is squeezed dry.

7. I do not understand why some people in explaining rarefaction have preferred to say that it arises from an increase in magnitude instead of considering the analogy of a sponge. Certainly when air or water are rarefied we do not see the pores between the particles of these bodies nor the method by which they become larger, nor the body which fills them. But it is much less reasonable to invent something unintelligible to give an apparent and nonsensical explanation of the rarefaction of a body than to deduce from the fact that it *is* rarefied the existence of pores or interstices between its particles, which are enlarged and filled with some other body. And it is not difficult to believe that rarefaction takes place as I have described, although we cannot detect the body which fills the pores by our senses; for there is absolutely no reason why we must believe that we ought to perceive by our senses every body around us, and we see that it is very easy to explain the situation in this way and impossible to understand it in any other. For lastly, as it seems to me, there is a manifest contradiction in supposing that a body could be increased by a magnitude or extension which it did not have before without by the same means being increased by a new, extended substance or a new body; because it is not possible to imagine that magnitude or extension can be added to a body except by adding something large and extended, as will appear more clearly by what follows.

8. The reason for this is that magnitude does not differ from what is large or number from what is numbered. That is to say, it is possible for us to conceive the nature of an extended object occupying a space of ten feet without taking notice of the *measure*

ten feet (because the nature of this object is the same in the parts as in the whole); and we can think of the number ten or of a magnitude contained in ten feet without thinking of a specific object, because the idea we have of the number ten is the same whether we consider the number of ten feet or any other ten, and we can even conceive a magnitude of ten feet without thinking about any particular body, although we cannot conceive it without something extended. At the same time it is clear that it is not possible to remove any part of such a magnitude or extension without in the same way subtracting as much from the object, and conversely it is impossible to subtract from the object without similarly subtracting as much from the magnitude or extension.

9. Even those who discourse differently upon this subject do not, I think, have any other conception than the above. For when they distinguish between substance, extension and magnitude, either they understand nothing by the word "substance," or they have only a confused idea of incorporeal substance which they ascribe to corporeal substance and leave to extension the true concept of this material substance, calling it an "accident," and this inappropriately in that it is easy to detect that their words do not at all fit their thoughts.

10. Space or intrinsic place differs from the body that comprises it only in our minds. For truly the extension in length, breadth, and depth that constitutes *space* constitutes *body*. The difference between them is only that we ascribe to a *body* a particular extension. This, as we conceive, exactly changes place with it as it is moved about, whereas we ascribe to a *space* an extension of so general and vague a character that when the body occupying a certain space is removed, we do not think moving the extension belonging to that space, because we suppose that the same extension remains so long as it has the same magnitude, shape, and position with respect to the particular external bodies in reference to which we fix it.

11. But it will be easy to perceive that the extension which constitutes the nature of a body also constitutes the nature of a space, so that there is no more difference between them than that between the nature of genus or species and the nature of an individual. To understand the true idea of a body better, let us examine a stone and remove from it everything that does not belong to the nature of the body. Thus, first remove hardness, for if the

stone is reduced into powder it will lose its hardness without ceasing for all that from being a body; next, remove color, for we have all sometimes seen stones transparent to the point of having no color; remove weight, for we see that fire, though very light, is yet material; remove cold, heat, and all other similar properties, for we never think they are in the stone or that the stone will change its nature whether it seems warm or cold. Having thus examined the stone we see that our real and distinct idea of it is only of a substance which has extension in length, breadth, and depth. Now this is the very notion we have of space, both that filled with body and that called void.

12. True, there is a difference in the way we think about the stone and space—or if we remove a stone from the space or place in which it was, we conceive that we remove its extension, for we regard them as inseparable. At the same time we consider the extension of the place where the stone was as persisting, although this place is now filled by wood, water, air, or some other body, or even is apparently empty; for we conceive that extension in general may be common to stones, wood, water, air, and all other bodies, even to the vacuum if there is any, provided it retains the same magnitude, shape, and position with respect to the external bodies which determine this space.

13. The reason for this is that *place* and *space* do not signify anything truly different from the body said to be in a place; they merely denote its magnitude, shape, and position with respect to other bodies. To determine this position it is necessary to observe other bodies considered immovable; but we may say, in regard to the different bodies we take note of, that the same thing at one and the same time changes its place and does not do so. For example, consider a man seated in the stern of a ship being carried out of the harbor by the wind only in relation to the ship; to us it seems that the man does not change his place, because we see him remaining always in the same position with respect to the parts of his ship. But in relation to the neighboring shores, it seems that this same man is constantly changing his place as he recedes from them and approaches others. And, moreover, we imagine the earth as turning on its axis from west to east exactly as far as the ship travels from east to west, it will once more seem that the man seated in the stern does not change his place, as we now determine place by certain immovable points which we conceive of in the

heavens. But if we think that there is absolutely no point in the universe which is truly at rest (and it will appear from what follows that this can be demonstrated), we shall conclude that nothing in the world has a fixed and constant place unless we fix it there in our thoughts.

14. Yet *place* and *space* have different names because *place* signifies particularly *position* while *space* signifies *magnitude* and *shape*. For we say that something takes the *place* of another although it may not have exactly the same size or shape, so that we do not conceive that it occupies the same *space*; and when the *position* is changed, we say that the *place* is also changed, although size and shape remain the same. Thus if we say that something is *in* a place, we think only of its being placed somewhere in respect to certain other bodies; but if we add that it occupies some space or place, we think besides of some definite size and shape which it fills exactly.

15. Thus we make no distinction between space and extension in length, breadth, and depth. But sometimes we take place to be inherent in the object, sometimes to be external to it. Inherent place is identical with space, but external place is either the surface immediately surrounding the object (and it must be noted that by "surface" is not meant any part of the object surrounded, but only the boundary between the object which surrounds and which is surrounded, which is only an aspect of them) or rather surface in general, which is no more part of one object than of another and which always appears the same so long as it has the same size and shape. For even though we see a body which surrounds another body move elsewhere with its surface, we do not ordinarily say that the surrounded body for all that has changed its place so long as it retains its position relative to other bodies which we regard as fixed. Thus we say that a boat does not move when it is carried downward by the flow of the river, but is driven backward by the wind with an exactly equal force so that it does not change its position with respect to the banks, even though we see that the surface surrounding it continually changes.

16. As for a vacuum, in the philosophical sense—that is, a space devoid of all substance—it is obvious that there is no such space anywhere, since the extension of a space or inherent place is no different from the extension of a body. And as it is from the single fact that a body has extension in length, breadth, and depth that

we conclude it to be a substance, since we cannot conceive of "nothing" having extension, we must conclude the same of a space conceived empty; that is, that as it has extension, it must necessarily also possess substance.

17. But when we take this word in its usual meaning and say that a place is empty, it is accepted that we do not intend to say that there is nothing at all there, but only that there is nothing there of what we think ought to be there. Thus, since a pitcher is made to hold water, we say that it is empty when it contains only air, and that a fishpond without fish has nothing in it, though it is full of water, and a ship is empty when instead of its usual cargo it is only laden with sand (to help it resist the force of the wind); in the same way we say a space is empty when it contains nothing detectable to the senses, although it contains created matter and extended substance. For we commonly take note of bodies near us only when they occasion such strong impressions upon our sense organs that we can detect them thereby. If, then, instead of re-collecting the proper meaning of the words "empty or "nothing," we think of a space in which our senses detect nothing as containing no created matter, we fall into as gross an error as if saying that a pitcher is empty when it contains only air were to lead us to judge that air is not a thing or substance.

18. We have almost all committed this error early in life, for seeing no necessary connection between the container and the thing contained, it appeared that God could remove the whole of that contained body and preserve the container in its former state, without the necessity for another body's taking the place of that which was removed. We can correct this totally false opinion by noting that, though there is no necessary connection between the container and its contents, there is an absolutely necessary connection between the concave shape of the container and the extension that must be contained within this concavity. So much so that it is no more repugnant to sense to conceive a mountain without a valley than such a concavity without its contained extension, or that extension without something extended, because nothing, as has already been noted many times, can have no extension. If it is asked what would happen if God did remove all the contained body without permitting any other body to enter, the answer is, that the sides of the container would be so close that they would be in contact. For two bodies must be in contact when

there is nothing between them, since there would be a contradiction if these two bodies were separated so that there was some distance between them, and yet there was nothing in this distance, distance being an attribute of extension which cannot exist without some extended body.

19. Having seen that the nature of corporeal substance or body consists only in its being something extended, and that its extension does not differ at all from that attributed to empty space, it is easy to perceive that it is no way possible for any of its parts to occupy more space at one time than at another. Thus it can only be rarefied in the manner described above, and there can be no more matter or body in a container full of gold or lead or any other hard and heavy body than when it is only full of air and appears empty. For the magnitude of a piece of matter is not determined by its weight or solidity, as already noted, but only the extension, which is always the same for a given container.

20. It is also very easy to understand that there can be no atoms or indivisible particles of bodies, such as some philosophers have imagined. However small these particles may be taken to be, since they necessarily possess extension, we must think that every one of them must be capable of division into two or more smaller ones, from which it follows that they are divisible. And if we clearly and distinctly know that a thing may be divisible, we must conclude that it is divisible; for if we conclude otherwise we are contradicting what we know. Even if we were to imagine that God had reduced some particle to such an extreme minuteness that it could not be divided further, we would not be entitled to conclude that it was indivisible. For when God had made that particle so small that no creature could divide it, He still could not deprive Himself of the power to divide it, since it is impossible for God to diminish His omnipotence, as has already been noted. This is why we say that the smallest possible extended particle can always, by its very nature, be divided.

21. We know also that this world (or the extended matter which makes up the universe) has no limit. For, imagine such a limit where we will, we can still conceive space indefinitely extending beyond it; this we can not only imagine, but perceive to be so, namely, containing an indefinitely extended body. For our idea of extension on any space whatever is the true idea we should have of the definition of a body.

22. Finally, it is not difficult to infer from all this that the heavens and earth are made of the same material and that even if there were an infinity of worlds they would still be made of the same matter. Whence it follows that there cannot be other worlds, because we clearly perceive that matter (whose only characteristic property is extension) now occupies all the imaginable space where these other worlds might exist and we do not know how to conceive an idea of any other kind of matter. . . .

54. To determine the nature of solid and fluid bodies, we must first take into account the evidence of our senses, since these are sensible qualities, and this teaches us that it consists only in the fact that the particles of fluid bodies so easily give up their place that they offer no resistance to our hands in an encounter, and that on the contrary the particles of solid bodies are so closely united that they cannot be separated without force, which then breaks their union. If we inquire further into the reason why some bodies yield their place without resistance and why others do not do so, we can only find that bodies already prepared for motion in no way prevent the places they are about to leave themselves from being occupied by other bodies. But those at rest cannot be chased out of their places without an external force to cause this change. Whence it follows that a body is fluid when it is divided into many particles, each of which moves individually and in its own way, and that it is solid when all its particles touch one another and are in no state to move away from each other.

55. And I can imagine no cement fitter to unite the particles of solid bodies than *rest*. For what kind of a thing is it? It cannot be more material, for since all these particles are bodies, there is no more reason why they should be united by another body than by themselves. It cannot be a property which is not rest, because there is no quality which differs more from the motion that can separate these particles than the rest which belongs to them. And apart from bodies and their properties what other things are there?

56. As for fluid bodies, although we do not observe the motion of their particles, because they are too small, we can yet know of its existence by several effects. The chief effect arises from the fact that air and water corrupt many other substances and that the particles of which these fluids are composed can produce no corporeal action such as corruption without real motion. I shall later explain what causes there are for the motion of these parts. . . .

61. It is easy to understand by the previous discussion that a solid body at rest between the particles of a fluid body which surrounds it on all sides is so balanced that the slightest force can push it from side to side, even if it is thought to be large, and this whether the force is external or arises because the surrounding fluid body is moving in a certain direction. So rivers flow to the sea and air towards the sunset when the east wind blows. For in such a case the solid body surrounded by a fluid on all sides is carried with it. . . .

62. And even if we pay strict attention to the true nature of motion—which is, properly speaking, the transference of the moving body from the neighborhood of other bodies with which it is in contact, while this transference is equally applicable to all the contiguous bodies—although we do not normally say that both bodies move, we yet know that it is not correct to say that only the solid body moves when, being surrounded on all sides by a fluid, it follows the fluid's path; or that if it had sufficient force to resist it could prevent itself from being moved, for it stays much closer to the surrounding particles if it follows the path of the fluid than if it does not do so. So one must not say that a solid body moves when it is thus carried along by a fluid body.

63. Whence it comes about that there are bodies so solid that they cannot be divided by our hands, although they are smaller than these. If it is true that the particles of solid bodies are not united by any cement and that there is absolutely nothing preventing their separation except the relative rest between them, as stated above, and if it is also true that a body in motion, however slow, always possesses sufficient force to move another smaller body which is at rest . . . then one may ask why we cannot break with our bare hands a nail or any other piece of iron smaller than they are, when each half of the nail may be considered as a body at rest relative to the other half and thus it seems the nail ought to be capable of division by our hands which are larger than it, since motion consists in the moving body's being separated from its contiguous bodies. It must be noted that our hands are very soft and thus share the nature of fluids more than of solids, which is why their particles are not sufficiently agitated against the body we wish to divide. For as the half of a nail can be taken for a body, since it can be separated from its other half, so the part of our hand which touches that half of the nail, and which is much

smaller than the whole hand, can be taken as another body, since it can be separated from the other parts of the hand. And as it can be separated from the rest of the hand more easily than a part of the nail from the rest of the nail, and as we feel pain at such an act, we cannot break a nail with our hands. But if we take a hammer or file or shears or some other similar instrument and if we so employ it as to apply the force of our hand against the part of the body we wish to divide, which must be smaller than the part of the instrument applied against it, we can overcome the solidity even of a large body.

Part III.
The Visible World

48. These things having been considered, let us, in order to begin to see what effects can deduced from the laws of nature, reflect that all the matter of which the world is composed was at the beginning divided into many equal particles, but these particles could not at first have been all round, because many balls joined together could not make up a solid and continuous body like this universe in which, as I have shown, there can be no vacuum. But whatever shape these particles had before, they must have become round in the course of their various circular motions. And since the force with which they were moved at the beginning was large enough to separate one from another, this force still continuing afterwards was, without doubt, great enough to smooth off all their angles successively whenever they met, for it would require less for this effect than it did for the other. And from this one fact, that all the angles of a body are thus smoothed off, it is easy to conceive that it is round.

49. Now as there can be no empty space in any part of the universe and as the particles of matter, being round, are prevented from joining together so tightly as to leave no little intervals or recesses between them, it necessarily follows that these recesses are filled with some other particles of matter, which must be extremely flexible so as to change their shape at any moment to conform to that of the places they enter into. Hence we must believe that what comes off the angles of material particles as they are rounded off by striking one against another is so flexible and acquires such a great speed that the impetuousness of its motion can divide it into innumerable particles which, having no predetermined size or

shape, easily fill up all the little angles or recesses through which the other particles of matter cannot pass.

50. It must be noted that, inasmuch as what comes off as shavings of the material particles when they are rounded off is more flexible, it can also be more easily set in motion and, as a result, whittled down or divided into yet smaller particles than before. This follows because the smaller a body, the larger the surface in proportion to the quantity of matter, and the large surface makes for more numerous encounters with other bodies, which in turn make an effort to move it or divide it, at the same time that its small amount of matter makes it less resistant to their force.

51. It must also be noted that although what thus derives from the whittling away of the particles as they are rounded off has no motion except what comes from these particles, it nevertheless must move much more quickly, because while these particles travel by straight and open ways, they constrain these shavings or dust which lies between them to travel by other, narrower and more twisted ways. In the same way, in compressing a bellows pretty slowly the air is made to go out quite quickly because the hole by which the air makes its escape is straight. And I have already proved that there is necessarily a part of matter which moves quickly and divides into an infinity of little particles, so that all the circular and nonuniform motions in the world can take place without rarefaction or any vacuum, and I do not belive that anything more suitable for this purpose can be imagined that what I have just described.

52. Thus we can say that we have already found two different forms of matter which may be taken to be the forms of the two first elements of the visible world. The first is that of those shavings which must have been separated from the other particles of matter when they were made round and which move with such speed that the force of their motion alone is sufficient to ensure that, meeting with other bodies, they will be crushed and divided by them into an infinite number of little particles whose shape is such as always exactly to fill up all the recesses which they find about other bodies. The other is that of all the remainder of matter whose particles are round and very small in comparison with the bodies we see on earth, but nevertheless possess a determinate quantity, insomuch as they can be divided into other much smaller particles. And we shall find hereafter yet a third form of

some parts of matter—that is, of those which because of their size and shape cannot be so easily set in motion as the ones discussed above. I shall try to show that all the bodies of this visible world are composed of these three forms to be found in matter, as if of three different elements. Thus the sun and the fixed stars have the form of the first of these elements; the heavens that of the second; and the earth with the planets and comets that of the third. . . .

Part IV.

The Earth

1. Although I do not at all wish everyone to be convinced that the bodies of which the visible world is made up were ever produced in the way I have described above, as I have already warned the reader, yet I must here retain the same hypothesis to explain terrestrial objects, so that I may clearly show (as I hope) that by this means easily intelligible and exact causes for all the things observed there may be be deduced. Further, since the same cannot be done by any other hypothesis that may be devised, we may reasonably conclude that, although the world was not made initially in this way but was directly created by God, yet everything in it does not fail to be *now* of the same nature as it would have been had it been so made.

2. Let us therefore now conceive that this earth of ours was formerly composed of the matter of the first element only, which occupied the center of one of the four vortices contained in the space which we call the first heaven, so that it did not differ in any way from the sun except in being smaller. But little by little (as we conceive) the less subtle particles of this matter, attaching themselves together, assembled on the surface and made the clouds and other thicker and darker bodies (like the spots which are seen constantly to be produced and soon dissipated on the surface of the sun), and these dark bodies were also dissipated a little after their production, so that the remaining particles, being heavier than those of the first two elements and having the form of the third, were piled in confusion around this earth and, surrounding it on all sides, made a substance very like the air we breath. Then when that air become thicker and much heavier, the dark bodies which continued to form on the surface of the earth could not be so easily destroyed as before, so that they slowly covered and obscured it completely. And perhaps several layers of such bodies

were piled upon one another, all of which so decreased the force of the containing vortex that it was entirely destroyed, and the earth, with the air and the dark bodies surrounding it, descended towards the sun to the place where it now is. . . .

14. Now when the earth, then composed of three distinct regions, descended toward the sun, this could not cause much change to the two lower, central regions, but did so much to the outermost that it was obliged to divide into two different bodies, then into three, four, and more.

15. I shall try to explain here how these bodies must have been produced. But first I must say something about three or four of the chief actions which contributed to this production. The first consists of the motion of the particles of the celestial matter taken in general; the second, of what is called weight; the third, of light; and the fourth, of heat. By "the motion of the particles of celestial matter taken in general," I understand their continual agitation, which is so great that it not only suffices to cause them to make a compete rotation about the sun each year and another each day around the earth, but also to move them in several other ways. And since once they have set out in some direction they always continue in a straight line as far as possible, it follows that, as they are intermingled with the particles of the third element which compose all the bodies of this outermost region of the earth, they produce various effects, of which I shall mention here the three chief.

16. The first is, that they render transparent all those fluid bodies composed of particles of the third element sufficiently small and so little crowded together that the particles of the second element can go around them on all sides. For in thus going between the particles of these bodies and having enough power to make them move, they do not fail to make passages which are quite straight in all directions, or at least passages equally suitable for transmitting the action of light and so making the body transparent. Thus we learn by experience that every fluid on earth which is both pure and composed of sufficiently small particles is transparent. . . . And all solid bodies are transparent that have been made from some transparent fluid whose particles have come to rest little by little one against another without anything being mixed with them that could change their order. Conversely, all bodies are opaque or dark whose particles have been united by

some outside force not subject to the motion of the celestial matter. For though there may remain in such bodies many pores through which the particles of the second element can pass, at the same time since these pores are stopped up or interrupted in various places they cannot transmit the action of light. . . .

18. The second effect produced by the agitation of the subtle matter in terrestrial bodies, chiefly in fluid ones, is that when two or more kinds of particles in these bodies are mingled in confusion, either it separates them or it sorts them out and distributes them equally in all parts of the body and so purifies it and makes every drop exactly like every other. The reason for this is, that sliding on all sides between unequal terrestrial particles it continually pushes those which by their size or shape or position get more in the way than the others, until it has so changed their position that they are uniformly spread through the body, and so fitted in with the others that it no longer impedes the motion of the subtle matter; or, if it cannot arrange them in this way, it separates them completely and makes a different body of them. . . .

19. The third effect of this celestial matter is to make all drops of fluid quite round whenever they are entirely surrounded by air or any other fluid whose nature is sufficiently different to prevent their mixing up together. . . .

20. The second action of which I have undertaken to speak here is that which makes bodies heavy, which has a close connection with what makes drops of water round. For it is the same subtle matter which, by the mere fact of moving equally on all sides around a drop of water, pushes all the particles of the surface toward the center of the drop that, by the mere fact of moving around the earth, also pushes toward the earth all the bodies we call heavy, among which are the particles . . .

22. Since there is no empty space around the earth and since it does not itself possess the power of turning every twenty-four hours about its axis, but is carried around by the celestial matter which surrounds it and which penetrates through all its pores, it should be regarded as a body without motion. Similarly, the celestial matter would not be thought either heavy or light if it had no other motion than that which makes it rotate in twenty-four hours with the earth, but as it has many more motions not required to produce this effect, its other motions are employed either to turn more quickly than the earth in the same direction, or to make

other motions in other directions. And these cannot be in such straight lines as if the earth were not in the way, not only does the subtle matter [ether] try to make the earth round or spherical, as with a drop of water but also has more power of rising away from the center about which it turns than the particles of the earth do, which makes it light compared to them. . . .

23. And it must be noted that the power the celestial matter possesses of getting farther form the center of the earth can have its effect only if the particles which rise take the place of some terrestrial particles which fall at the same time. . . . Thus each heavy body is not pushed toward the center of the earth by all the celestial matter which surrounds it, but only by those particles of this matter which rise into its place at the same time that it falls and which consequently are exactly as large as it is. . . .

28. As for light, which is the third action which we have to consider here, I think I have already sufficiently explained its *nature*, it only remains to note that although all the rays come in the same direction from the sun and do nothing except exert pressure in a straight line on the bodies which they encounter, nevertheless they cause different motions in the particles of the third element of which the outermost region of the earth is com-posed, because these particles, having other causes of motion, do not always present themselves to the rays in the same fashion. . . .

29. It is a similar agitation of the particles of terrestrial bodies that is called *heat* (whether it be excited by the light of the sun or by any other cause), chiefly when it is larger than usual and can move the nerves in our hands enough to be felt, for this appella-tion "heat" is related to the sense of touch. And here it may be noted why heat produced by light remains in terrestrial bodies even when the light has gone until some other cause removes it—it consists only in the motion of the particles of the body and when they have once been put into such motion they remain in motion according to the laws of nature until this motion can be trans-ferred to some other body.

30. It should be noted also that the terrestrial particles thus agitated by the sun's light agitate others underneath themselves, and these in turn agitate others lower still, and so on, so that although the sun's rays do not pass farther than the outer surface of opaque terrestrial bodies at the same time, its heat reaches to

the lowest particles of the third element, which composes the second or middle region.

31. Finally, this agitation of the particles of terrestrial bodies is ordinarily the reason why they occupy more space than when they are at rest or are less agitated. The reason of this is that, their shapes being irregular, they fit together better when they keep the same relative positions than when their motion makes them change positions. And hence heat rarefies almost all terrestrial bodies, some more than others according to their various shapes and the arrangements of their particles—so that there are some which heat condenses, because their particles are better arranged and can come closer together when agitated than when not. Snow and ice are examples of this. . . .

133. Hitherto I have tried to explain the nature and chief properties of air, water, earth, and fire, since these are the substances most commonly found everywhere in this sublunary region in which we dwell; hence they are called the four elements. But there is yet another substance—the magnet—which may be said to be more widely distributed than any of these four, since the whole mass of the earth is a magnet, and since we know of no place into which we can go where its power cannot be detected. So, not wishing to overlook what is most common to earth, I must now explain the magnet. Let us return to what was said earlier about the grooved or screwed particles of the first element of this visible world, and applying here to the earth what is there said about a certain star, consider that there are in the earth's central regions certain pores or little pipes parallel to its axis through which the grooved particles pass freely from one pole to the other, and that these pipes are so hollow and so well arranged for the shape of the grooved particles that those which accept particles coming from the South Pole cannot accept those coming from the North Pole and vice versa, because they are threaded in opposite directions . . . After these grooved particles have traversed the whole earth from one half to the other, following lines parallel to its axis, some return through the surrounding air toward the half by which they entered and thus passing reciprocally from earth to air and from air to earth, make up a kind of vortex [which in turn explains attraction].

207. Because I do not wish to trust too much in myself, I do

not assert anything here and I submit my opinions to the judgment of the wise and the authority of the Church; I even beg my readers not to have complete faith in what they shall find written here, but only to examine it, and not to accept anything what the strength and evidence of reason compels them to believe.

32. Boyle's Corpuscular Philosophy

Of the Excellency and Grounds of the Corpuscular or Mechanical Philosophy

By embracing the corpuscular or mechanical philosophy, I am far from supposing with the Epicureans that atoms accidentally meeting in an infinite vacuum were able, of themselves, to produce a world and all its phenomena: nor do I suppose, when God had put into the whole mass of matter an invariable quantity of motion, he needed do no more to make the universe; the material parts being able, by their own unguided motions, to throw themselves into a regular system. The philosophy I plead for reaches but to things purely corporeal; and distinguishing between the first origin of things and the subsequent course of nature, teaches that God indeed gave motion to matter; but that, in the beginning, he so guided the various motion of the parts of it as to contrive them into the world he designed they should compose; and established those rules of motion, and that order amongst things corporeal, which we call the laws of nature. Thus the universe being once framed by God and the laws of motion settled and all upheld by his perpetual concourse and general providence; the same philosophy teaches, that the phenomena of the world are physically produced by the mechanical properties of the parts of matter, and, that they operate upon one another according to mechanical laws. 'Tis of this kind of corpuscular philosophy, that I speak.

And the first thing that recommends it is the intelligibleness or clearness of its principles and explanations. Among the peripa-

SOURCE: Robert Boyle, *The Excellency and Grounds of the Mechanical Hypothesis* (London, 1674), taken from Peter Shaw's abridgement (1725), II, 187-96.

tetics[1] there are many intricate disputes about matter, privation, substantial forms, their educations, etc. And the chymists are puzzled to give such definitions, and accounts, of their hypostatical principles[2] as are consistent with one another, and to some obvious phenomena: and much more dark and intricate are their doctrines about the Archeus, Astral Beings, and other odd notions; which perhaps, have in part occasioned the darkness and ambiguity of their expressions, that could not be very clear, when the conceptions were obscure. And if the principles of the Aristotelians and chymists are thus obscure, it is not to be expected that the explications made by the help of such principles only should be intelligible. And, indeed, many of them are so general and slight, or otherwise so unsatisfactory, that, granting their principles, 'tis very hard to understand or admit their applications of them to particular phenomena. And, methinks, even in some of the more ingenious and subtle of the peripatetic discourses, the authors, upon their superficial and narrow theories, have acted more like painters than philosophers; and only shown their skill in making men fancy they see castles, cities, and other structures, that appear solid, magnificent, and extensive; when the whole piece is superficial, artificially made up of colours, and comprized within a frame. But, as to the corpuscular philsophy, men do so easily understand one another's meaning, when they talk of local motion, rest, magnitude, shape, order, situation, and contexture, of material substances; and these principles afford such clear accounts of those things, that are rightly deduced from them alone; that, even such peripatetics or chymists, as maintain other principles, acquiesce in the explications made by these, when they can be had; and seek no further: though, perhaps, the effect be so admirable, as to make it pass for that of a hidden form, or an occult quality. Those very Aristotelians, who believe the celestial bodies to be moved by intelligences, have no recourse to any peculiar agency of theirs to account for eclipses: and we laugh at those East Indians who, to this day, go out in multitudes, with some instruments, to relieve the distressed luminary; whose loss of light, they

1. Aristotelians, especially the reactionary formalists of the sixteenth and seventeenth centuries.
2. Essential principles or elements. The reference is to the followers of Paracelsus and van Helmont. The "Archeus" is a vital spirit responsible for both chemical and physiological reactions.

fancy, proceeds from some fainting fit; out of which it must be roused. For no intelligent man, whether chymist or perpatetic, flies to his peculiar principles, after he is informed that the moon is eclipsed, by the interposition of the earth betwixt her, and it; and the sun, by that of the moon, betwixt him and the earth. And, when we see the image of a man cast into the air by a concave spherical speculum; though most men are amazed at it, and some suspect it to be no less than an effect of witchcraft, yet he who is skilled enough in catoptrics will, without consulting Aristotle or Paracelsus or flying to hypostatical principles or substantial forms, be satisfied that the phenomenon is produced by rays of light reflected and made to converge according to optical and mathematical laws.

I next observe that there cannot be fewer principles than the two grand ones of our philosophy, matter and motion; for matter alone, unless it be moved, is wholly unactive; and, whilst all the parts of a body continue in one state, without motion, that body will not exercise any action, or suffer any alteration; though it may, perhaps, modify the action of other bodies that move against it.

Nor can we conceive any principles more primary than matter and motion: for either both of them were immediately created by God; or, if matter be eternal, motion must either be produced by some immaterial supernatural agent; or it must immediately flow, by way of emanation, from the nature of the matter it appertains to.

There cannot be any physical principles more simple than matter and motion; neither of them being resoluble into any other thing.

The next thing which recommends the corpuscular principles is their extensiveness. The genuine and necessary effect of the strong motion of one part of matter against another is either to drive it on, in its entire bulk, or to break and divide it into particles of a determinate motion, figure, size, posture, rest, order or texture. The two first of these, for instance, are each of them capable of numerous varieties: for the figure of a portion of matter may either be one of the five regular geometrical figures, some determinate species of solid figures, or irregular, as the grains of sand, feathers, branches, files etc. And, as the figure, so the motion of one of these particles may be exceedingly diversified, not only by the determination to a particular part of the world but by several other things: as by the almost infinitely different degrees of celerity; by the manner of its progression, with or without rotation, etc. and more yet by the line wherein it moves; as circular, elliptical, parabolical, hyperbolical, spiral, etc. For, as later geometricians have shown that these curves may be compounded of several motions, that is, described by a body whose motion is mixed, and results from two or more simple motions; so, how many more curves may be made by new compositions, and re-compositions of motion, is not easy to determine.

Now, since a single particle of matter, by virtue of only two mechanical properties that belong to it, may be diversified so many ways; what a vast number of variations may we suppose capable of being produced by the compositions, and re-compositions of myriads of single invisible corpuscles, that may be contained and concreted in one small body; and each of them be endued with more than two or three of the fertile, universal principles above-mentioned? And the aggregate of those corpuscles may be further diversified by the texture resulting from their convention into a body; which, as so made up, has its own magnitude, shape, pores, and many capacities of acting and suffering, upon account of the place it holds among other bodies, in a world constituted like ours: so that, considering the numerous diversifications that compositions and re-compositions may make of a small number, those who think the mechanical principles may serve, indeed, to account for the phenomena of some particular part of natural philosophy, as statics, the theory of planetary motions etc. but prove unapplicable to all the phenomena of things corporeal seem to imagine, that by putting together the letters of the alphabet one may, indeed, make up all the words to be found in Euclid or Virgil, or in the Latin or English language, but that they can by no means supply words to all the books of a great library; much less, to all the languages in the world.

There are other philosophers, who, observing the great efficacy of magnitude, situation, motion, and connection in engines are willing to allow those mechanical principles a great share in the operations of bodies of a sensible bulk and manifest mechanism; and, therefore, to be usefully employed, in accounting for the effects and phenomena of such bodies: though they will not admit that these principles can be applied to the hidden transactions among the minute particles of bodies; and, therefore, think it necessary to refer these to what they call nature, substantial forms, real qualities, and the like unmechanical agents. But this is not necessary: for the mechanical properties of matter are to be found, and the laws of motion take place, not only in the great masses and the middle-sized lumps, but in the smallest fragments of matter: a less portion of it being as much a body as a greater, must as necessarily as the other have its determinate bulk and figure. And whoever views sand through a good microscope will easily perceive that each minute grain has as well its own size and shape as a rock or a mountain. Thus too, when we let fall a large stone, and a pebble, from the top of a high building, they both move comformable to the laws of acceleration, in heavy descending bodies: and the rules of motion are observed, not only in cannon-bullets, but in small shot; and the one strikes down a bird, according to the same laws, as the other batters a wall. And though nature works with much finer materials, and employs more curious contrivances, than art; yet an artist, according to the quantity of the matter he employs, the exigency of the design he undertakes, and the magnitude and shape of the instruments he uses, is able to make pieces of work of the same nature or kind, of extremely different bulks where yet the like art, contrivance, and motion may be observed. Thus a smith who, with a hammer and other large instruments, can, out of masses of iron, forge great bars or wedges to make strong and ponderous chains to secure streets and gates may, with lesser instruments, make smaller nails, and filings, almost as minute as dust; and with yet finer tools, make links wonderfully light and slender. And therefore, to say that though in natural bodies, whose bulk is manifest and their structure visible, the mechanical principles may be usefully admitted but are not to be extended to such portions of matter, whose parts and texture are invisible, is like allowing that the laws of mechanism may take place in a town-clock, and not in a pocket-watch: or, because the terraqueous globe is a vast magnetical body, one should affirm that magnetical laws are not to be expected manifest in a small spherical piece of loadstone; yet experience shows us that, notwithstanding the immense disproportion betwixt these two spheres, the terella[3] as well as the earth, hath its poles, equa-

---

3. William Gilbert's spherical loadstone.

tor, and meridians; and in several other magnetical properties resembles the terrestrial globe.

When, to solve the phenomena of nature, agents are made use of which, though they involve no contradiction in their notions, as many think substantial forms and real qualities do, yet are such that we conceive not how they operate to produce effects; such agents I means, as the soul of the world, the universal spirit, the plastic power etc., the curiosity of an inquisitive person is not satisfied hereby; who seeks not so much to know what is the general agent that produces a phenomenon, as by what means, and after what manner, it is produced. Sennertus,[4] and other physicians, tell us of diseases which proceed from incantation; but sure, it is very trivial to a sober physician, who comes to visit a patient reported to be bewitched, to hear only that the strange symptoms he meets with, and would have an account of, are produced by a witch or the devil; and he will never be satisfied with so short an answer, if he can by any means reduce those extravagant symptoms to any more known and stated diseases; as epilepsies, convulsions, hysteric fits, etc. and if he cannot, he will confess his knowledge of this distemper to come far short of what might be expected and attained in other diseases, wherein he thinks himself bound to search into the morbific matter; and will not be satisfied, till he can, probably, deduce from that, and the structure of the human body, and other concurring physical causes, the phenomena of the malady. And it would be of little satisfaction to one who desires to understand the causes of the phenomena in a watch, and how it comes to point at and strike the hours to be told that a certain watch-maker so contrived it: or, to him who would know the true causes of an echo, to be answered that it is a man, a vault, or a wood, that makes it.

I come now to consider that which I observe most alienates other sects from the mechanical philosophy; viz. a supposition, that it pretends to have principles so universal and mathematical that no other physical hypothesis can be tolerated by it.

This I look upon as an easy, indeed but an important mistake: for the mechanical principles are so universal, and appliable to so

4. Daniel Sennert (1572-1637), a widely read German writer on medical and
   chemical subjects, and an early exponent of a corpuscular chemistry
   (though he was also an Aristotelian).

many purposes, that they are rather fitted to take in, than to exclude, any other hypothesis founded on nature. And such hypotheses, if prudently considered, will be found, as far as they have truth on their side, to be either legitimately deducible from the mechanical principles or fairly reconcileable to them. For such hypotheses will, probably, attempt to account for the phenomena of nature, either by the help of a determinate number of material ingredients, such as the tria prima of the chymists,[5] or else by introducing some general agents, as the Platonic soul of the world, and the universal spirit, asserted by some chymists; or, by both these ways together.

Now, the chief thing that a philosopher should look after, in explaining difficult phenomena, is not so much what the agent is or does as, what changes are made in the patient, to bring it to exhibit the phenomena proposed; and by what means, and after what manner, those changes are effected. So that the mechanical philosopher being satisfied, one part of matter can act upon another, only by virtue of local motion, or the effects and consequences thereof; he considers, if the proposed agent be not intelligible and physical, it can never physically explain the phenomena; and if it be intelligible and physical, it will be reducible to matter and some or other of its universal properties. And the indefinite divisibility of matter, the wonderful efficacy of motion, and the almost infinite variety of coalitions and structures that may be made of minute and insensible corpuscles being duly weighed; why may not a philosopher think it possible to make out, by their help, the mechanical possibility of any corporeal agent, how subtle, diffused, or active soever, that can be solidly proved to have a real existence in nature? Though the Cartesians are mechanical philosophers, yet their subtle matter, which the very name declares to be a corporeal substance, is, for ought I know, little less diffused through the universe, or less active in it, than the universal spirit of some chymists; not to say the world soul of the Platonists. But whatever be the physical agent, whether it be inanimate, or living, purely corporeal, or united to an intellectual substance; the above-mentioned changes, wrought in the body made to exhibit the phenomena, may be effected by the same, or the like means; or after the same, or the like manner: as,

5. The Paracelsan elements of salt, sulphur, and mercury.

for instance, if corn be reduced to meal, the materials and shape of the mill-stones and their peculiar motion and adaptation will be much of the same kind; and, to be sure, the grains of corn will suffer a various attrition, and comminution in their passage to the form of meal, whether the corn be ground by a watermill, or a windmill, a horsemill, or a handmill; that is, a mill, whose stones are turned by inanimate, by brute, or by rational agents. And if an angel himself should work a real change in the nature of a body, 'tis scarce conceivable to men how he could do it without the assistance of local motion; since, if nothing were displaced, or otherwise moved than before it is hardly conceivable how it should be, in itself, different from what it was before.

But if the chymists, or others, who would deduce a compleat natural philosophy from salt, sulphur, and mercury, or any determined number of ingredients of things, would well consider what they undertake, they might easily discover that the material parts of bodies can reach but to a few phenomena of nature, whilst these things [ingredients] are considered but as quiescent things, whence, they would find themselves to suppose them active; and that things purely corporeal cannot but by means of local motion, and the effects that may result from it, be very variously shaped, sized, and combined parts of matter: so that the chymists must leave the greatest part of the phenomena of the universe unexplained, by means of the ingredients of bodies, without taking in the mechanical and more comprehensive properties of matter, especially local motion. I willingly grant that salt, sulphur, and mercury, or some substances analogous to them, are obtainable, by the action of the fire, from a very great many dissipable bodies here below. Nor do I deny that in explaining several phenomena of such bodies, it may be of use to a naturalist to know and consider that as sulphur, for instance, abounds in the body proposed, it may be, thence, probably argued that the qualities usually attending that principle, when predominant, may be also upon its account found in the body that so largely partakes of it. But, though chymical explications are, sometimes, the most obvious, yet they are not the most fundamental and satisfactory: for the chymical ingredient itself, whether sulphur, or any other must owe its nature and other qualities to the union of insensible particles, in a convenient size, shape, motion, or rest, and texture; all which are but mechanical properties of convening corpuscles. And this may be illustrated by what happens in artificial fire-works. For, though in most of those sorts, made either for war, or re-creation, gun-powder be a principal ingredient; and many of the phenomena may be derived from the greater or less proportion wherein it enters the compositions: yet there may be fire-works made without gun-powder, as appears by those of the ancient Greeks and Romans. And gun-powder owes its aptness to fire, and to be exploded, to the mechanical texture of more simple portions of matter, nitre, charcoal, and sulphur. And sulphur itself, though it be by many chymists mistaken for an hypostatical [essential] prin-ciple, owes its inflammability to the union of still more simple and primary corpuscles; since chymists confess that it had an inflamma-ble ingredient: and experience shows that it very much abounds with an acid and uninflammable salt and is not destitute of a terrestrial part. It may, indeed, be here alleged that the productions of chymical analyses are simple bodies; and, upon that account, irresoluble; but that several substances, which chymists call the salts, sulphurs, or mercuries of the bodies that afford them, are not simple and homogeneous is demonstrable. Nor is their not being easily dissipable, or resoluble, a clear proof of their not being made up of more primitive portions of matter. For com-pounded bodies may be as difficultly resoluble as most of those that chymists obtain by the fire: witness common greenglass, which is far more durable, and irresoluble, than many of those which pass for hypostatical substances. And some enamels will, for several times, even vitrify in the forge, without losing their nature or often so much as their colour: yet, enamel consists of salt, powder of pebbles, or sand, and calcined tin; and, if not white, usually of some tinging metal or mineral. But how indestructible soever the chymical principles are supposed, several of the operations ascribed to them will never be made appear without the help of local motion: were it not for this, we can but little better solve the phenomena of many bodies by knowing what ingredients compose them than we can explain the opera-tions of a watch by knowing of how many and of what metals, the balance, the wheels, the chain, and other parts consist; or than we can derive the operations of a windmill from barely knowing that it is made up of wood, stone, canvas, and iron. And here let me add that it would not at all overthrow the corpuscularian hypothesis, though, either by more exquisite purifications or by some other operations, than the usual analysis by fire, it should appear that the material principles of mixed bodies are not the tria prima of the vulgar chymists; but, either substances of another nature, or fewer in number; or, if it were true that the Helmontians had such a resolving menstruum as their master's alkahest,[6] by which he affirms that he could reduce stones into salt, of the same weight with the mineral; and bring both that salt, and all other mixed and tangible bodies, into insipid water. For whatever be the number or qualities of the chymical principles, if they really exist in nature, it may very possibly be shown that they are made up of insensible corpuscles, of determinate bulks and shapes: and, by the various coalitions and textures of such corpuscles, many material ingredients may be composed, or made to result. But though the alkahestical reductions, newly mentioned, should be admitted, yet the mechanical principles might well be accommodated even to them. For the solidity, taste, etc. of salt may be fairly accounted for by the stiffness, sharpness, and other mechanical properties of the minute particles whereof salt consists: and if, by a farther action of the alkahest, the salt, or any other solid body, be reduced into insipid water, this also may be explained by the same principles; supposing a farther comminution of its parts, and such an attrition as wears off the edges and points that enabled them to strike briskly upon the organ of taste: for as to fluidity and firmness, they, principally, depend upon two of our grand principles, motion and rest. And 'tis certain that the agitation, or rest, and the looser contact, or closer cohesion of the particles, is able to make the same portion of matter at one time a firm and at another a fluid body. So that, though future sagacity and industry of chymists should obtain, from mixed bodies, homogeneous substances, different in number, nature, or both, from their vulgar salt, sulphur, and mercury; yet the corpuscular philosophy is so general and fertile as to be fairly reconcilable to such a discovery; and also so useful, that these new material principles will, as well as the old tria prima, stand in need of the more universal principles of the corpuscularians; especially of local motion. And, indeed, whatever elements or ingredients men have pitched upon; yet, if they take not in the mechanical properties of matter, their principles are so deficient that I have observed both the materialists and chymists

6. A universal solvent.

not only leave many things unexplained, to which their narrow principles will not extend; but, even in the particulars they presume to give an account of, they either content themselves to assign such common and indefinite causes as are too general to be satisfactory; or, if they venture to give particular causes, they assign precarious or false ones, liable to be easily disproved by circumstances, or instances, whereto their doctrines will not agree. The chymists, however, need not be frightened from acknowledging the prerogative of the mechanical philosophy, since that may be reconcilable with the truth of their own principles, so far as they agree with the phenomena they are applied to: for these more confined hypotheses may be subordinate to those more general and fertile principles; and there can be no ingredient assigned that has a real existence in nature but may be derived, either immediately or by a row of compositions, from the universal matter, modified by its mechanical properties. For if with the same bricks, differently put together and ranged, several bridges, vaults, houses, and other structures may be raised merely by a various contrivance of parts of the same kind; what a great variety of ingredients may be produced by nature from the various coalitions and contextures of corpuscles, that need not be supposed, like bricks, all of the same size and shape; but to have, both in the one and the other, as great a variety as could be wished for? And the primary and minute concretions that belong to these ingredients may, without opposition from the mechanical philosophy, be supposed to have their particles so minute and strongly coherent that nature of herself scarce ever tears them asunder. Thus mercury and gold may be successively made to put on a multitude of disguises; and yet so retain their nature as to be reducible to their pristine forms.

From hence it is probable if, besides rational souls, there be any immaterial substances, such as the heavenly intelligences, and the substantial forms of the Aristotelians, that are regularly to be numbered among natural agents; their way of working being unknown to us, they can only help to constitute and effect things, but will very little help us to conceive how things are effects; so that, by whatever principles natural things are constituted, 'tis by the mechanical principles that their phenomena must be clearly explained. For instance though we grant, with the Aristotelians, that the planets are made of a quintessential matter and moved by angels or immaterial intelligences; yet, to explain the stations, progressions and retrogradations, and other phenomena of the planets, we must have recourse either to excentrics, epicycles, etc. or to motions, made in elliptical, or other peculiar lines; and, in a word, to theories wherein the motion, figure, situation and other mathematical, or mechanical properties are chiefly employed. But if the principles proposed be corporeal, they will then be fairly reducible or reconcilable to the mechanical principles; these being so general and fertile that, among real material things, there is none but may be derived from or reduced to them. And when the chymists shall show that mixed bodies owe their qualities to the predominance of any one of their three grand ingredients, the corpuscularians will show that the very qualities of this or that ingredient flow from its peculiar texture, and the mechanical properties of the corpuscles that compose it. And to affirm that because the chemical furnaces afford a great number of uncommon productions, and phenomena, that there are bodies or operations amongst things purely corporeal not derivable from or reconcilable to the principles of mechanical philosophy is to say, because there are many and various hymns, pavanes, threnodies, courants, gavottes, sarabands, etc. in a music book, many of the tunes, or notes have no dependence on the scale of music; or as if because excepting rhomboids, squares, pentagons, chiliagons, and numerous other polygons, one should affirm there are some rectilineal figures not reducible to triangles, or that have properties which overthrow Euclid's doctrine of triangles and polygons.

I shall only add that as mechanical principles and explanations, where they can be had, are, for their clearness, preferred by materialists themselves; so the sagacity and industry of modern naturalists and mathematicians, having happily applied them to several of those difficult phenomena which before were referred to occult qualities it is probable that when this philosophy is more scrutinized and farther improved, it will be found applicable to the solution of still more phenomena of nature. And 'tis not always necessary that he who advances an hypothesis in astronomy, chymistry, anatomy, etc. be able, a priori, to prove it true, or demonstratively to show that the other hypothesis proposed about the same subject must be false; for as Plato said that the world is God's epistle to mankind; and might have added, in his own way, that it was written in mathematical characters; so, in the physical explanations of the parts of the system of the world, methinks there is somewhat like what happens when men conjecturally frame several keys to read a letter written in ciphers. For though one man, by his sagacity, finds the right key, it will be very difficult for him either to prove, otherwise than by trial, that any particular word is not such as 'tis guessed to be by others, according to their keys; or to show, a priori, that theirs are to be rejected and his to be preferred; yet, if due trial being made, the key he proposes be found so agreeable to the characters of the letter, as to enable one to understand them, and make coherent sense of them, its suitableness to what it should decipher is, without either confutations or foreign positive proofs, alone sufficient to make it accepted as the right key of that cipher. Thus, in physical hypotheses, there are some that, without falling foul upon others, peacably obtain the approbation of discerning men only by their fitness to solve the phenomena for which they were devised, without thwarting any known observation or law of nature; and therefore, if the mechanical philosophy shall continue to explain corporeal things, as it has of late, 'tis scarce to be doubted but that in time unprejudiced persons will think it sufficiently recommended, by its being consistent with itself and applicable to so many phenomena of nature.

33. Newton on Attraction

Conclusion

Hitherto I have explained the System of this visible world, as far as concerns the greater motions which can easily be detected. There are however innumerable other local motions which on account of the minutenesss of the moving particles cannot be detected, such as the motions of the particles in hot bodies, in fermenting bodies, in putrescent bodies, in growing bodies, in the organs of sensation and so forth. If any one shall have the good fortune

SOURCE: Isaac Newton, *Principia*, The Conclusion; from A. Rupert Hall and Marie Boas Hall, *Unpublished Scientific Papers of Isaac Newton* (Cambridge: The University Press, 1962), pp. 334-44.

to discover all these, I might almost say that he will have laid bare the whole nature of bodies so far as the mechanical causes of things are concerned. I have least of all undertaken the improvement of this part of philosophy. I may say briefly, however, that nature is exceedingly simple and conformable to herself. Whatever reasoning holds for greater motions, should hold for lesser ones as well. The former depend upon the greater attractive forces of larger bodies, and I suspect that the latter depend upon the lesser forces, as yet unobserved, of insensible particles. For, from the forces of gravity, of magnetism and electricity it is manifest that there are various kinds of natural forces, and that there may be still more kinds is not to be rashly denied. It is very well known that greater bodies act mutually upon each other by those forces, and I do not clearly see why lesser ones should not act on one another by similar forces. If spirit of vitriol (which consists of common water and an acid spirit) be mixed with Sal Alkali or with some suitable metallic powder, at once commotion and violent ebullition occur.[1] And a great heat is often generated in such operations. That motion and the heat thence produced argue that there is a vehement rushing together of the acid particles and the other particles, whether metallic or of Sal Alkali; and the rushing together of the particles with violence could not happen unless the particles begin to approach one another before they touch one another. The force of whatever kind by which distant particles rush towards one another is usually in popular speech, called an attraction. For I speak loosely when I call every force by which distant particles are impelled mutually towards one another, or come together by any means and cohere, an attraction. Moreover, if the solution of a metal in spirit of vitriol be distilled, that spirit which by itself will ascend at the heat of boiling water does not ascend before the material is heated almost to incandescence, being held down by the attraction of the metal. So also spirit of nitre (which is composed of water and an acid Spirit) violently unites with salt of tartar;[2] then, although the spirit by itself can be distilled in a gently heated bath, nevertheless it cannot be separated from the salt of tartar except by a vehement fire. And

1. Sulphuric acid mixed with potassium carbonate will bubble violently as carbon dioxide is released.
2. Nitric acid and potassium carbonate react to produce potassium nitrate (KNO_3) and carbon dioxide.

similar reasoning applies to all sufficiently strong acids and the bodies dissolved in them. In the same way if oil of vitriol be mixed with common water, a vehement heat arises from the impetuous rushing together of the particles, and then the united particles so cohere that water, although by itself very volatile, cannot be wholly distilled off before a large part of the oil is driven to ascend by the heat of the water at boiling point, or rather more. And by a similar cohesion the phlegmatic water and the acid spirit, of which aqua fortis is composed, hold on to each other and distill over together.[3] And the acid spirit in butter of antimony,[4] because it is copious and by itself extremely volatile, cohering with the metallic particles takes them with it in distillation, so that both ascend together by a not very violent heat in the form of a fusible salt. In the same way the spirits of urine and salt by cohering compose Sal Ammoniac;[5] and that salt by cohering with other bodies carries up the particles of these in sublimation. And Sal Akali liquefies by attracting vapour from the air, and retains the attracted liquor so firmly that the two can scarcely be separated by distillation.

And because all the particles of a compound are greater than the component particles, and larger are agitated with greater difficulty, so the particles of sal ammoniac are less volatile than the smaller particles of the spirits of which they are composed. So gold, which is the most fixed of all bodies, seems to consist of compound particles, not all of which, on account of their massiveness, can be carried up by the agitation of heat, and whose component parts cohere with one another too strongly to be separated by that agitation alone. That the particles of water and spirits, however, are most subtle and small of all, and for that reason exceedingly volatile, is consonant with reason. And the acid spirit of salts rendered more subtle in mercury sublimate divides Antimony and metals already sublimed in a fresh sublimation, and in dividing volatilizes, better than the same spirit by itself distilled from these same salts.

Moreover, when bodies dissolved in acids are precipitated by salt of tartar, the precipitation is probably caused by the stronger

3. Newton is, like Boyle, trying to speak of the composition of bodies not in terms of elements but of simple substances (see below). Aqua fortis is strong nitric acid.
4. Butter of antimony (so called from its appearance) is antimony chloride.
5. Ammonium chloride (NH_4CL). Spirit of urine is an ammoniacal solution.

attraction by which the salt of tartar draws those acid spirits from the dissolved bodies to itself. For if the spirit does not suffice to retain them both, it will cohere with that which attracts more strongly. Thus also the acid spirit in mercury sublimate,[6] acting on metals, leaves the mercury. That spirit in butter of antimony coalesces with water poured on to it, and allows the antimony abandoned by it to be precipitated. And the acid spirit, joined with common water in aqua fortis and spirit of vitriol, by acting on metals dissolved in those menstruums,[7] leaves the water and allows the water to ascend by itself with a merely gentle heat, whereas before it could be separated from the spirit by distillation. And spirit of vitriol, meeting with the fixed particles of salt of nitre, looses the spirit of nitre which was formerly joined to those fixed particles, so that that latter spirit can be more easily distilled than before.

It is probable that the particles of homogenous bodies, whether solid or fluid, cohere by a mutual attraction, and thence it is that whenever the particles of quicksilver in a Torricellian Tube are everywhere contiguous to each other and to the tube, the quicksilver can be suspended and sustained to a vertical height of 40, 50, 60 inches and more. It is agreeable to reason, however, that bodies, like their cohering particles (which according to the variation in their magnitudes, shapes and forces move either by gliding between one another more or less easily, or by receding from each other in some way) should be fluid or solid, soft or hard, ductile or elastic and more or less apt to be liquefied by fire.

And just as magnetic bodies repel as well as attract each other, so also the particles of bodies can recede from each other by certain forces. Water does not mix with oil, or lead with iron or copper, on account of the repulsion between the particles. Mercury penetrates gold, silver, tin and lead, but it does not penetrate wood and bladders and salts and rock; water on the contrary penetrates wood, bladders, salts and rock but does not penetrate metals: the cause of this is not the greater or lesser subtlety of the parts of water or mercury, but is, I suspect, to be attributed to the mutual attraction of particles of a like kind, less to the repulsion of likes. Just as the magnet is endowed with a double force, the one of gravity and the other magnetic, so there are various forces of the

6. Mercuric chloride.
7. Solutions.

same particles arising from various causes. There must be one force whereby particles of oil attract each other, and another whereby they repel the particles of water. The particles of metals attract the particles of the acids which dissolve them, and the particles composed of them both swimming in the water flee from each other. For although their specific gravity is greater than that of water, and for that reason they ought to fall to the bottom, nevertheless by fleeing from each other they spread uniformly throughout the water in which they swim, even though its volume is over a thousand times greater than theirs. Conceive the metallic particles on account of their fixity to be larger than the particles of acid and let them have a double force. The first force is an attractive one and is the stronger, but it quickly decreases with distance from the particle; the second is a repulsive force which decreases more slowly and on that account extends more widely. Since there will be a certain limit within which the attractive force will be stronger than the repulsive, beyond this limit the repulsive force will prevail. Each and every particle [of metal] attracts so many particles of acid as can remain within that limit and will retain them about itself, and together with them composes a particle of salt, and it will repel the other particles, whether acid or metallic, outside that limit. And although it is possible to make the particles come together by evaporation of the water and finally join together into crystals (because of the narrowness of the space in which they are contained) and that regularly on account of the regular arrangement between [the particles] which they had before when swimming separately in the water; yet they cohere in such crystals by the attractive force of the acid particles, now in contact with each other. If, however, the crystals are immersed in sufficient springwater, and if the forces by which the [crystalline] particles are attracted to each other are not much greater than the forces by which they attracted towards the particles of water, the former particles are easily driven off from the crystals when agitated by a gentle motion; and then, being beyond [the limit of] the forces of mutual attraction, they flee from each other, and by that repulsion are diffused throughout the liquid. So also the particles of liquid and solid bodies attract each other when contiguous, but when separated by heat or fermentation they flee from each other and constitute vapours, exhalations, and air. And on the other hand coming together by the action of cold or fermentation they cohere and revert into liquids and compact bodies. For that remarkable Philosopher, the Honourable Mr. Boyle, teaches us that true air arises from the fermentation of dense bodies and returns into them.[8] And this, in connection with the violent expansion of branched particles by an elastic force (as is commonly supposed), is difficult to understand. Indeed, particles of air in the narrow cavities of tubes flee from the glass, so that air seems to be more rare there than in free spaces and if the lower orifice of the tube is immersed in stagnating water it ascends quickly (as a certain person[9] has ingeniously remarked following an hypothesis of his own) and fills the cavity from which air tries to recede. And from the similar ascent of liquids in the small passages of sponges and other bodies it is learnt that the particles of air endeavour to recede from all bodies universally. Whence it is that air when confined in a bladder does not escape through its pores even when compressed although water very easily permeates them. And in the same manner we see why the separated parts of hard bodies can only with great difficulty be brought together so that they touch each other completely. Whence they do not cohere as they did before. I placed on a bi-convex objective lens from a fifty-foot telescope a plane glass of less weight, and in the middle of the glasses about the point where they almost touched there appeared certain coloured circles.[10] Then lightly pressing upon the glass, so that it came closer to the lower one, there emerged new coloured circles in the middle of the former circles; and then increasing the pressure there emerged in the middle of the white circle a sort of spot darker than the other colours. this spot was the indication of the contact of the pieces of glass. For the surfaces of the glasses which elsewhere reflected light from themselves, reflected almost no light at all at that spot, but offered very little hindrance to the unimpaired passage of light just if the glasses were continuous there, and being joined together by fusion were deprived of intermediate surfaces. Therefore the upper glass could be brought by its weight into contact with the lower one. That was effected by the greater force of the hands pressing down. Indeed, with all the force with which I compressed those glasses I could scarcely make them touch at all. For on removing my hands they did not cohere

8. See further, p. 289, for Boyle's work in pneumatic chemistry.
9. Boyle.
10. I.e., Newton's rings.

in the least. The matter may be better understood from this experiment. I covered a fairly large [soap] bubble, such as boys blow in play, with a glass vessel so that the external air should not by its motion disturb the colours appearing in the bubble. The top of the bubbles was ringed round with concentric circles of different colours; these circles [as the bubble became thinner] widened out; they crept down the sides towards the lower part of the bubble while new circles arose towards the top. At length a whiteness appeared towards the top and soon in the midst of it a spot of dark blue colour in whose centre a black spot then appeared and afterward in its centre a blacker spot which hardly reflected the light at all, and then at last the bubble usually burst. These colours are of the same kind and have the same origin as the colours between glasses. For the watery skin of the bubble between the internal and external air serves in this case to exhibit colours just as the layer of air between the two glasses does. Just as when the layer of air was made thinner by compression of the glasses new colours arose until the black spot appeared, so when the skin of the soap-bubble becomes thinner at the top by the continual flowing down of the upper parts to the lower part, new colours appear successively and ultimately the black spot is observed. And as the skin of the soap-bubble is thinnest at the top where the black spot appears but notwithstanding this is not wholly lacking in thickness, so from the black spot in the middle of the glasses it cannot be concluded that the layer of intermediate air near the spot is destitute of all thickness. Therefore separate and distinct bodies vehemently avoid mutual contact. And thence it happens that a fly walks upon water without wetting its feet, and that particles of dry powder cannot be brought together by a compressive force so that they cohere as the parts of a solid body do. But if water is poured on so that [the particles] compose a continuous body, and then the water is evaporated, they are brought into contact and through that contact they form a body by cohesion; when it is dried by fire so that the fluid softening it is driven off, it is converted into a firm and hard body.

I have briefly set these matters out, not in order to make a rash assertion that there are attractive and repulsive forces in bodies, but so that I can give an opportunity to imagine further experiments by which it can be ascertained more certainly whether they exist or not. For if it shall be settled that they are true [forces] it will remain for us to investigate their causes and properties diligently, as being the true principles from which, according to geometric reasoning, all the more secret motions of the least particles are no less brought into being than are the motions of greater bodies which as we saw in the foregoing [books] derived from the laws of gravity. For through the forces by which contiguous bodies cohere, and distant ones seek to separate from one another, these particles will not collect together in the composition of natural bodies like a heap of stones, but they coalesce into the form of highly regular structures almost like those made by art, as happens in the formation of snow and salts. Undoubtedly, following the laws of geometry they can be formed into very long and elastic rods, and by the connection of the rods into retiform particles, and by the composition of these into greater particles, and so at length into perceptible bodies. And such bodies suffer light to pass through them every way, and there can be great differences of density between them. Thus gold is nineteen times denser than water yet it is not destitute of pores through which mercury can penetrate, and light (if the gold be dissolved in aqua regia and reduced into crystals) very freely be transmitted. Bodies of this kind easily receive the motion of heat by means of the free vibrations of the elastic particles and they will conserve [heat] a long time through that motion if it is slow and long-lasting. Their particles coalesce in new ways and by means of the attractive forces of contiguous ones they come together more densely: for which reason that rare substance water can be transformed by continued fermentation into the more dense substances of animals, vegetables, salts, stones and various earths. And finally by the very long duration of the operation be coagulated into mineral and metallic substances. For the matter of all things is one and the same, which is transmuted into countless forms by the operations of nature, and more subtle and rare bodies are by fermentation and the processes of growth commonly made thicker and more condensed. By the same motion of fermentation bodies can expel certain particles, which thereupon by their repulsive forces are caused to recede from each other violently; if they are denser, they constitute vapours, exhalations and air; if on the other hand they are very small they are transformed into light. These last undoubtedly adhere more strongly, since bodies do not shine save by a vehement heat. After they are separated they recede from bodies more violently, then in passing through other bodies sometimes they are attracted towards them, sometimes repelled; and by attraction they are certainly refracted, and sometimes reflected as I explained above;[11] by repulsion they are always reflected. Bodies which reflect all incident light will shine on account of the abundance of reflected light, as do mercury and white metals. Others which abound in rather large translucent particles disseminated through uniform material of a different density, according to the size of the particles, more copiously reflect rays of one kind or of some other kind, whence they will appear coloured. For that the various colours of bodies arise from various sizes of refelcting particles appears from the preceding soap-bubble experiment, where the pellucid skin of the bubble reflected various colours according to variations in its thickness.

Bodies which expel particles of light because of vehement heat are fire. For I think that burning bodies are distinguished from non-burning ones only by heat and light; and they are of two kinds, vapours and solid bodies. For just as *ignes fatui,* which are cold shining vapours, differ from shining rotten wood, so flames, which are vapours and exhalations made hot and shining by the vehement agitation of heat, are distinguished from coals and other dense bodies whether solid or fluid which likewise glow because of vehement heat. And just as glowing metal differs from non-glowing, so glowing coal and glowing vapour are distinguished from non-glowing. For I seem to understand that flame is nothing else but a glowing vapour from the fact that bodies do not feed flame unless they copiously emit a sulphureous vapour, and that if a [lighted] lamp is brought close to such a vapour it ignites it, and by the propagation of the fermentation it immediately converts all the material of the vapour into flame, as may be found with the vapour of spirit of wine and with the vapour of a recently extinguished candle. Accordingly, glowing bodies seem to be distinguished from non-glowing ones solely by the motion of their internal parts. For bodies grow hot by motion and always shine by intense heat. Thus iron grows warm under the hammer, and can be so agitated by hammering that at length it glows. The axles of waggons often burst into flame from the motion of the wheels. Particles of steel when rubbed or struck with flint (as someone [Boyle] has in-

11. *Principia,* prop. XCVI (pp. 230 - 31 in F. Cajori's edition).

geniously noted) glow from the stroke; and closely packed hay grows warm from the motion of the vapours, and sometimes takes fire from the warmth. In coals and ignited vapours, however, heat seems to be excited and conserved by the action of a sulphureous spirit. For fire can hardly burn and be supported without fatty and sulphureous matter; with the addition of sulphur it generally becomes intense. For the fume of sulphur abounds in an acid spirit, which makes the eyes smart, and when condensed under the bell runs down as a corrosive liquid of the same kind as spirit of nitre and oil of vitriol. These only differ by the phlegm in the spirit, and when mixed with other bodies whether dry or fluid excite heat in them, and not infrequently vehement heat. Therefore spirit of sulphur meeting with the particles of coals and fumes heats them till they glow; for the encounter of hot bodies is the more vehement. And inasmuch as air abounds in sulphureous spirits, it also makes ignited matter grow hot and because of the subtlety of the spirit is required for the maintenance of fire. Whence I suspect that the heat of the Sun may be conserved by its own sulphureous atmosphere. However the volatile parts of ignited bodies are thrown off by their heat, while the fixed parts remain in the form of ashes. That sulphureous spirit meeting with the fixed parts of vegetable salts looses their spirits, however, and united with those fixed parts forms that fixed salt which is usually found in ashes and is called Sal Alkali, and by virtue of that sulphureous spirit runs *per deliquium*.[12] For such are the properties of oil and spirit of vitriol. They attract a humour from the air and meeting with the fixed particles of common salt and salt of nitre (which are both vegetable salts), they loose their spirits and expel them. But nitre deflagrated in the open air with sulphur or with any sulphureous body is converted into Sal Alkali, whereas if nitre alone is mixed with clay and distilled in a closed glass, no Sal Alkali is extracted from the *caput mortuum*.[13] What has been said before is confirmed by gunpowder, which is composed of nitre, sulphur, and charcoal. The powdered charcoal readily takes fire and ignites the powdered sulphur. The acid fume of the sulphur invades the nitre and by this encounter makes it hot, and looses its spirit, which spirit driven out by heat and converted into vapour expands itself violently. If Sal Alkali be mixed in due proportion

12. Deliquesces. A "sulphureous spirit" is a volatile, inflammable component.
13. Residuum.

with it, and the powder dried by fire to drive off all the humour
that this salt commonly attracts to itself, the powder now de-
flagrates with a more violent explosion after the manner of fulmi-
nating gold. For the sulphureous spirit meets more willingly and
violently with Sal Alkali than with salt of nitre, and by that meeting agitates all the material more and makes it grow more intensely
hot. Indeed, it is to be believed that the culinary fire is excited and
maintained at a slow and gradual rate by the material causes of fire
which act more vigorously in these compositions, for those causes
abound less in it. But ignited metals, because they do not abound
with volatile parts which are turned into fumes by heat, nor do
they have passages wide enough to admit sulphureous spirit into
their internal parts, so they do not nourish flame, nor do they by
themselves conserve the heat of ignition, nor are they easily turned
into ashes; except that zinc, as that imperfect and volatile metal is
called, by emitting a copious fume does nourish a flame, and Mr.
Boyle has proved the same for copper divided and made more
subtle by mercury sublimated which could emit a fume. So also
stones are not consumed by burning, nor (except those bitumin-
ous ones which emit a copious sulphureous vapour) do they conserve fire.

34. Chemical Atomism

On Chemical Synthesis

When any body exists in the elastic state,[1] its ultimate particles
are separated from each other to a much greater distance than in
any other state; each particle occupies the center of a compara-
tively large sphere, and supports its dignity by keeping all the rest,
which by their gravity, or otherwise are disposed to encroach up
it, at a respectful distance. When we attempt to conceive the *number* of particles in an atmosphere, it is something like attempting

SOURCE: John Dalton, *A New System of Chemical Philosophy,* part I (London, 1808), pp. 211-16.

1. I.e., is a gas.

to conceive the number of stars in the universe; we are confounded with the thought. But if we limit the subject, by taking a given volume of any gas, we seem persuaded that, let the divisions be ever so minute, the number of particles must be finite; just as in a given space of the universe, the number of stars and the planets cannot be infinite.

Chemical analysis and synthesis go no farther than to the separation of particles one from another, and to their reunion. No new creation or destruction of matter is within the reach of chemical agency. We might as well attempt to introduce a new planet into the solar system, or to annihilate one already in existence; as to create or destroy a particle of hydrogen. All the changes we can produce, consist in separating particles that are in a state of cohesion or combination, and joining those that were previously at a distance.

In all chemical investigations, it has justly been considered an important object to ascertain the relative *weights* of the simples[2] which constitute a compound. but unfortunately the inquiry has terminated here; whereas from the relative weights in the mass, the relative weights of the ultimate particles or atoms of the bodies might have been inferred, from which their number and weight in various other compounds would appear, in order to assist and to guide future investigations, and to correct their results. Now it is one great object of this work, to show the importance and advantage of ascertaining *the relative weights of the ultimate particles, both of simple and compound bodies, the number of simple elementary particles which constitute one compound particle, and the number of less compound particles which enter into the formation of one more compound particle.*

If there are two bodies, A and B, which are disposed to combine, the following is the order in which the combinations may take place, beginning with the most simple:

> 1 atom of A + 1 atom of B = 1 atom of C, binary.
> 1 atom of A + 2 atoms of B = 1 atom of D, ternary.
> 2 atoms of A + 1 atom of B = 1 atom of E, ternary.
> 1 atom of A + 3 atoms of B = 1 atom of F, quaternary.
> 3 atoms of A + 1 atom of B = 1 atom of G, quaternary.
>   etc., etc.

2. Simple substances.

The following general rules may be adopted as guides in all our investigations respecting chemical synthesis.

1st. When only one combination of two bodies can be obtained, it must be presumed to be a *binary* one, unless some cause appear to the contrary.

2d. When two combinations are observed, they must be presumed to be a *binary* and a *ternary*.

3d. When three combinations are obtained, we may expect one to be a *binary*, and the other two *ternary*.

4th. When four combinations are obtained, we may expect one to be a *binary*, two *ternary*, and one *quaternary*, etc.

5th. A *binary* compound should always be specifically heavier than the mere mixture of its two ingredients.

6th. A *ternary* compound should be specifically heavier than the mixture of a binary and a simple, which would, if combined, constitute it; etc.

7th. The above rules and observations equally apply, when two bodies, such C and D, D and E, etc. are combined.

From the application of these rules, to the chemical facts already well ascertained, we deduce the following conclusions; 1st. That water is a binary compound of hydrogen and oxygen, and the relative weights of the two elementary atoms are as 1:7, nearly; 2d. That ammonia is a binary compound of hydrogen and azote,[3] and the relative weights of the two atoms are as 1:5, nearly; 3d. That nitrous gas is a binary compound of azote and oxygen, the atoms of which weigh 5 and 7 respectively; that nitric acid is a binary or ternary compound according as it is derived, and consists of one atom of azote and two of oxygen, together weighing 19;[4] that nitrous acid is a binary compound of nitric acid and nitrous gas, weighing 31; that oxynitric acid is a binary compound of nitric acid and oxygen, weighing 26; 4th. That carbonic oxide is a binary compound, consisting of one atom of charcoal, and one of oxygen, together weighing nearly 12; that carbonic acid is a ternary compound, (but sometimes binary) consisting of one atom of charcoal,

3. The early (Lavoisieran, and still used in France) name for nitrogen, meaning "without life."

4. It was not yet known that all acids contain hydrogen, a point which only began to be understood after Davy's discovery of the composition of hydrochloric acid (HCL) in 1810. It had previously been thought that oxygen was the essential ingredient of all acids.

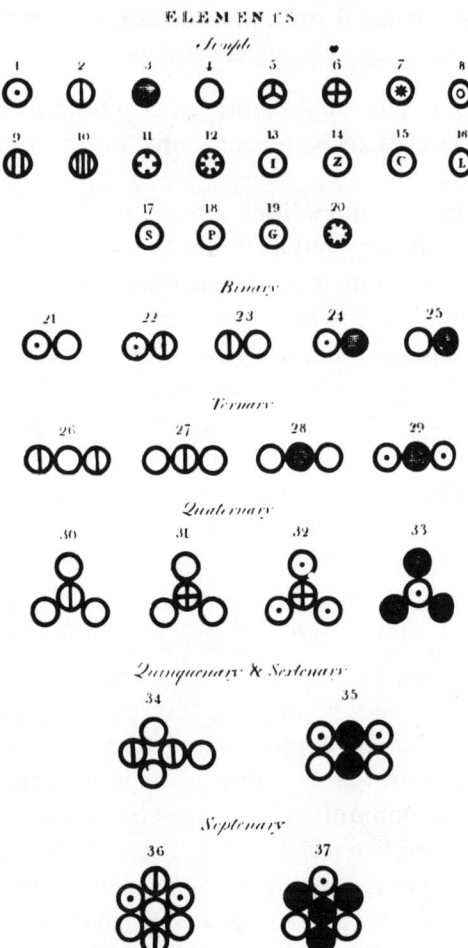

12. Chemical combinations.
From John Dalton, *A New System of Chemical Philosophy*

This plate contains the arbitrary marks or signs chosen to represent the several chemical elements or ultimate particles.

1. Hydrog.; its rel. weight	1	11. Strontites	46
2. Azote	5	12. Barytes	68
3. Carbon or charcoal	5	13. Iron	38
4. Oxygen	7	14. Zinc	56
5. Phosphorus	9	15. Copper	56
6. Sulphur	13	16. Lead	95
7. Magnesia	20	17. Silver	100
8. Lime	23	18. Platina	100
9. Soda	28	19. Gold	140
10. Potash	42	20. Mercury	167

21. An atom of water or steam, composed of 1 of oxygen and 1 of hydrogen, retained in physical contact by a strong affinity, and supposed to be surrounded by a common atmosphere of heat; its relative weight =   8
22. An atom of ammonia, composed of 1 of azote and 1 of hydrogen   6
23. An atom of nitrous gas, composed of 1 of azote and 1 of oxygen   12
24. An atom of olefiant gas, composed of 1 of carbon and 1 of hydrogen   6
25. An atom of carbonic oxide composed of 1 of carbon and 1 of oxygen   12
26. An atom of nitrous oxide, 2 azote + 1 oxygen   17
27. An atom of nitric acid, 1 azote + 2 oxygen   19
28. An atom of carbonic acid, 1 carbon + 2 oxygen   19
29. An atom of carburetted hydrogen, 1 carbon + 2 hydrogen   7
30. An atom of oxynitric acid, 1 azote + 3 oxygen   26
31. An atom of sulphuric acid, 1 sulphur + 3 oxygen   34
32. An atom of sulphuretted hydrogen, 1 sulphur + 3 hydrogen   16
33. An atom of alcohol, 3 carbon + 1 hydrogen   16
34. An atom of nitrous acid, 1 nitric acid + 1 nitrous gas   31
35. An atom of acetous acid, 2 carbon + 2 water   26
36. An atom of nitrate of ammonia, 1 nitric acid + 1 ammonia + 1 water   33
37. An atom of sugar, 1 alcohol + 1 carbonic acid   35

Enough has been given to show the method; it will be quite unnecessary to devise characters and combinations of them to exhibit to view in this way all the subjects that come under investigation; nor is it necessary to insist upon the accuracy of all these compounds, both in number and weight; the principle will be entered into more particularly hereafter, as far as respects the individual results. It is not to be understood that all those articles marked as simple substances, are necessarily such by the theory; they are only necessarily of such weights. Soda and Potash, such as they are found in combination with acids, are 28 and 42 respectively in weight; but according to Mr. Davy's very important discoveries, they are metallic oxides; the former then must be considered as composed of an atom of metal, 21, and one of oxygen, 7; and the latter, of an atom of metal, 35, and one of oxygen, 7. Or, soda contains 75 per cent metal and 25 oxygen; potash, 83.3 metal and 16.7 oxygen. It is particularly remarkable, that according to the above-mentioned gentleman's essay on the Decomposition and Composition of the fixed alkalies, in the Philosophical Transactions (a copy of which essay he has just favored me with) it appears that "the largest quantity of oxygen indicated by these experiments was, for potash 17, and for soda, 26 parts in 100, and the smallest 13 and 19."

and two of oxygen, weighing 19; etc., etc. In all these cases the weights are expressed in atoms of hydrogen, each of which is denoted by unity.

In the sequel, the facts and experiments from which these conclusions are derived, will be detailed; as well as a great variety of others from which are inferred the constitution and weight of the ultimate particles of the principal acids, the alkalis, the earths, the metals, the metallic oxides and sulphurets,[5] the long train of neutral salts, and in short, all the chemical compounds which have hitherto obtained a tolerably good analysis. Several of the conclusions will be supported by original experiments.

From the novelty as well as importance of the ideas suggested in this chapter, it is deemed expedient to give plates, exhibiting the mode of combination in some of the more simple case. A specimen of these [fig. 12] accompanies this first part. The elements or atoms of such bodies as are conceived at present to be simple, are denoted by a small circle, with some distinctive mark; and the combinations consist in the juxtaposition of two or more of these; when three or more particles of elastic fluids are combined together in one, it is to be supposed that the particles of the same kind repel each other, and therefore take their stations accordingly.

5. Sulphides.

Chemical Theories

---

35. The Purpose of Early Chemistry

The Author's Preface

Authors who have treated of chemistry have had very different discoveries and apprehensions, and thence it is that they have disagreed very much in their writings. Those who have applied themselves unto the high chemistry,[1] and have penetrated into its greatest mysteries, have contented themselves with the knowledge thereof; and though it may seem that they wrote with an intent to be understood, yet they have penned things so obscurely, that they gave us no grounds to question whether they uttered realties, or have given us phantoms for bodies, and thorns for fruits: Others who have not soared so high, have notwithstanding had some good skill therein, and have themselves discovered some preparations, which have made them considerable to posterity; but some of these have sought their own satisfaction also, and taken pleasure to perplex men's minds, and to cast them into labyrinths, without giving them any means of retreating thence.

Others, much less capable, have yet attained some small light, but not knowing all, nor having wrought themselves what they write, and desiring for all that, to pass for sufficient artists in a profession they had learned by halves only, they made their imagination pass for certain truths, whose falsehood and imperfection the practise hath frequently discovered.

Finally, others, which deserve not the name of chemists, but rather of ignorant fireblowers, working by copies or stolen receipts, which they commonly take in some contrary sense, and having consumed their own or others' money in a ridiculous

SOURCE: Christophe Glaser, *Traité de la chymie* (Paris, 1663); the text is that of the anonymous English translation, *The Compleat Chymist, or A New Treatise of Chymistry* (London, 1677).

1. That is, alchemy.

labour, have made many others accompany them in their fortunes, engaging the vulgar by promises of enriching them in certain practices, whereby they have reduced the best metal into smoke, unless perhaps they passed some part of it through their own hands, which is not the least of their operations. Hence it is that I wonder not, that many have declaimed against such authors, and against chemistry itself; not having understood truly the good things which it contains. As for myself, who profess to say nothing but what I know, and to write nothing but what I have done, I purpose only in this little treatise to publish a short and easy method for the happy attainment of all the most necessary preparations of chemistry. Those who take the pains to read and well consider it, shall observe therein nothing tedious, superfluous or defective in any point which ought to be known: and though indeed the preparations of all things cannot be found therein, yet sufficient examples thereof will be had from it. And though it was very difficult to comprise in this little tract all the discoveries which I have made in this Profession, yet I have not concealed any manual operation, and have sincerely discovered all the circumstances necessary to make a good artist, and by practice to attain the greatest knowledge thereof. I set down no preparation but what I have made and thoroughly experienced, and what any one following the rules I have prescribed, may do after me. I speak very succinctly of the theory, but I say so much therein as is necessary for directions to the preparations, and you may find in a few words the substance of many great books.

I confine myself to operations on minerals, vegetables, and animals and proceed therein orderly, and forget nothing that is necessary. I am persuaded that the experience of all that which I have advanced, will manifest my communicative freeness unto all, and that they will be well pleased with the care that I have taken therein. Unto which I have thought myself so much the more obliged by the choice which Mr. Vallot, his Majesty's chief and most worthy physician, hath made of me, to make the public chemical lectures and preparations in the Royal Garden.[2] For this cause I have desired to manifest as well by writing as work, that I

2. Glaser was appointed one of the two professors at the Royal Botanic Garden in Paris (now the (Jardin des Plantes) in 1660. The professors at this time lectured on practical chemistry—the preparation of chemical drugs; they nearly all, like Glaser, turned their lectures into textbooks.

have no other design than to acknowledge the honor he hath done me, by giving satisfaction to the public, according to this inclination by all ways which are possible for me.

Of the Usefulness of Chemistry

They that have any true knowledge of this noble art, are without doubt fully persuaded of the usefulness of it; for it is the key which alone can unlock to all naturalists the door of Nature's secrets; by reducing things to their first principles; by giving to them new forms; and by imitating Nature in all its productions and physical alterations. Without it physicians would be at a stand how to penetrate into the many fermentations, effervescences, distillations and other operations, which are performed in the body of man, and are the immediate cause of many grievous distempers;[3] to which art the same physicians must be beholden for the remedy as well as the knowledge of the disease; for we must own that chemistry does furnish us with the more effectual medicines for the more inveterate and obstinate affections, and often supply the failings and deficiencies of those of the vulgar pharmacy. Surgeons likewise cannot be without chemistry, nor can with good success undertake the cure of all diseases pertaining to their art without chemical remedies, and knowledge of their operations: and it is impossible that apothecaries should make their compositions like true artists, if they knew not how to preserve the principal virtue of their ingredients, and separate the pure from the impure and heterogeneous in natural commixtures, as unprofitable to their intention, which is not learned without the aid of this noble and excellent art. Finally, most of the ingenuous mechanic arts are beholden to this: painters have from it their most lively and glorious colors; engravers cannot work without the assistance of these corrosive spirits prepared by chemistry; dyers cannot exalt their colors without the instructions of chemists, one might allege an infinity of other such examples, which might prove the necessity of this art; but we omit them for brevity's sake.

3. This interpretation of disease as the result of chemical imbalance, to be therefore cured by chemical means, derives ultimately from Paracelsus, though most fully developed a century later by J. B. van Helmont, by whom Glaser was influenced.

Of the Subject Matter of Chemistry and Its Functions

This art is of a very great extent, since it embraces for its subject the bodies of the three families, animal, vegetable and mineral; which by fire it reduces into different substances, which the philosophers call *first principles* and do constitute five of them, of which three are active, and two passive.[4] The active are, the spirit which is called Mercury, the oil which is called Sulphur, and the salt, which has no other name. The passive principles are the water or phlegm, and the earth or terrestrial part. These names are bestowed upon them for the likeness they have to common Mercury, Sulphur and Salt, and the elementary water and earth: the Mercury appears to us in the resolution of bodies, in the form of a most aerial, subtle liquor; the Sulphur is apparent to our smell and taste, by which we distinguish it from the insipid and inodorous Phlegm, which sometimes ascends with it; and it appears to us in the form of a penetrating, inflammable oil; the salt remains joined to the body of the earth till it be extracted by elevation. Now while these principles remain in the body of the mixt [combination of substances], those that are active are confounded with the passive, so that their virtue is hidden, and, as it were, buried; but chemistry coming to separate them, purifies each by itself, then unites them again, to make of them bodies much purer, and more active than they were before.

4. An extension of the common theories of the three Paracelsan principles (salt, sulphur, and mercury) and the four Aristotelian elements (earth, air, fire, water). These are not, like the modern elements, merely the last product of chemical analysis; they are rather the necessary ingredients of all bodies, so that all bodies must be ultimately analyzable into these elements or principles—five in number for Glaser.

36. *The New Chemistry and the New Physics*

---

*Some Specimens of an Attempt to Make Chymical Experiments
Useful to Illustrate
the Notions of the Corpuscular Philosophy*

THE PREFACE

Giving an account of the two following treatises, and proposing the desirableness of a good intelligence betwixt the Corpuscularian Philosophers and the Chymists.

There are many learned men, who being acquainted with chymistry but by report, have from the illiterateness, the arrogance and the impostures of too many of those, that pretend skill in it, taken occasion to entertain so ill an opinion, as well of the art, as of those that profess it, that they are apt to repine, when they see any person, capable of succeeding in the study of solid philosophy, addict himself to an art they judge so much below a philosopher, and so unserviceable to him: nay, there are some, that are troubled when they see a man, acquainted with other learning, countenance by his example sooty empirics, and a study, which they scarce think fit for any but such, as are unfit for the rational and useful parts of physiology. I now take notice of these things, because they gave occasion to the two following treatises. For perceiving divers years ago, that some learned men of the temper above described thought it strange (if not amiss also) that one, of whose studies they were pleased to have too favourable an expectation, should spend upon chymical trials (to which I then happened to be invited by the opportunity of some furnaces and leisure) much of those endeavours, which they seemed to think might be far more usefully employed, than upon such empty and

SOURCE: Robert Boyle, *Certain Physiological Essays* (1661), as printed in Thomas Birch, *The Works of the Honourable Robert Boyle* (6 vols., London, 1772) I, 354-50.

deceitful study: perceiving this, I say, I thought it not amiss to endeavour to manifest, that without seeking after the elixir, that alchymists generally hope and toil for (but which they, that knew me, knew to be not at all in my aim) I did not in the prosecution of chymical trials do any thing either without an end, or unsuitable to the design I had of attempting to promote men's knowledge of the works of nature, as well as their power over them. In order to do this, I did not think it enough to show, that by an insight into chymistry one may be enabled to make some meliorations (I speak not of transmutations) of mineral and metalline bodies, and many excellent medicines for the health of men, besides divers other preparations of good use in particular trades, and in several occurrences of human life: I did not, I say, think it enough to do this, because that though this might suffice to evince that a rational man might without losing his time employ some of it to understand and promote chymistry; yet this would scarce suffice to manifest it to be useful to philosophy. And therefore there seemed requisite some specimens which might show that chymical experiments might be very assistant even to the speculative naturalist in his contemplations and inquiries.

But against my attempting anything of this nature, three difficulties opposed themselves. The first was the want of leisure, in regard I was already pre-engaged to write of other subjects, and to prosecute some experiments, whose event I was concerned to know. Another impediment was, that for other reasons elsewhere mentioned, and chiefly to keep my judgment as unprepossessed as might be with any of the modern theories of philosophy, till I were provided of experiments to help me to judge of them, I had purposely refrained from acquainting myself thoroughly with the entire system of either the Atomical, or the Cartesian, or any other whether new or revived philosophy; and therefore I could scarce be fit to show how chymical experiments might illustrate their doctrines. And thirdly, some of those learned men, for whom I was to write, more favouring the Epicurean; and others (though but a few) being more inclinable to the Cartesian opinions, it seemed very difficult to gratify by the same composures persons of differing persuasions.

But as to the first of these discouragements, since my pre-engagements to other themes were not unknown to those, for whom I was to write, it might reasonably be presumed they would over-look such unaccurateness as should appear imputable to haste: and besides, some such subject might be chosen to write of, as would conveniently admit enlargements and additions, according as my leisure should afterwards serve me to annex them.

On the occasion of the second impediment, I remembered, that having divers years before read the lives of the Atomical, among other philosophers, in Diogenes Laertius; and having sometimes occasionally heard mention of divers Epicurean and Cartesian notions, and having hence framed to myself some general, though but imperfect, idea of the way of philosophizing my friends esteemed; I thought I might, without a more particular explicit inquiry into it, say something to illustrate some notions of it, by making choice of such, as, being of the more simple and obvious, did not require skill in the more mysterious points of the hypothesis they belonged to. . . .

37. Chemical Indicators

To Determine Whether an Acid or an Alkaline Salt Is Present in a Solution or Solid

The distinction of salts into acid or volatile and fixed or alkaline may possibly be so useful in natural philosophy as to render it an acceptable corollary of the preceding experiment [not given here], from thence to deduce a method of trying which, or whether any one of those salts is predominant in chemical liquors. We have already shown a way, by means of the tincture of Lignum Nephriticum, or of syrup of violets, to discover whether a salt proposed be acid or not; yet we can thereby only find in general, that particular salts belong not to the tribe of acids: but cannot determine whether they belong to the tribe of the urinous; (under which I comprehend all those volatile salts of animal, or other substances, that are contrary to acids) or to that of alkalies: for both these salino-sulphureous salts will restore the blue colour to

SOURCE: Robert Boyle, *Experiments and Considerations Touching Colours* (London, 1664), from the abridgement by Peter Shaw (1725), II, 86-88.

that tincture, and turn the syrup of violets green. The present experiment, therefore, commodiously supplies this deficiency. I found, that all those chemical salts I thought fit to make trial of, would, if they were of a lixivious [alkaline] nature, make, with sublimate dissolved in fair water, an orange-tawney precipitate; but if they were of an urinous nature, the precipitate would be white and milky. So that having always at hand some syrup of violets, and a solution of sublimate; I can, by the help of the former, immediately discover whether a proposed salt, or saline body, be of an acid nature; if it be, I need inquire no further; but if not, I can very easily and readily distinguish between the other two kinds of salts, by the white, or orange-colour, immediately produced by letting fall a few drops or grains of the salt to be examined, into a spoonful of the clear solution of sublimate. Thus, for example, it has been supposed, that when sal-ammoniac, mixed with an alkali, is forced from it by the fire in close vessels, the volatile salt thence obtained, is but a more fine and subtle sort of sal-ammoniac; which, 'tis presumed, this operation only purifies more exquisitely than common solutions, filtrations, and coagulations. But this opinion may easily be proved erroneous, as by other arguments, so particularly by our method of distinguishing the tribe of salts. For the saline spirit of sal-ammoniac, as it is in many other manifest qualities very like the spirit of urine; so, like that, it will instantly turn syrup of violets to a lovely green; a solution of good verdigrease into an excellent azure; and make the solution of sublimate deposite a white precipitate. Therefore, in most experiments, where I only design to produce a sudden change of colour; I scruple not to use spirit of sal-ammoniac, when at hand, instead of spirit of urine; as indeed it seems chiefly to consist of volatile urinous salt. And by this way of examining chemical liquors, we may not only, in general, conclude affirmatively, but, in some cases, negatively.

Thus, since spirit of wine, and, as far as I have tried, the chemical oils will not turn syrup of violets red, or green; nor the solution of sublimate, white or yellow, I infer, with probability, that either they are destitute of salt, or have such as belongs not to either of the three grand families mentioned.

And thus, upon examining the spirit of oak, or of such like substances, forced over the retort, I found by this means, amongst others, that those chemists are much mistaken, who make it a simple liquor, and one of their hypostatical principles; for, not to mention what phlegm it may have, a few drops of one of this sort of spirits, mixed with a large proportion of syrup of violets, turned it purplish; by the affinity of which colour to redness, I conjectured that the spirit contained some acid corpuscles, and accordingly found, that it would destroy the blueness of the nephritic tincture; and that being put upon corals, it would corrode them like common spirit of vinegar, and other acid liquors. And to examine whether there were not a great part of the liquor of other than an acid nature; having separated the sour part from the rest, we concluded, the remaining part, though it had a strong taste as well as smell, to be of a nature different from that of any of our three sorts of salts; for it did as little as spirit of wine, and chemical oils, alter the colour of syrup of violets, and a solution of sublimate: whence we also inferred, that the change made of that syrup into a purple colour, was effected by the vinegar, that proved one of the ingredients of the liquor which usually passes for a simple, or uncompounded spirit.

And by the same way we may examine and discover many changes that are produced in bodies, either by nature or art; for both are able, by changing the texture of some substances, to qualify them to operate after a new manner, upon the syrup, or solution. Thus I have discovered, that there are factitious bodies, which, though they run as readily as salt of tartar, belong, in other respects, not to the family of alkalies, much less to that of volatile or acid salts. Perhaps, too, I know a way of making a highly operative saline body, that shall neither change the colour of syrup of violets, nor precipitate the solution of sublimate. And I can, likewise, conceal the liquors wherewith I make such changes and colours as those lately mentioned, by quite altering the texture of some ordinary chemical productions.

38. *The New Gases*

Of Dephlogisticated Air and of
the Constitution of the Atmosphere

The contents of this section will furnish a very striking illustration of the truth of a remark, which I have more than once made in my philosophical writings, and which can hardly be too often repeated as it tends greatly to encourage philosophical investigations; viz. that more is owing to what we call chance, that is, philosophically speaking, to the observation of events arising from unknown causes, than to any proper design, or pre-conceived theory in this business. This does not appear in the works of those who write synthetically upon these subjects, but would, I doubt not, appear very strikingly in those who are the most celebrated for their philosophical acumen, did they write analytically and ingenuously.

For my own part, I will frankly acknowledge that, at the commencement of the experiments recited in this section, I was so far from having formed any hypothesis that led to the discoveries I made in pursuing them that they would have appeared very improbable to me had I been told of them; and when the decisive facts did at length obtrude themselves upon my notice, it was very slowly, and with great hesitation that I yielded to the evidence of my senses. And yet, when I re-consider the matter, and compare my last discoveries relating to the constitution of the atmosphere with the first, I see the closest and the easiest connection in the world between them, so as to wonder that I should not have been led immediately from the one to the other. That this was not the case, I attribute to the force of prejudice, which, unknown to ourselves, biases not only our judgments, properly so called, but even the perceptions of our senses: for we may take a maxim so strongly for granted, that the plainest evidence of sense will not entirely change, and often hardly modify our persuasions; and the more ingenious a man is, the more effectually he is entangled in

SOURCE: Joseph Priestley, *Experiments and Observations on Different Kinds of Air* (London, 1775), II, 29-61.

his errors; his ingenuity only helping him to deceive himself, by evading the force of truth.

There are, I believe, very few maxims in philosophy that have laid firmer hold upon the mind, than that air, meaning atmospherical air (free from various foreign matters, which were always supposed to be dissolved, and intermixed with it) is a simple elementary substance, indestructable and unalterable, at least as much so as water is supposed to be. In the course of my enquiries, I was, however, soon satisfied that atmospherical air is not an unalterable thing; for that the phlogiston with which it becomes loaded from bodies burning in it, and animals breathing it, and various other chemical processes, so far alters and depraves it, as to render it altogether unfit for inflammation, respiration and other purposes to which it is subservient;[1] and I had discovered that agitation in water, the process of vegetation, and probably other natural processes, by taking out the superfluous phlogiston, restore it to its original purity. But I own I had no idea of the possibility of going any farther in this way, and thereby procuring air purer than the best common air. I might, indeed, have naturally imagined that such would be air that should contain less phlogiston than the air of the atmosphere; but I had no idea that such a composition was possible.

It will be seen in my last publication, that, from the experiments which I made on the marine acid air,[2] I was led to conclude, that common air consisted of some acid (and I naturally inclined to the acid that I was then operating upon) and phlogiston; because the union of this acid vapour and phlogiston made inflammable air;[3] and inflammable air, by agitation in water, ceases to be inflammable, and becomes respirable.[4] And though I could never make it quite so good as common air, I thought it very probable that vegetation, in more favourable circumstances than any in which I could apply it, or some other natural process, might render it more pure.

1. This is a clear and simple statement of the function of phlogiston as conceived at this time (and indeed never relinquished by Priestley): that it was given off in combustion and respiration and that substances which supported either of these processes were necessarily rich in phlogiston.
2. Hydrochloric acid gas (HCl).
3. Hydrogen.
4. Priestley here is somewhat confused; hydrogen does not react with water, but it is probable that his original "inflammable air" was mixed with ordinary air to a certain extent.

Upon this, which no person can say was an improbable supposition, was founded my conjecture, of volcanos having given birth to the atmosphere of this planet, supplying it with a permanent air, first inflammable, then deprived of its inflammability by agitation in water, and farther purified by vegetation.

Several of the known phenomena of the nitrous acid[5] might have led me to think, that this was more proper for the constitution of the atmosphere than the marine acid: but my thoughts had got into a different train, and nothing but a series of observations, which I shall now distinctly relate, compelled me to adopt another hypothesis, and brought me, in a way of which I had then no idea, to the solution of the great problem, which my reader will perceive I have had in view ever since my discovery that the atmospherical air is alterable, and therefore that it is not an elementary substance, but a composition viz. what this composition is, or what is the thing that we breathe, and how is it to be made from its constituent principles.

At the time of my former publication, I was not possessed of a burning lens of any considerable force; and for want of one, I could not possibly make many of the experiments that I had projected, and which, in theory, appeared very promising. I had, indeed, a mirror of force sufficient for my purpose. But the nature of this instrument is such, that it cannot be applied, with effect, upon substances that are capable of being suspended, or resting on a very slender support. It cannot be directed at all upon any substance in the form of powder, nor hardly upon anything that requires to be put into a vessel of quicksilver; which appears to me to be the most accurate method of extracting air from a great variety of substances, as was explained in the Introduction to this volume. But having afterwards procured a lens of twelve inches diameter, and twenty inches focal distance, I proceeded with great alacrity to examine, by the help of it, what kind of air a great variety of substances, natural and factitious, would yield, putting them into . . . vessels . . . which I filled with quicksilver and kept inverted in a basin of the same. Mr. Warltire, a good chymist, and lecturer in natural philosophy, happening to be at that time in Calne, I explained my views to him, and was furnished by him with many substances, which I could not otherwise have procured.

5. Probably the modern nitrogen dioxide (NO_2).

With this apparatus, after a variety of other experiments, an account of which will be found in its proper place, on the 1st of August, 1774, I endeavoured to extract air from mercurius calcinatus per se[6] and I presently found that, by means of this lens, air was expelled from it very readily. Having got about three or four times as much as the bulk of my materials, I admitted water to it, and found that it was not imbibed by it. But what surprised me more than I can well express, was, that a candle burned in this air with a remarkably vigorous flame, very much like that enlarged flame with which a candle burns in nitrous air, exposed to iron or liver of sulphur; but as I had got nothing like this remarkable appearance from any kind of air besides this particular modification of nitrous air, and I knew no nitrous acid was used in the preparation of mercurius calcinatus, I was utterly at a loss how to account for it.

In this case, also, though I did not give sufficient attention to the circumstance at that time, the flame of the candle, besides being larger, burned with more splendor and heat than in that species of nitrous air;. and a piece of red-hot wood sparkled in it, exactly like paper dipped in a solution of nitre, and it consumed very fast; an experiment which I had never thought of trying with nitrous air.

At the same time that I made the above mentioned experiment, I extracted a quantity of air, with the very same property, from the common red precipitate, which being produced by a solution of mercury in spirit of nitre,[7] made me conclude that this peculiar property, being similar to that of the modification of nitrous air above mentioned, depended upon something being communicated to it by the nitrous acid; and since the mercurius calcinatus is produced by exposing mercury to a certain degree of heat, where common air has access to it, I likewise concluded that this substance had collected something of nitre, in that state of heat, from the atmosphere.

This, however, appearing to me much more extraordinary than it ought to have done, I entertained some suspicion that the mercurius calcinatus, on which I had made my experiments, being

6. Mercuric oxide, HgO; here Priestley was decomposing it to get pure oxygen.
7. Mercury treated with nitric acid would form mercuric oxide, as Priestley says.

bought at a common apothecary's, might in fact, be nothing more than red precipitate; though, had I been anything of a practical chymist, I could not have entertained any such suspicion. However, mentioning this suspicion to Mr. Warltire, he furnished me with some that he had kept for a specimen of the preparation, and which, he told me, he could warrant to be genuine. This being treated in the same manner as the former, only by a longer continuance of heat, I extracted much more air from it than from the other.

This experiment might have satisfied any moderate sceptic; but, however, being at Paris in the October following, and knowing that there were several eminent chymists in that place, I did not omit the opportunity by means of my friend Mr. Magellan, to get an ounce of mercurius calcinatus prepared by Mr. Cadet, of the genuineness of which there could not possibly be any suspicion; and at the same time, I frequently mentioned my surprise at the kind of air which I had got from this preparation to Mr. Lavoisier, Mr. le Roy, and several other philosophers, who honoured me with their notice in that city; and who, I dare say, cannot fail to recollect the circumstance.

At the same time, I had no suspicion that the air which I had got from the mercurius calcinatus was even wholesome, so far was I from knowing what it was that I had really found, taking it for granted, that it was nothing more than such kind of air as I had brought nitrous air to be by the processes above mentioned; and in this air I have observed that a candle would burn sometimes quite naturally, and sometimes with a beautiful enlarged flame, and yet remain perfectly noxious.

At the same time that I had got the air above mentioned from mercurius calcinatus and the red precipitate, I had got the same kind from red lead or minium.[8] In this process that part of the minium on which the focus of the lens had fallen, turned yellow. One third of the air, in this experiment, was readily absorbed by water, but, in the remainder, a candle burned very strongly, and with a crackling noise.

That fixed air is contained in red lead I had observed before; for I had expelled it by the heat of a candle, and had found it to be very pure. . . I imagine it requires more heat than I then used to expel any of the other kind of air.

8. Lead oxide, Pb_3O_4.

This experiment with red lead confirmed me more in my suspicion, that the mercurius calcinatus must get the property of yielding this kind of air from the atmosphere, the process by which that preparation, and this of red lead is made, being similar. As I never make the least secret of anything I observe, I mentioned this experiment also, as well as those with mercurius calcinatus, and the red precipitate, to all my philosophical acquaintance at Paris, and elsewhere, having no idea, at that time, to what these remarkable facts would lead.

Presently after my return from abroad, I went to work upon the mercurius calcinatus, which I had procured from Mr. Cadet; and, with a very moderate degree of heat, I got from about one fourth of an ounce of it, an ounce-measure of air, which I observed to be not readily imbibed, either by the substance itself from which it had been expelled (for I suffered them to continue a long time together before I transferred the air to any other place) or by water, in which I suffered this air to stand a considerable time before I made any experiment upon it.

In this air, as I had expected, a candle burned with a vivid flame, but what I observed new at this time (Nov. 19), and which surprised me no less than the fact I had discovered before, was that, whereas a few moments' agitation in water will deprive the modified nitrous air of its property of admitting a candle to burn it in; yet, after more than ten times as much agitation as would be sufficient to produce this alteration in the nitrous air, no sensible change was produced in this. A candle still burned in it with a strong flame; and it did not, in the least, diminish common air, which I have observed that nitrous air, in this state, in some measure, does.

But I was much more surprised, when, after two days, in which this air had continued in contact with water (by which it was diminished about one twentieth of its bulk) I agitated it violently in water about five minutes, and found that a candle still burned in it as well as in common air. The same degree of agitation would have made phlogisticated nitrous air fit for respiration indeed, but it would certainly have extinguished a candle.

These facts fully convinced me, that there must be a very material difference between the constitution of the air from mercurius calcinatus, and that of phlogisticated nitrous air, notwithstanding their resemblance in some particulars. But though I did not doubt that the air from mercurius calcinatus was fit for respiration, after being agitated in water, as every kind of air without exception, on which I had tried the experiment, had been, I still did not suspect that it was respirable in the first instance, so far was I from having any idea of this air being, what it really was, much superior, in this respect, to the air of the atmosphere.

In this ignorance of the real nature of this kind of air, I continued from this time (November) to the 1st of March following; having, in the mean time, been intent upon my experiments on the vitriolic acid air above recited, and the various modifications of air produced by spirit of nitre, an account of which will follow. But in the course of this month, I not only ascertained the nature of this kind of air, though very gradually, but was led by it to the complete discovery of the constitution of the air we breathe.

Till this 1st of March, 1775, I had so little suspicion of the air from mercurius calcinatus, etc., being wholesome, that I had not even thought of applying to it the test of nitrous air; but thinking (as my reader must imagine I frequently must have done) on the candle burning in it after long agitation in water, it occurred to me at last to make the experiment; and putting one measure of nitrous air to two measures of this air, I found, not only that it was diminished, but that it was diminished quite as much as common air, and that the redress of the mixture was likewise equal to that of a similar mixture of nitrous and common air.[9]

After this I had no doubt but that the air from mercurius calcinatus was fit for respiration, and that it had all the other properties of genuine common air. But I did not take notice of what I might have observed, if I had not been so fully possessed by the notion of there being no air better than common air, that the redness was really deeper, and the diminution something greater than common air would have admitted.

Moreover, this advance in the way of truth, in reality, threw me back into error, making me give up the hypothesis I had first formed, viz. that the mercurius calcinatus had extracted spirit of nitre from the air; for I now concluded, that all the constituent

9. "Nitrous air" absorbed oxygen from atmospheric air and hence diminished the volume. The effect was no less marked, obviously, with pure oxygen.

parts of the air were equally, and in their proper proportion, imbibed in the preparation of this substance, and also in the process of making red lead. For at the same time that I made the above mentioned experiment on the air from mercurius calcinatus, I likewise observed that the air which I had extracted from red lead, after the fixed air was washed out of it, was of the same nature, being diminished by nitrous air like common air: but, at the same time, I was puzzled to find that air from the red precipitate was diminished in the same manner, though the process for making this substance is quite different from that of making the two others. But to this circumstances I happened not to give much attention.

I wish my reader be not quite tired with the frequent repetition of the word surprise, and others of similar import; but I must go on in that style a little longer. For the next day I was more surprised than ever I had been before, with finding that, after the above mentioned mixture of nitrous air and the air from mercurius calcinatus had stood all night (in which time the whole diminution must have taken place; and, consequently, had it been common air, it must have been made perfectly noxious, and entirely unfit for respiration or inflammation) a candle burned in it, and even better than in common air.

I cannot, at this distance of time, recollect what it was that I had in view in making this experiment; but I know I had no expectation of the real issue of it. Having acquired a considerable degree of readiness in making experiments of this kind, a very slight and evanescent motive would be sufficient to induce me to do it. If, however, I had not happened, for some other purpose, to have had a lighted candle before me, I should probably never have made the trial; and the whole train of my future experiments relating to this kind of air might have been prevented.

Still, however, having no conception of the real cause of this phenomenon, I considered it as something very extraordinary; but as a property that was peculiar to air extracted from these substances, and adventitious; and I always spoke of the air to my acquaintance as being substantially the same thing with common air. I particularly remember my telling Dr. Price,[10] that I was

10. Richard Price, a Nonconformist Minister and writer; he was a friend of Priestley's and shared many of his liberal political and theological views. The mouse mentioned later in the sentence would have shown the new air to be as respirable as common air.

myself perfectly satisfied of its being common air, as it appeared to be so by the test of nitrous air; though for the satisfaction of others, I wanted a mouse to make the proof quite complete.

On the 8th of this month I procured a mouse, and put it into a glass vessel, containing two ounce-measures of the air from mercurius calcinatus. Had it been common air, a full grown mouse, as this was, would have lived in it about a quarter of an hour. In this air, however, my mouse lived a full half hour; and though it was taken out seemingly dead, it appeared to have been only exceedingly chilled, for, upon being held to the fire, it presently revived, and appeared not to have received any harm from the experiment.

By this I was confirmed in my conclusion, that the air extracted from mercurius calcinatus etc., was at least *as good* as common air; but I did not certainly conclude that it was any *better*; because, though one mouse would live only a quarter of an hour in a given quantity of air, I knew it was not impossible but that another mouse might have lived in it half an hour; so little accuracy is there in this method of ascertaining the goodness of air; and indeed I have never had recourse to it for my own satisfaction, since the discovery of that most ready, accurate, and elegant test that nitrous air furnishes. But in this case I had a view to publishing the most generally satisfactory account of my experiments that the nature of the thing would admit of.

This experiment with the mouse, when I had reflected upon it some time, gave me so much suspicion that the air into which I had put it was better than common air, that I was induced, the day after, to apply the test of nitrous air to a small part of that very quantity of air which the mouse had breathed so long; so that, had it been common air, I was satisfied it must have been very nearly, if not altogether, as noxious as possible, so as not to be affected by nitrous air; when, to my surprise again, I found that though it had been breathed so long, it was still better than common air. For after mixing it with nitrous air, in the usual proportion of two to one, it was diminished in the proportion of 4½ to 3½; that is, the nitrous air had made it two ninths less than before, and this in a very short space of time; whereas I had never found that, in the longest time, any common air was reduced more than one fifth of its bulk by any proportion of nitrous air, nor more than one fourth by any phlogistic process whatever. Thinking of this extraordinary fact upon my pillow, the next morning I put another measure of nitrous air to the same mixture, and, to my utter astonishment, found that it was farther diminished to almost one half of its original quantity. I then put a third measure to it; but this did not diminish it any farther: but however, left it one measure less than it was even after the mouse had been taken out of it.

Being now fully satisfied that this air, even after the mouse had breathed it half an hour, was much better than common air, and having a quantity of it still left, sufficient for the experiment, viz. an ounce-measure and a half, I put the mouse into it; when I observed that it seemed to feel no shock upon being put into it, evident signs of which would have been visible, if the air had not been very wholesome; but that it remained perfectly at its ease another full half hour, when I took it out quite lively and vigorous. Measuring the air the next day, I found it to be reduced from 1½ to $^2/_3$ of an ounce-measure. And after this, if I remember well (for in my register of the day I only find it noted, that it was "considerably diminished" by nitrous air) it was nearly as good as common air. It was evident, indeed, from the mouse having been taken out quite vigorous that the air could not have been rendered very noxious.

For my farther satisfaction I procured another mouse, and putting it into less than two ounce measures of air extracted from mercurius calcinatus and air from red precipitate (which, having found them to be of the same quality, I had mixed together) it lived three quarters of an hour. But not having had the precaution to set the vessel in a warm place, I suspect that the mouse died of cold. However, as it had lived in the same quantity of common air, and I did not expect much accuracy from this kind of test, I did not think it necessary to make any more experiments with mice.

Being now fully satisfied of the superior goodness of this kind of air, I proceeded to measure that degree of purity, with as much accuracy as I could, by the test of nitrous air, and I began with putting one measure of nitrous air to two measures of this air, as if I had been examining common air, and now I observed that the diminution was evidently greater than common air would have suffered by the same treatment. A second measure of nitrous air reduced it to two thirds of its original quantity, and a third measure to one half. Suspecting that the diminution could not proceed much farther, I then added only half a measure of nitrous air, by which it was diminished still more; but not much, and another half measure made it more than half of its original quantity; so that, in this case, two measures of this air took more than two measures of nitrous air, and yet remained less than half of what it was. Five measures brought it pretty exactly to its original dimensions.

At the same time, air from the red precipitate was diminished in the same proportion as that from mercurius calcinatus, five measures of nitrous air being received by two measures of this without any increase of dimensions. Now as common air takes about one half of its bulk of nitrous air, before it begins to receive any addition to its dimensions from more nitrous air, and this air took more than four half-measures before it ceased to be diminished by more nitrous air, and even five half-measures made no addition to its original dimensions, I conclude that it was between four and five times as good as common air. It will be seen that I have since procured air better than this, even between five and six times as good as the best common air that I have ever met with.

Being now fully satisfied with respect to the nature of this new species of air, viz. that, being capable of taking more phlogiston from nitrous air, it therefore originally contains less of this principle; my next enquiry was, by what means it comes to be so pure, or philosophically speaking, to be so much dephlogisticated; and since the red lead yields the same kind of air with mercurius calcinatus, though mixed with common air, and is a much cheaper material, I proceeded to examine all the preparations of lead, made by heat in the open air, to see what kind of air they would yield, beginning with the grey calx, and ending with litharge.[11]

The red lead which I used for this purpose yielded a considerable quantity of dephlogisticated air, and very little fixed air; but to what circumstance in the preparation of this lead, or in the keeping of it, this difference is owing, I cannot tell. I have frequently found a very remarkable difference between different specimens of red lead in this respect, as well as in the purity of the air which they contain. This difference, however, may arise in a great measure, from the care that is taken to extract the fixed air from it. In this experiment two measures of nitrous air being put to one measure of this air, reduced it to one third of what it was at

11. Lead dioxide (PbO_2) and lead monoxide (PbO), respectively.

first, and nearly three times its bulk of nitrous air made very little addition to its original dimensions; so that this air was exceedingly pure, and better than any that I had procured before.

The preparation called massicot[1,2] (which is said to be a state between the grey calx and the red lead) also yielded a considerable quantity of air, of which about one half was fixed air, and the remainder was such, that when an equal quantity of nitrous air was put to it, it was something less than at first; so that this air was about twice as pure as common air.

I thought it something remarkable, that in the preparations of lead by heat, those before and after these two, viz. the red lead and massicot, yielded only fixed air. I would also observe by the way, that a very small quantity of air was extracted from lead ore by the burning lens. The bulk of it was easily absorbed by water. The remainder was not affected by nitrous air, and it extinguished a candle.

I got a very little air by the same process from the grey calx of lead, of precisely the same quality with the former. That part of it which was not affected by nitrous air extinguished a candle, so that both of them may be said to have yielded fixed air, only with a larger portion than usual of that part of it which does not unite with water.

Litharge (which is a state that succeeds the red lead) yielded air pretty readily; but this also was fixed air. That which was not absorbed by water, was not affected by nitrous air.

Much more than I had any opportunity of doing remains to be done in order to ascertain upon what circumstances, in these preparations of lead, the quality of the air which they contain, depends. It can only be done by some person who shall carefully attend to the processes, so as to see himself in what manner they are made, and examine them in all their different states. I very much wished to have attempted something of this kind myself, but I found it impossible in my situation. However, I got Dr. Higgins[1,3] (who furnished me with several preparations that I could not easily have procured elsewhere) to make me a quantity of red lead, that I might, at least, try it when fresh made, and after

12. A yellow form of litharge, less stable than the ordinary red form.
13. Bryan Higgins (c. 1737-1820), uncle of the better known William Higgins. Priestley attended some of Bryan Higgins' lectures and bought chemicals from him.

keeping it some time in different circumstances; and though by the help of this preparation, I did not do the thing that I expected, I did something else, much more considerable.

This fresh made red lead had a yellowish cast, and had in it several pieces entirely yellow. I tried it immediately, in the same manner in which I had made the preceding experiments, viz. with the burning lens in quicksilver, and found that it yielded very little air, and with great difficulty; requiring the application of a very intense heat. With an equal quantity of nitrous air, a part of this air was reduced to one half of its original bulk, and 3½ measures saturated it. The air, therefore, was very pure, and the quantity that it yielded being very small, it proved to be in a very favourable state for ascertaining on what circumstances its acquiring this air depended.

My object now was to bring this fresh made red lead, which yielded very little air, to that state in which other red lead had yielded a considerable quantity; and taking it, in a manner, for granted, in consequence of the reasoning intimated above, that red lead must imbibe from the atmosphere some kind of acid, in order to acquire that property, I took three separate half-ounces of this fresh made red lead, and moistened them till they made a kind of paste, with each of the three mineral acids, viz. the vitriolic, the marine and the nitrous, and as I intended to make the experiments in a gun-barrel, lest the iron should be too much affected by them, I dried all these mixtures, till they were perfectly hard; then pulverizing them, I put them separately into my gun-barrel, filled up to the mouth with pounded flint, which I had found by trial to yield little, or no air when treated in this manner. I had also found that no quantity of air sufficient to make an experiment could be procured from an equal quantity of this red lead by this process.

Those portions of the red lead which had been moistened with the vitriolic and marine acids became white; but that which had been moistened with the nitrous acid, had acquired a deep brown colour. The mixtures with the nitrous and marine acids dried pretty readily but that with the vitriolic acid was never perfectly dry; but a great part of it remained in the form of softish paste.

Neither the vitriolic nor the marine acid mixtures gave the least air when treated in the manner above mentioned; but the moment that the composition into which the nitrous acid had entered became warm, air began to be produced; and I received the produce in quicksilver. About one ounce-measure was quite transparent, but presently after it became exceedingly red; and being satisfied that this redness was owing to the nitrous acid vapour having dissolved the quicksilver, I took no more than two ounce-measures in this way, but received all the remainder, which was almost two pints, in water. Far the greatest part of this was fixed air, being readily absorbed by water, and extinguishing a candle. There was, however, a considerable residuum, in which the flame of a candle burned with a crackling noise, from which I concluded that it was true dephlogisticated air.

In this experiment I had moistened the red lead with spirit of nitre several times, and had dried it again. When I repeated this experiment, I moistened it only once with the same acid, when I got from it not quite a pint of air; but it was almost all of the dephlogisticated kind, about five times as pure as common air. N.B. All the acids made a violent effervescence with the red lead.

Though there was a difference in the result of these experiments, which I shall consider hereafter, I was now convinced that it was the nitrous acid which the red lead had acquired from the air, and which had enabled it to yield the dephlogisticated air, agreeable to my original conjecture. Finding also, as will be seen in the following section, that the same kind of air is produced by moistening with the spirit of nitre any kind of earth[14] that is free from phlogiston, and treating it as I had done the red lead in the last mentioned experiment, there remained no doubt in my mind, but that atmospherical air, or the thing that we breathe, consists of the nitrous acid and earth, with so much phlogiston as is necessary to its elasticity; and likewise so much more as is required to bring it from its state of perfect purity to the mean condition in which we find it.

For this purpose I tried, with success, flowers of zinc, chalk, quick-lime, slaked-lime, tobacco-pipe clay, flint and Muscovy talc, with other similar substances, which will be found to comprise all the kinds of earth that are essentially distinct from each other, according to their chymical properties. A particular account of the processes with these substances, I reserve for another section; thinking it sufficient in this to give a history of the discovery, and

14. Priestley was still, like his contemporaries, inclined to use the ancient "elements" to denote the generalized characteristics of bodies. As the next section shows, he is thinking here of inert minerals.

a general account of the nature of this dephlogisticated air, with this general inference from the experiment, respecting the constitution of the atmosphere.

I was the more confirmed in my idea of spirit of nitre and earth constituting respirable air, by finding, that when any of these matters, on which I had tried the experiment, had been treated in the manner above mentioned, and they had thereby yielded all the air that could be extracted from them by this process; yet when they had been moistened with fresh spirit of nitre, and were treated in the same manner as before, they would yield as much dephlogisticated air as at the first. This may be repeated till all the earthy matter be exhausted. It will be sufficient to recite one or two facts of this kind from my register.

April 18, I took the remains of the fresh made red lead, out of which a great quantity of dephlogisticated air had been extracted, and moistening about three quarters of an ounce of it a second time with spirit of nitre, I got from it about two pints of air, all of which was nearly six times as pure as common air. This air was generated very fast and the glass tube through which it was transmitted was filled with red fumes; the nitrous acid, I suppose, prevailing in the composition of the air, but being absorbed by the water in which it was afterwards received.

In this, and many other processes, my reader will find a great variety in the purity of the air procured from the same substances. But this will not be wondered at, if it be considered that a small quantity of phlogistic matter, accidentally mixing with the ingredients for the composition of this air, depraves it. It will also be unavoidably depraved, in some measure, if the experiment be made in a gun-barrel, which I commonly made use of, as was generally the case, it was sufficiently exact for my purpose, on account of its being the easiest, and in many respects, the most commodious process.

The reason of this is, that if the produce of air be not very rapid, there will be time for the phlogiston to be disengaged from the iron itself, and to mix with the air. Accordingly I have seldom failed to find, that when I endeavoured to get all the air I possibly could from any quantity of materials, and received the produce at different times (as for my satisfaction I generally did) the last was inferior in purity to that which came first. Not unfrequently it was phlogisticated air; that is, air so charged with phlogiston, as to be perfectly noxious; and sometimes, as the reader will find in the next section, it was even nitrous air.

On the same account it frequently happened, that when I used a considerable degree of heat, the red lead which I used in these experiments would be changed into real lead, from which it was often very difficult to get the gun-barrel perfectly clear.

A good deal will also depend upon the ingredients which have been used in the gun-barrel in preceding experiments: for it is not easy to get such an instrument perfectly clean from all the matters that have been put into it: and though it may be presumed, in general, that every kind of air will be expelled from such ingredients by making the tube red hot; yet matters containing much phlogiston, as charcoal, etc., will not part with it in consequence of the application of heat, unless there be at hand some other substance with which it may combine. Though, therefore, a gun-barrel, containing such small pieces of charcoal as cannot be easily wiped out of it, be kept a long time in a red heat, and even with its mouth open; yet if it be of a considerable length, some part of the charcoal may remain unconsumed, and the effect of it will be found in the subsequent experiment. Of this I had the following very satisfactory proof.

Being desirous to show some of my friends the actual production of dephlogisticated air, and having no other apparatus at hand, I had recourse to my gun-barrel; but apprized them, that having used it the day before to get air from charcoal, with which it had been filled for that purpose, though I had taken all the pains I could to get it all out, yet so much would probably remain, that I could not depend on the air I should get from it being dephlogisticated; but that it would probably be of an inferior quality, and perhaps even nitrous air. Accordingly, having put into it a mixture of spirit of nitre and red lead (being part of a quantity which I had often used before for the same purpose) dried, and pounded, I put it into the fire, and received the air in water.

The first produce, which was about a pint, was so far nitrous, that two measures of common air, and one of this, occupied the space of little more than two measures; that is, it was almost as strongly nitrous as that which is produced by the solution of metals in spirit of nitre. The second pint was very little different from common air, and the last produce was better still, being more than twice as good as common air. If, therefore, any person shall propose to make dephlogisticated air, in large quantities, he should have an apparatus appropriated to that purpose; and the greatest care should be taken to keep the instruments as clear as possible from all phlogistic matter, which is the very bane of purity with respect to air, they being exactly plus and minus to each other.

The hypothesis maintained in this section viz. that atmospherical air consists of the nitrous acid and earth, suits exceedingly well with the facts relating to the production of nitre; for it is never generated but in the open air, and by exposing to it such kinds of earth as are known to have an affinity with the nitrous acid; so that by their union common nitre may be formed.

Hitherto it has been supposed by chymists, that this nitrous acid, by which common nitre is formed, exists in the atmosphere as an extraneous substance, like water, and a variety of other substances, which float in it, in the form of effluvia; but since there is no place in which nitre may not be made, it may, I think, with more probability be supposed, according to my hypothesis, that nitre is formed by a real decomposition of the air itself, the bases that are presented to it having, in such circumstances, a nearer affinity with the spirit of nitre than that kind of earth with which it is united in the atmosphere.

My theory also supplies an easy solution of what has always been a great difficulty with chymists, with respect to the detonation of nitre. The question is, what becomes of the nitrous acid in this case? The general, I believe the universal, opinion now is, that it is destroyed, that is, that the acid is properly decomposed, and resolved into its original elements, which Stahl[15] supposed to be earth and water. On the other hand, I suppose that, through the common properties of the acid, as combined with water, disappear, it is only in consequence of its combination with some earthy or inflammable matter, with which it forms some of the many species of air, into the composition of which this wonderful acid enters. It may be common air, it may be dephlogisticated air, or it may be nitrous air, or some of the other kinds, of which an account will be given in a subsequent section. That it should really be the nitrous acid, though so much disguised by its union with earthy, or other matters, will not appear extraordinary to any

15. G. E. Stahl, the early eighteenth-century proponent of the phlogiston theory.

person who shall consider how little the acid of vitriol is apparent in common sulphur.

With respect to mercurius calcinatus, and red lead, their red colour favours the supposition of their having extracted spirit of nitre from the air.

39. New French Chemistry

Of all the phenomena of animal life none is more striking or more worthy of the attention of scientists and naturalists than those which accompany respiration. If, on the one hand, we know little of the purpose of this peculiar function, we know, on the other, that it is so essential to life that it cannot be suspended for any length of time without exposing the animal to the danger of imminent death.

Air, as everyone knows, is the agent or, more precisely, the subject of respiration; but, at the same time, every sort of air, or more generally every sort of elastic fluid, is not suitable for supporting it, and there are large numbers of airs which animals cannot breathe without dying as quickly as if they did not breathe at all.

The experiments of a number of scientists, and particularly those of Mr. Hales, began to shed some light on this important subject; recently, Mr. Priestley, in a paper published last year in London, has pushed back the limits of our understanding much farther and has sought to prove, by very ingenious and very precise experiments of a very novel kind, that animal respiration had the property of phlogisticating the air, like the calcination of metals and several other chemical processes, and that air did not cease to be respirable until it was supercharged and in some way saturated with phlogiston.

However probable the theory of this distinguished scientist

SOURCE: Translated by the editor from A. L. Lavoisier, *"Mémoires sur la Respiration des Animaux et sur les changements qui arrivent à l' air en passant par leur poumons"* published in *Mémoires de l' Académie des sciences,* Paris, 1777).

might appear at first sight, however numerous and well constructed the experiments upon which he tried to base it, I confess that I
found it contrary to so many phenomena that I believed I had the
right to question it. In consequence, I worked on a different plan
and found myself irrefutably led as a result of my experiments to
conclusions which are totally opposed to his. I shall not pause at
this point to discuss in detail each of Mr. Priestley's experiments,
nor to demonstrate how they are all really favorable to the theory
which I am going to develop in this memoir; I shall only refer to
my own and explain their consequences.

I enclosed in a suitable apparatus (which it would be difficult to
describe without a figure) fifty cubic inches of common air; I then
added four ounces of very pure mercury and proceeded to calcine
this mercury by maintaining it for twelve days at a degree of heat
just below that necessary to make it boil.

Nothing remarkable occurred during the first day: the mercury,
though not boiling, was continually evaporating and covering the
inside of the apparatus with little drops, at first very small, later
increasing in size until, when large enough, they fell back into the
bottom of the flask. On the second day I began to see some small
red particles floating on the surface of the mercury; these in a few
days increased in size and number. At the end of twelve days when
I removed the fire and let the flask cool, I observed that the air
contained in it had diminished by eight or nine cubic inches, that
is to say, by about one-sixth of its volume; at the same time a
pretty large quantity, I estimate about forty-five grains in weight,
of mercury precipitated *per se* (calx of mercury) had been
formed.[1]

The air thus diminished did not react with limewater;[2] it extinguished flames; it killed rapidly animals placed in it; it gave
virtually no red vapors with nitrous air, nor was it noticeably
decreased in volume by this latter.[3] In a word, it was an absolutely
mephitic state.

It is known from Mr. Priestley's experiments and those I have
made myself that mercury precipitated *per se* is nothing but a
combination of mercury with about one-twelfth of its weight of

1. Mercuric oxide.
2. I.e., it contained no carbon dioxide.
3. These tests in fact showed that it contained no free oxygen.

an air which is much better and more respirable (if it is permissible to use such an expression) than common air. Now the preceding experiment appears to prove that the mercury in undergoing calcination has absorbed the best and most respirable part of the air, leaving only the mephitic or nonrespirable part. The following experiment further confirmed this truth.

I carefully collected the forty-five grains of mercury formed during the preceding calcination; I put them into a very small glass retort whose neck, in a double curve, fitted under a bell jar filled with water, and I proceeded to reduce them without adding anything else. By this operation I recovered very nearly the same quantity of air which had been absorbed during the calcination, that is, between about eight and nine cubic inches. When I recombined these eight or nine cubic inches with the air vitiated by the calcination of the mercury, I returned that pretty exactly to the state in which it had been before the calcination, that is, to being ordinary air. This air, thus reconstituted, no longer extinguished flames, no longer killed animals breathing it, and was almost as much diminished in volume by nitrous air as is atmospheric air.

This is an example of the firmest possible kind of proof attainable in chemistry with the decomposition and recomposition of air; from this it follows: (1) that five-sixths of the air we breathe is (as I have already stated in a previous paper) in a mephitic state, that is, one incapable of supporting the respiration of animals, or the illumination and combustion of bodies; (2) that the remainder, that is to say, only one-fifth of the whole volume of the atmospheric air, is fit for respiration; (3) that, in calcination, metallic mercury absorbs the healthful part of the air, leaving behind only the mephitic part; (4) that by combining these two parts of the air which have been separated (the respirable and the mephitic parts), air like that of the atmosphere is reconstituted.

These preliminary facts concerning the calcination of metals will lead to simple conclusions concerning animal respiration; and, since air which has served for some time to sustain that vital function has much in common with that in which metals have been calcined, knowledge about one naturally may be applied to the other.

I put a house sparrow under a bell jar filled with ordinary air and plunged this in a bowl full of mercury. The empty part of the jar was thirty-one cubic inches in volume. The bird for the first few moments seemed quite unaffected, only a little sleepy; at the end of a quarter of an hour it became disturbed, its respiration labored and quick and, from this point on, these difficulties increased continually until finally it died at the end of fifty-five minutes with a sort of convulsive motion. In spite of the bird's own heat, which during the first moments necessarily expanded the air contained in the bell jar, there was a noticeable decrease in the total volume; this decrease was about a fortieth at the end of the first quarter of an hour but, instead of increasing thereafter, it was a little less at the end of half an hour; and after the bird's death, when the air under the bell jar had acquired the temperature of the surroundings, the decrease was found to be no more than a sixtieth.

This air, which had been respired by a living creature, had become very different from atmospheric air; it precipitated limewater; it extinguished a candle; it was no longer diminished by nitrous air; a new bird introduced in it lived only a few moments. In short, it was entirely mephitic and, in that respect, appeared quite similar to what remained after the calcination of mercury.

However, a more thorough examination made me perceive two very striking differences between these two airs, that is, between that which had sustained calcination and that breathed by the sparrow: first, the decrease of volume was much less in this latter than in the former; second, respired air precipitated limewater whereas the air which had served for calcination did nothing of the sort.

This difference between these two airs and, on the other hand, the close analogy which they presented on many points made me suppose that two causes are involved in respiration, of which I probably as yet understood only one; and, to clear up my conjectures on this point, I made the following experiment.

I made twelve inches of air vitiated by respiration pass into a glass bell jar full of mercury and plunged into mercury, and I introduced into it a thin layer of fixed caustic alkali;[4] I would have made use of limewater for the purpose, but the volume of

4. Potassium hydroxide, which reacted with the carbon dioxide in the "vitiated air" to give potassium carbonate, which naturally effervesced with acids in producing carbon dioxide again.

that would necessarily have been so much too large as to be fatal to the success of the experiment.

The effect of the caustic alkali was to occasion a diminution in the volume of this air of nearly one-sixth; at the same time the alkali lost part of its causticity, acquiring the property of effervescing with acids and crystallizing under the bell jar in very regular rhomboids. Now these are properties which are only known to be communicated to alkali when it is combined with the kind of air or gas known under the name of fixed air and which I shall henceforth call aerial chalky acid [carbon dioxide]. From this it follows that air vitiated by respiration contains nearly a sixth part of an aerial acid, exactly like that which is elicited from chalk.

Air thus deprived of its fixable part by caustic alkali was far from having thus become restored to the state of common air; on the contrary, it was much closer to air which had served in the calcination of mercury, or rather it was exactly the same thing. Like it, it killed animals and extinguished lights; indeed, every experiment which I made to compare these two airs only demonstrated to me that there was not the least difference between them.

Now air which has served in the calcination of mercury is nothing else, as shown above, except the mephitic residue of atmospheric air, whose eminently respirable part has combined with the mercury in the course of the calcination. Hence air which has served in respiration, when it has been deprived of its portion of aerial chalky acid [fixed air, CO_2] is equally nothing but common air deprived of its respirable part. Indeed, having combined this air with about a quarter its volume of eminently respirable air, derived from mercuric calx, I restored it to its original state and thus rendered it just as fit for either respiration or combustion as common air, just as I had done with air vitiated by the calcination of metals.[5]

It follows from these experiments that in order to restore air vitiated by respiration to the state of common and respirable air, two results must be obtained; first, it is necessary to remove from

5. If oxygen from mercuric oxide is added to nitrogen, the mixture can be used for respiration or combustion again. Below Lavoisier correctly states that air used for respiration or combustion can be restored by removing the carbon dioxide and adding oxygen. How confusing this simple conclusion was at the time is indicated by his subsequent arguments.

this air (either by means of lime or of a caustic alkali) the portion of aerial chalky acid which it contains; second, return to it a quantity of eminently respirable or phlogisticated air [oxygen] equal to that which it lost. Respiration, as a necessary consequence, reverses these two results and, in this respect, I find myself led to two equally probable conclusions between which the experiments have not yet permitted me to choose.

From what has just been shown, it is possible to conclude that respiration produces one of two results—either the portion of eminently respirable air contained in atmospheric air is converted into aerial chalky acid [carbon dioxide] in passing through the lungs; or else there is an exchange in this organ so that, on the one hand, eminently respirable air is absorbed and, on the other, the lung returns in its place a very nearly equal quantity by volume of aerial chalky acid.

The first of these views is supported by an experiment which I have already described to the Academy. In a paper read at the public meeting at Easter, 1775, I showed that eminently respirable air could be entirely converted into aerial chalky acid by the addition of powdered charcoal, and I shall prove in other papers that this conversion can be made in various other ways. It is thus possible that respiration has this same property and that eminently respirable air which has entered the lungs leaves it as aerial chalky acid; but, on the other hand, strong analogies seem to militate in favor of the second opinion and lead to the belief that a portion of eminently respirable air remains in the lungs and combines with the blood. It is known that eminently respirable air has the property of giving a red color to bodies, especially to metallic substances with which it combines; mercury, lead, and iron furnish examples of this. These metals form with eminently respirable air calxes [oxides] of a fire-red color: the first known under the name of mercury precipitated *per se* or red precipitated mercury; the second under the name of minium; the third under that of colcothar. The same effects, the same phenomena, are found, as will be seen, in both the calcination of metals and the respiration of animals; all the circumstances are the same right down to the color of the residues. Is it not then possible to conclude by induction that the red color of blood is due to the combination of eminently respirable air with it, or, more precisely, as I shall show in a later paper, to the combination of the *base* of eminently respirable air with an animal fluid in the same fashion as the red color of red precipitated mercury and of minium is due to the combination of the *base* of the same air with a metallic substance? Although Mr. Priestley and modern authors concerned with this subject have not reached the same conclusion, I venture to say that there is hardly one of their experiments which does not seem to tend to establish it. For they, and especially Mr. Priestley, have proved that the blood is not red or crimson except when it is in continuous contact with atmospheric or eminently respirable air; that it becomes black in aerial chalky acid, in nitrous air, in inflammable air, in all the airs which are not fit for respiration, and in the pneumatic machine; while, on the contrary, it reacquires its red color when it is again put into contact with air and above all with eminently respirable air; and that this restoration of color is always accompanied by a diminution in the volume of the air. Now, does it not follow from all these facts that eminently respirable air has the property of combining with the blood and that it is this combination which constitutes its red color? Besides, whichever of these two opinions one adopts, either that the respirable portion of the air combines with the blood or that this portion changes into aerial chalky acid in passing through the lung, or even, as I am being led to believe, that both of these results occur during the act of respiration, it may, from the facts alone, be regarded as proved:

1. That respiration acts only on the portion of "pure air" or eminently respirable air contained in atmospheric air, and that the residue, that is to say, the mephitic part, is a purely passive medium, which enters the lungs only to leave in very nearly the same state as it enters, without change or alteration.

2. That the calcination of a metal in a fixed quantity of air only continues, as I have already proclaimed several times, up to the point when the portion of true air, of eminently respirable air which it contains, is used up and combined with the metal.

3. That besides, if an animal is shut up in a given quantity of air, it will die when it has absorbed or converted into aerial chalky acid the greater part of the respirable portion of air, and when this latter is reduced to the state of mephitic air.

4. That the kind of mephitic air remaining after the calcination of metals differs not at all (according to all my experiments) from that remaining after animal respiration, always provided that this

latter has been deprived, by lime or caustic alkali, of its fixable part; that is, of the aerial chalky acid which it contained. And that these two mephitic airs can be substituted one for another in all experiments, and that they can both be restored to the state of atmospheric air by a quantity of eminently respirable air equal to that they have lost. A new proof of this last truth is that if the quantity of true or eminently respirable air contained in a given quantity of atmospheric air is increased or diminished, the quantity of metal that can be calcined in it is increased or diminished in the same proportion and, up to a point, so is the length of time during which animals can live in it.

The limits which I have set myself in this paper have not permitted me to record many other experiments which support the theory here set out. . .these experiments, as I believe, will shed new light not only on the respiration of animals but also on combustion, operations which are much more closely connected than might be thought at first sight.

Chronology

<table>
<tr><td>1543</td><td>Copernicus' De Revolutionibus
Vesalius' Fabrica
Death of Copernicus</td></tr>
<tr><td>1572</td><td>The new star in Cassiopeia</td></tr>
<tr><td>1577</td><td>Tycho Brahe's comet</td></tr>
<tr><td>1601</td><td>Death of Tycho Brahe
Gilbert's De Magnete</td></tr>
<tr><td>1609</td><td>Publication of Kepler's first two laws</td></tr>
<tr><td>1610</td><td>Publication of Galileo's telescopic discoveries</td></tr>
<tr><td>1628</td><td>Publication of Harvey's discovery of the circulation of the blood</td></tr>
<tr><td>1632</td><td>Galileo's Dialogue leads to his trial</td></tr>
<tr><td>1637</td><td>Descartes' Discours de la méthode</td></tr>
<tr><td>1642</td><td>Death of Galileo
Birth of Newton</td></tr>
<tr><td>1644</td><td>Descartes' Principia Philosophia</td></tr>
<tr><td>1645</td><td>Meetings of scientific men in London lead to foundation of Royal Society</td></tr>
<tr><td>1657</td><td>Accademia del Cimento</td></tr>
<tr><td>1657</td><td>Publication of invention of air pump</td></tr>
<tr><td>1660</td><td>First formal meeting leading to Royal Society's charter
Publication of Boyle's first air pump experiments</td></tr>
<tr><td>1666</td><td>Foundation of Académie royale des sciences
Newton's "wonderful year"</td></tr>
<tr><td>1672</td><td>Newton's first paper on light and colors</td></tr>
<tr><td>1673</td><td>Huygens' Horologium Oscillatorium</td></tr>
<tr><td>1687</td><td>Newton's Principia</td></tr>
<tr><td>1690</td><td>Huygens' Traité de la lumière</td></tr>
<tr><td>1691</td><td>Death of Boyle</td></tr>
<tr><td>1699</td><td>Reorganization of Académie royale des sciences</td></tr>
<tr><td>1704</td><td>Newton's Opticks</td></tr>
</table>

1727   Hales' *Vegetable Staticks*
         Death of Newton
1756   Black's announcement of a nonatmospheric gas
1772   Lavoisier begins work on combustion
1774   Priestley discovers oxygen
1781   Herschel discovers a new planet
1789   Lavoisier's *Traité élémentaire de chimie*
1791   Galvani's account of "animal electricity"
1794   Death of Lavoisier
1800   Foundation of Royal Institution
         Publication of Volta's discovery of the electric battery
1807   Thomas Thomson publishes first account of Dalton's atomic
             theory
         Davy prepares sodium and potassium by electricity

Bibliography

For general histories of science during this period, see especially Herbert Butterfield, *The Origins of Modern Science* (London, 1949, and later paperback editions); A. R. Hall, *The Scientific Revolution* (1st edn., London, 1954, 2nd edn., London; both in Beacon Press paperbacks), with detailed bibliography; Marie Boas, *The Scientific Renaissance 1450–1630* (London and New York, 1962), supplemented by A. Rupert Hall, *From Galileo to Newton (1630–1720)* (London and New York, 1963), both with bibliographies.

There are numerous modern reprints and translations of original sources, though by no means everything printed here is readily available. The only complete translation of Copernicus' *De Revolutionibus* is in the Great Book Series; the Preface and Book I have been partially printed many times, the best translation being that by J. F. Dobson and Selig Brodetsky, Number 10 of *Occasional Notes* of the Royal Astronomical Society (London, 1947). There is also Edward Rosen, *Three Copernican Treatises* (New York, 1939, 1959). Virtually all of Galileo is available in English; especially recommended is Stillman Drake (ed.), *Discoveries and Opinions of Galileo* (Anchor Books). There are Dover reprints of Gilbert's *De Magnete*, of Hooke's *Micrographia*, and of Newton's *Opticks* and *Principia.* Bacon's works exist in numerous editions, Descartes' only in excerpts. Harvey's *De Motu Cordis* has several translations and many editions. Power's *Experimental Philosophy* (1966), Boyle's *Experiments and Considerations Touching Colours* (1964), and the *Saggi* (1964), in Richard Waller's translation, are available from the Johnson Reprint Corp., New York. More selections of Boyle's works are to be found in M. B. Hall (ed.), *Robert Boyle on Natural Philosophy* (Indiana University paperback).

For sixteenth-century astronomy there is F. R. Johnson, *Astronomical Thought in Renaissance England* (Baltimore, 1937) and J. L. E. Dreyer, *Tycho Brahe* (1890, now in paperback). Arthur

Koestler, *The Watershed* (1959, now in paperback) is a biographical study of Kepler. There are readily available biographies of Bacon, Boyle, Harvey, Hooke, Lavoisier, Newton, and Priestley.

Index

70 71 72 73   12 11 10 9 8 7 6 5 4 3 2 1

DOCUMENTARY HISTORY OF WESTERN CIVILIZATION
Edited by Eugene C. Black and Leonard W. Levy

ANCIENT AND MEDIEVAL HISTORY OF THE WEST

Morton Smith: ANCIENT GREECE

A. H. M. Jones: A HISTORY OF ROME THROUGH THE FIFTH CENTURY
Vol. I: The Republic
Vol. II: The Empire

Deno Geanakoplos: BYZANTINE EMPIRE

Marshall W. Baldwin: CHRISTIANITY THROUGH THE THIRTEENTH CENTURY

Bernard Lewis: ISLAM THROUGH SULEIMAN THE MAGNIFICENT

David Herlihy: HISTORY OF FEUDALISM

William M. Bowsky: RISE OF COMMERCE AND TOWNS

David Herlihy: MEDIEVAL CULTURE AND SOCIETY

EARLY MODERN HISTORY

Hanna H. Gray: CULTURAL HISTORY OF THE RENAISSANCE

Florence Edler de Roover: MONEY, BANKING,
AND COMMERCE, THIRTEENTH THROUGH SIXTEENTH CENTURIES

V. J. Parry: THE OTTOMAN EMPIRE

Ralph E. Giesey: EVOLUTION OF THE DYNASTIC STATE

J. H. Parry: THE EUROPEAN RECONNAISSANCE: Selected Documents

Hans J. Hillerbrand: THE PROTESTANT REFORMATION

John C. Olin: THE CATHOLIC COUNTER REFORMATION

Orest Ranum: THE CENTURY OF LOUIS XIV

Thomas Hegarty: RUSSIAN HISTORY THROUGH PETER THE GREAT

Marie Boas Hall: NATURE AND NATURE'S LAWS

Barry E. Supple: HISTORY OF MERCANTILISM

Arthur J. Slavin: IMPERIALISM, WAR, AND DIPLOMACY, 1550-1763

Herbert H. Rowen: THE LOW COUNTRIES

C. A. Macartney: THE HABSBURG AND HOHENZOLLERN DYNASTIES
IN THE SEVENTEENTH AND EIGHTEENTH CENTURIES

Lester G. Crocker: THE AGE OF ENLIGHTENMENT

Robert and Elborg Forster: EUROPEAN SOCIETY IN THE EIGHTEENTH CENTURY